Rogemar S. Mamon • Robert J. Elliott
Editors

Hidden Markov Models in Finance

Further Developments and Applications, Volume II

Editors
Rogemar S. Mamon
Department of Statistical & Actuarial
   Sciences
University of Western Ontario
London, ON, Canada

Robert J. Elliott
School of Mathematics
University of Adelaide
Adelaide, Australia

ISSN 0884-8289 ISSN 2214-7934 (electronic)
ISBN 978-1-4899-7441-9 ISBN 978-1-4899-7442-6 (eBook)
DOI 10.1007/978-1-4899-7442-6
Springer New York Heidelberg Dordrecht London

Library of Congress Control Number: 2007921976

© Springer Science+Business Media New York 2014
This work is subject to copyright. All rights are reserved by the Publisher, whether the whole or part of the material is concerned, specifically the rights of translation, reprinting, reuse of illustrations, recitation, broadcasting, reproduction on microfilms or in any other physical way, and transmission or information storage and retrieval, electronic adaptation, computer software, or by similar or dissimilar methodology now known or hereafter developed. Exempted from this legal reservation are brief excerpts in connection with reviews or scholarly analysis or material supplied specifically for the purpose of being entered and executed on a computer system, for exclusive use by the purchaser of the work. Duplication of this publication or parts thereof is permitted only under the provisions of the Copyright Law of the Publisher's location, in its current version, and permission for use must always be obtained from Springer. Permissions for use may be obtained through RightsLink at the Copyright Clearance Center. Violations are liable to prosecution under the respective Copyright Law.
The use of general descriptive names, registered names, trademarks, service marks, etc. in this publication does not imply, even in the absence of a specific statement, that such names are exempt from the relevant protective laws and regulations and therefore free for general use.
While the advice and information in this book are believed to be true and accurate at the date of publication, neither the authors nor the editors nor the publisher can accept any legal responsibility for any errors or omissions that may be made. The publisher makes no warranty, express or implied, with respect to the material contained herein.

Printed on acid-free paper

Springer is part of Springer Science+Business Media (www.springer.com)

Preface

Since the publication of our coedited monograph *Hidden Markov Models (HMM) in Finance* by Springer in 2007, there has been substantial research in many areas of finance which employ HMM. It is the objective of this edited volume to present some updates on the current state of the field. The book brings together several articles which explore the most recent developments of HMMs and their applications to financial modelling. The main themes in this collection, combining new theoretical advances and applications, are improvements of the EM algorithm for HMMs, interest rates, foreign exchange, insurance, option pricing, trading and parameter estimation within some extended HMM frameworks.

The lead paper 'Robustification of an Online EM Algorithm for Modelling Asset Prices Within an HMM', by C. Erlwein-Sayer and P. Ruckdeschel, considers the robustification of the EM algorithm in conjunction with Elliott's pioneering work (1994) on exact adaptive filters for Markov chains observed in Gaussian noise. This innovation aims to tackle the presence of outliers in financial data and preserves confidence in the results of the HMM estimation procedure.

The succeeding three articles showcase various approaches in modelling interest rates under HMM-driven regime-switching frameworks. In their paper 'Stochastic Volatility or Stochastic Central Tendency: evidence from a Hidden Markov Model of the Short-Term Interest Rate', C.A. Wilson and R.J. Elliott show the importance of stochastic volatility over stochastic central tendency in capturing the evolution of interest rates. It features an iterative procedure in determining the likelihood function; numerical maximisation was used to find maximum likelihood estimates. Wu and Zeng put forward a model with an analytically simple representation of the Markov regime shifts but which is capable of handling the stylised features of the yield curve in their paper 'An Econometric Model of the Term Structure of Interest Rates Under Regime-Switching Risk'. The efficient method of moments was utilised in their empirical examination of US data within a two-factor version of their proposed model. Continuing the theme of interest-rate modelling, L. Steinrücke, R. Zagst and A. Swishchuk extended the LIBOR market model to incorporate sudden market shocks, structural breaks and changes in economic climate. Their paper 'The LIBOR market model: a Markov-switching jump diffusion extension' includes model calibration to real data illustrating potential usability for practitioners.

'Exchange Rates and Net Portfolio Flows: a Markov-Switching Approach', by F. Menla, F. Spagnolo and N. Spagnolo, is a paper which describes the usefulness of Markov-switching tools to investigate the finer structure of the foreign exchange rate market. In particular, it probes the effects of bond and equity portfolio flows to exchange rate dynamics. Covering major currencies against the US dollar, it is a far-reaching empirical study that explores the relationship between portfolio flows and fluctuations in the exchange rate changes.

In recent years, there has been a proliferation of finsurance business due to the significant growth in the creation of derivatives that blend the characteristics of financial and insurance products. These instruments require new methods to deal with their complexity. P. Azimzadeh, P.A. Forsyth and K.R. Vetzal examine the valuation and hedging of guaranteed lifelong withdrawal and death benefits contracts in their paper 'Hedging Costs for Variable Annuities Under Regime Switching'. A general approach is constructed which enables utility-based pricing and other factors to be taken into account and yields a system of partial differential equations (PDEs). The system is solved using an implicit method for a large class of utility functions that govern the withdrawal behaviour of the policyholders.

D. Nguyen, G. Yin and Q. Zhang analyse an optimal trading strategy assuming a bull-bear regime-switching market. Their paper 'A Stochastic Approximation Approach for Trend-Following Trading' delves into a buy–sell strategy that maximises expected return. The optimality of such a strategy is achieved by determining threshold levels through a stochastic approximation algorithm. This circumvents the need to solve Hamilton–Jacobi–Bellman-type equations.

Some contributions to the ubiquitous and popular problem of option pricing under an HMM setting are contained in two papers included in this monograph. TK Siu's article 'A Hidden Markov-Modulated Jump Diffusion Model for European Option Pricing' is concerned with a two-step valuation procedure. It consists of using filtering methods and the Esscher transform to derive an integro-partial differential equation satisfied by the price of a European option. The paper 'An Exact Formula for Pricing American Exchange Options with Regime Switching' by L. Chan provides the valuation of exchange options of American style when the parameters of the underlying variable are governed by an HMM. A homotopy analysis-based method is utilised to calculate the value of an American option in this framework.

The last two papers highlight the estimation of model parameters under HMM frameworks with enhanced flexibility of modelling other features of observed market data. The emphasis is on their potential implementation in financial derivative valuation, risk management and asset allocation. X. Xi and R. Mamon's paper 'Parameter Estimation in a Weak Hidden Markov Model with Independent Drift and Volatility' gives the estimation procedure for the drift and volatility components in a model where they are independently influenced by two separate higher-order HMMs with different state spaces. Such a procedure enables the incorporation of memory in historical data in addition to occasional structural changes. Finally, another way to augment the capability of the HMM setting is to build an estimation scheme assuming that the signal model has a non-normal perturbation. This is discussed in the paper 'Parameter Estimation in a Regime-Switching Model with Non-normal Noise' by L. Jalen and R. Mamon. For this setting, the dynamic estimation of transition probabilities can still rely on recursively filtered estimates of quantities that are functions of the Markov chain. Concentrating on noise having a Student's t-distribution, it is demonstrated that the estimation of the other remaining parameters amounts to solving a system of nonlinear equations which can be readily accomplished by using modern computing software.

Acknowledgements Rogemar S. Mamon would like to express his appreciation for the financial support received from the Natural Sciences and Engineering Research Council (grant no. 341780-2012). Robert J. Elliott thanks the Australian Research Council and the Natural Sciences and Engineering Research Council for the support of his research. Particular thanks go to PhD candidate Ivy Zang (University of Western Ontario) for her technical help in producing the draft of this monograph consistent with Springer's style.

Both editors of this volume gratefully acknowledge the following dependable colleagues whose invaluable inputs were essential in the review process of each contribution:

Leunglung Chan (University of New South Wales, Australia)
Bujar Ghasi (University of Liverpool, UK)
Cody Hyndman (Concordia University, Montréal, Canada)
Kenneth Jackson (University of Toronto, Canada)
Michael Ludkovski (University of California at Santa Barbara, USA)
Nigel Meade (Imperial College, London, UK)
Marianito Rodrigo (University of Wollonggong, Australia)
Ken Siu (Cass Business School, City University, London, UK)
Anatoliy Swishchuk (University of Calgary, Canada)
Shu Wu (University of Kansas, USA)

Finally, we wish to thank Fred Hillier (Springer's international series in OR editor) and Matthew Amboy (associate editor of OR & MS Springer Science + Business Media) for their prompt and reliable assistance in making this project a success.

London, ON, Canada   Rogemar S. Mamon
Adelaide, Australia   Robert J. Elliott
November 2013

Contents

1 Robustification of an On-line EM Algorithm for Modelling Asset Prices Within an HMM 1
Christina Erlwein-Sayer and Peter Ruckdeschel
1.1 Introduction 1
1.2 Hidden Markov Model Framework for Asset Returns 3
1.3 Essential Steps in Elliott's Algorithm 4
    1.3.1 Change of Measure 4
    1.3.2 Filtering for General Adapted Processes............... 4
    1.3.3 Filter-Based EM Algorithm 7
    1.3.4 Summary of Algorithm 8
1.4 Outliers in Asset Allocation Problem 8
    1.4.1 Outliers in General 8
    1.4.2 Time-Dependent Context: Exogenous and Endogenous Outliers 9
    1.4.3 Evidence for Robustness Issue in Asset Allocation 10
1.5 Robust Statistics 12
    1.5.1 Concepts of Robust Statistics 12
    1.5.2 Our Robustification of the HMM: General Strategy 14
    1.5.3 Robustification of Step (0) 15
    1.5.4 Robustification of the E-Step 15
    1.5.5 Robustification of the (M)-Step 18
1.6 Implementation and Simulation 23
1.7 Conclusion.. 24
    1.7.1 Contribution of This Paper 24
    1.7.2 Outlook... 26
Appendix 27
References 29

2 Stochastic Volatility or Stochastic Central Tendency: Evidence from a Hidden Markov Model of the Short-Term Interest Rate 33
Craig A. Wilson and Robert J. Elliott
- 2.1 Introduction 33
- 2.2 The Model 36
- 2.3 Maximum Likelihood Estimation 37
- 2.4 The Likelihood Function 37
- 2.5 The Interest Rate Model 39
- 2.6 Data 40
- 2.7 Results 45
- 2.8 Conclusion 51
- References 52

3 An Econometric Model of the Term Structure of Interest Rates Under Regime-Switching Risk 55
Shu Wu and Yong Zeng
- 3.1 Introduction 55
- 3.2 The Model 58
  - 3.2.1 A Simple Representation of Markov Regime Shifts 58
  - 3.2.2 Other State Variables 59
  - 3.2.3 The Term Structure of Interest Rates 60
  - 3.2.4 Bond Risk Premiums Under Regime Shifts 61
  - 3.2.5 An Affine Regime-Switching Model 63
  - 3.2.6 The Effects of Regime Shifts on the Yield Curve 66
- 3.3 Empirical Results 68
  - 3.3.1 Data and Summary Statistics 68
  - 3.3.2 Estimation Procedure 70
  - 3.3.3 Discussions 74
- 3.4 Conclusion 80
- References 80

4 The LIBOR Market Model: A Markov-Switching Jump Diffusion Extension 85
Lea Steinrücke, Rudi Zagst, and Anatoliy Swishchuk
- 4.1 Introduction 85
- 4.2 Mathematical Preliminaries 87
- 4.3 The Log-Normal LIBOR Framework 90
  - 4.3.1 An Introduction to the LIBOR Market Model: The Log-Normal Dynamics 91
  - 4.3.2 Pricing of Caps and Floors in the Log-Normal LMM 93
- 4.4 The Markov-Switching Jump Diffusion (MSJD) Extension of the LMM 94
  - 4.4.1 Presenting the Extended Framework 95
  - 4.4.2 The Measure Changes and Its Consequences 97
- 4.5 Pricing in the MSJD Framework 100
  - 4.5.1 Determining the Characteristic Function of Y_{N-1} 101

		4.5.2	Determining the Characteristic Function of Y_j, $j = 1,\ldots,N-2$ 105

- 4.6 Calibration 105
  - 4.6.1 The Data 106
  - 4.6.2 Discussion of the Results of the Calibration 108
- 4.7 Conclusion 112
- References 114

5 Exchange Rates and Net Portfolio Flows: A Markov-Switching Approach 117
Faek Menla Ali, Fabio Spagnolo, and Nicola Spagnolo
- 5.1 Introduction 117
- 5.2 The Model 119
- 5.3 Data 120
- 5.4 Empirical Results 122
- 5.5 Conclusions 129
- References 130

6 Hedging Costs for Variable Annuities Under Regime-Switching 133
Parsiad Azimzadeh, Peter A. Forsyth, and Kenneth R. Vetzal
- 6.1 Introduction 134
- 6.2 Hedging Costs 136
  - 6.2.1 Derivation of the Pricing Equation 137
  - 6.2.2 Events 139
  - 6.2.3 Loss-Maximizing Strategies 141
  - 6.2.4 Regime-Switching 142
- 6.3 Optimal Consumption 143
  - 6.3.1 Utility PDE 144
  - 6.3.2 Events 144
  - 6.3.3 Consumption-Optimal Withdrawal 145
  - 6.3.4 Regime-Switching 146
  - 6.3.5 Hyperbolic Absolute Risk-Aversion 147
- 6.4 Numerical Method 147
  - 6.4.1 Homogeneity 147
  - 6.4.2 Localized Problem and Boundary Conditions 150
  - 6.4.3 Determining the Hedging Cost Fee 151
- 6.5 Results 151
  - 6.5.1 Loss-Maximizing and Contract Rate Withdrawal 151
  - 6.5.2 Consumption-Optimal Withdrawal 154
- 6.6 Conclusion 159
- Appendix 160
- References 165

7 A Stochastic Approximation Approach for Trend-Following Trading 167
Duy Nguyen, George Yin, and Qing Zhang
- 7.1 Introduction 167
- 7.2 Problem Formulation 168
- 7.3 Asymptotic Properties 171
- 7.4 Numerical Examples 176
- References 183

8 A Hidden Markov-Modulated Jump Diffusion Model for European Option Pricing 185
Tak Kuen Siu
- 8.1 Introduction 185
- 8.2 Hidden Regime-Switching Jump-Diffusion Market 188
- 8.3 Filtering Theory and Filtered Market 192
  - 8.3.1 The Separation Principle 192
  - 8.3.2 Filtering Equations 194
- 8.4 Generalized Esscher Transform in the Filtered Market ... 198
- 8.5 European-Style Option 203
- 8.6 Conclusion 207
- References 207

9 An Exact Formula for Pricing American Exchange Options with Regime Switching 211
Leunglung Chan
- 9.1 Introduction 211
- 9.2 Asset Price Dynamics 213
- 9.3 Problem Formulation 215
- 9.4 A Closed-Form Formula 219
- 9.5 Conclusion 224
- References 224

10 Parameter Estimation in a Weak Hidden Markov Model with Independent Drift and Volatility 227
Xiaojing Xi and Rogemar S. Mamon
- 10.1 Introduction 227
- 10.2 Modelling Background 230
- 10.3 Filters and Parameter Estimation 231
- 10.4 Numerical Implementation 235
- 10.5 Conclusion 239
- References 239

11 Parameter Estimation in a Regime-Switching Model with Non-normal Noise 241
Luka Jalen and Rogemar S. Mamon
- 11.1 Introduction 241

	11.2	Model Set Up 242
	11.3	Reference Probability Measure 243
	11.4	Recursive Estimation 244
	11.5	Parameter Estimation 246
		11.5.1 EM Algorithm and the Estimation of Transition Probabilities 247
		11.5.2 Student's t-Distributed Noise Term 248
	11.6	Numerical Application of the Filters 252
		11.6.1 Filtering Using Simulated Data 252
		11.6.2 Application of the Filters to Observed Market Data 258
	11.7	Conclusions 260
References 261		

List of Contributors

Parsiad Azimzadeh
Cheriton School of Computer Science, University of Waterloo, Waterloo, Canada

Leunglung Chan
School of Mathematics and Statistics, University of New South Wales, Sydney, Australia

Craig A. Wilson and Robert J. Elliott
Haskayne School of Business, Scurfield Hall, University of Calgary, Calgary, Canada
School of Mathematical Sciences, North Terrace Campus, University of Adelaide, Adelaide, Australia

Christina Erlwein-Sayer
Department of Financial Mathematics, Fraunhofer ITWM, Kaiserslautern, Germany

Parsiad Azimzadeh, Peter A. Forsyth, and Kenneth R. Vetzal
Cheriton School of Computer Science, University of Waterloo, Waterloo, Canada

Luka Jalen
MVG Equities, Nomura International, London, UK

Rogemar S. Mamon
Department of Statistical and Actuarial Sciences, University of Western Ontario, London, Canada

Faek Menla
Department of Economics and Finance, Brunel University, London, UK

Duy Nguyen
Department of Mathematics, University of Georgia, Athens, GA, USA

Peter Ruckdeschel
Department of Financial Mathematics, Fraunhofer ITWM, Kaiserslautern, Germany

Tak Kuen Siu
Cass Business School, City University London, London, UK

Faculty of Business and Economics, Department of Applied Finance and Actuarial Studies, Macquarie University, Sydney, Australia

Fabio Spagnolo
Department of Economics and Finance, Brunel University, London, UK

Nicola Spagnolo
Department of Economics and Finance, Brunel University, London, UK

Centre for Applied Macroeconomic Analysis (CAMA), Australian National University, Canberra, Australia

Lea Steinrücke
Berlin Doctoral Program in Economics and Management Science, Humboldt Universität Berlin, Berlin, Germany

Anatoliy Swishchuk
Department of Mathematics and Statistics, University of Calgary, Calgary, Canada

Parsiad Azimzadeh, Peter A. Forsyth, and Kenneth R. Vetzal
School of Accounting and Finance, University of Waterloo, Waterloo, Canada

Craig A. Wilson
Edwards School of Business, University of Saskatchewan, Sakatoon, Canada

Shu Wu
Department of Economics, University of Kansas, Lawrence, KS, USA

Xiaojing Xi
BMO Mortgage Central, Bank of Montreal, Ottawa, Canada

George Yin
Department of Mathematics, Wayne State University, Detroit, MI, USA

Rudi Zagst
Technische Universität München, Munich, Germany

Yong Zeng
Department of Mathematics and Statistics, University of Missouri at Kansas City, Kansas City, MO, USA

Qing Zhang
Department of Mathematics, University of Georgia, Athens, GA, USA

Biographical Notes

Editors

Rogemar S. Mamon is a professor of quantitative finance in the Department of Statistical and Actuarial Sciences at the University of Western Ontario, London, Ontario, Canada. Prior to joining Western, he held academic appointments at Brunel University, London, England; University of British Columbia; University of Waterloo; and University of Alberta. He spent short research or teaching visits at the University of Adelaide, Australia; Institute for Mathematics and its Applications, University of Minnesota, USA; CIMAT, México; Aarhus University, Denmark; Isaac Newton Institute for Mathematical Sciences, University of Cambridge, UK; Maxwell Institute for the Mathematical Institute, Edinburgh, Scotland; University of Wollongong, New South Wales, Australia; and University of Calabria, Italy. His publications have appeared in various peer-reviewed journals in statistics, applied mathematics, mathematical finance and economics. He is a coeditor of the *IMA Journal of Management Mathematics*, a fellow of the UK's Higher Education Academy, a fellow and chartered mathematician of the UK's Institute of Mathematics and its Applications and a chartered scientist of the Science Council in the UK.

Robert J. Elliott received his bachelor's and master's degrees from Oxford University, and his PhD and DSc from Cambridge University. He has held positions at Newcastle, Yale, Oxford, Warwick, Hull, Alberta, and visiting positions in Toronto, Northwestern, Kentucky, Brown, Paris, Denmark, Hong Kong and Australia. From 2001 to 2009, he was the RBC Financial Group Professor of Finance at the University of Calgary, Canada, where he was also an adjunct professor in both the Department of Mathematics and the Department of Electrical Engineering. Currently he is an Australian Professorial Fellow at the University of Adelaide. Professor Elliott has authored nine books and over 450 papers. His book with PE Kopp *Mathematics of Financial Markets* was published by Springer in 1999 and has been reprinted three times. The Hungarian edition was published in 2000, and the second edition was published in September 2004. An edition in China was published in 2010. Springer

Verlag published his book *Binomial Methods in Finance*, written with John van der Hoek, in the summer of 2005. He has also worked in signal processing, and his book with L. Aggoun and J. Moore on *Hidden Markov Models: Estimation and Control* was published in 1995 by Springer Verlag and reprinted in 1997. A revised and expanded edition was printed in 2008. His book with L. Aggoun *Measure Theory and Filtering* was published by Cambridge University Press in June 2004. His earlier book *Stochastic Calculus and Applications* was published by Springer in 1982 and a Russian translation appeared in 1986.

Contributors

Parsiad Azimzadeh is a PhD candidate at the University of Waterloo. His main area of interest is computational finance and numerical methods for partial differential equations. He holds an MMath and a BSc in computer science.

Leunglung Chan is a lecturer in statistics in the School of Mathematics and Statistics at the University of New South Wales. Prior to joining the University of New South Wales, he held postdoctoral fellowships at the University of Calgary, the University of Texas at Dallas and the University of Technology at Sydney. He has published research papers in academic journals, including the *Journal of Computational and Applied Mathematics, Applied Mathematics and Computation, The ANZIAM Journal, Annals of Finance, Journal of Stochastic Analysis and Applications, Applied Mathematical Finance, Asia-Pacific Financial Markets, International Journal of Theoretical and Applied Finance and Quantitative Finance*. He received his PhD from the University of Calgary, Canada.

Christina Erlwein-Sayer is a research associate in the Financial Mathematics group of the Fraunhofer Institute for Industrial Mathematics ITWM, Kaiserslautern, Germany. She received her PhD in financial mathematics on hidden Markov models in finance from Brunel University in London, UK, in 2008. She was awarded a Marie Curie Fellowship for Early Stage Researchers and worked within international research projects on financial mathematics. She published several papers on applications of HMMs in finance. Her coauthored article published in the *North American Actuarial Journal* was awarded the Best Paper Prize by the Society of Actuaries in 2010. Since 2008 she is affiliated to ITWM, where she works on various projects with the financial industry, ranging from modelling alternative investments to developing software concepts for statistical models. Her current research interests include portfolio optimisation in regime-switching models as well as news analytics.

Parsiad Azimzadeh, Peter A. Forsyth, and Kenneth R. Vetzal is a professor in the Cheriton School of Computer Science at the University of Waterloo. He is an associate editor of the *Journal of Computational Finance and Applied Mathematical*

Finance. From 2008 to 2013, he was the editor-in-chief of the *Journal of Computational Finance*. His current research is focused on numerical solution of optimal stochastic control problems in finance.

Luka Jalen is currently an equity derivatives quantitative analyst at Nomura International, London, and was previously in a quantitative role at Credit Suisse, London. He has a PhD from Brunel University in London, where his dissertation focused on filtering and applications of Markov models to financial and actuarial problems. Prior to that, he received his BSc and MSc degrees in applied mathematics from the University of Ljubljana, Slovenia. He coauthored research articles that were published in *Insurance: Mathematics and Economics, Applied Mathematics and Computation, Journal of Computational and Applied Mathematics, Operations Research Letters,* and *Mathematical and Computer Modelling*.

Faek Menla is currently a final-year PhD student at Brunel University, UK, and he will start as a lecturer in economics and finance at Brunel University, UK. His main areas of interest are international finance, open-economy macroeconomics, macro and financial econometrics, and financial economics. He holds an MSc in international money, finance and investment from Brunel University, UK. He has published an article in the *Journal of Defence and Peace Economics*.

Duy Nguyen is a recent PhD graduate in mathematics from the University of Georgia. His research interest is in mathematical finance. More specifically, he is interested in applying stochastic control theory in solving optimal asset trading problems.

Peter Ruckdeschel is a research associate in the financial mathematics group of the Fraunhofer Institute for Industrial Mathematics ITWM, Kaiserslautern, Germany. He received his PhD in statistics from the University Bayreuth in 2001 with an award-winning dissertation on Robust Kalman Filtering. Since then, he has published research on robust statistics and computational statistics and got his Habilitation in Statistics at Kaiserslautern University in 2011. He is a coauthor of 14 R packages available on CRAN and r-forge, in particular including R package robKalman. Since his affiliation to ITWM in 2008, he has been working in several industry projects covering various topics in fraud detection, parameter estimation in SDEs for pricing, computation and robustness aspects of operational risk, and assessing autocorrelation effects in determining a bank's risk-bearing ability.

Tak Kuen Siu is a professor of actuarial science in the Faculty of Actuarial Science and Insurance at the Cass Business School, City University London, UK. His research interests are mathematical finance, actuarial science, risk management, stochastic calculus, filtering and control, as well as their applications. He has published over 140 research papers. His research papers have been published in international peer-reviewed journals such as *SIAM Journal on Control and Optimization, Journal of Economic Dynamics and Control, IEEE Transactions on Automatic Control, Systems and Control Letters, Automatica, Insurance: Mathematics and*

Economics, ASTIN Bulletin, Scandinavian Actuarial Journal, North American Actuarial Journal, Quantitative Finance, Annals of Operations Research, Energy Economics, and Risk Magazine. He serves as a member of the editorial board in several international peer-reviewed journals such as *Stochastics* (formerly *Stochastics and Stochastics Reports*), *IMA Journal of Management Mathematics and Annals of Financial Economics.*

Fabio Spagnolo is a reader in finance at Brunel University, UK. He has published extensively in the area of econometrics and finance. His works appeared in *Economics Letters, Journal of Applied Econometrics, Journal of Econometrics, Journal of Time Series Analysis and Journal of Forecasting.* He was a coauthor of an article published in *Studies in Nonlinear Dynamics and Econometrics* that won the Arrow Prize for Senior Economists in 2006.

Nicola Spagnolo is a reader in economics and finance at Brunel University, UK. He is also research associate at the Centre for Applied Macroeconomics, Australia. His works appeared in *Economics Letters, Journal of Empirical Finance, Journal of Time Series Analysis and Studies in Nonlinear Dynamics and Econometrics.*

Lea Steinrücke studied mathematics at Technische Universität München (TUM), Germany. After obtaining her diploma in 2012, she joined the graduate school "Research Training Group 1659 – Interdependencies in the Regulation of Markets" and is currently pursuing her PhD in economics at Humboldt-Universität at Berlin, Germany. Her research focus is the regulation of financial markets and its effects on the real economy.

Anatoliy Swishchuk is a professor of mathematical finance at the University of Calgary, Canada. His research interests include modelling and pricing swaps, weather and energy derivatives, option pricing, regime-switching models, random evolutions and their applications, and stochastic models in finance with path-dependent history. He is the author of many research papers and 10 books. He received his PhD and DSc from the Institute of Mathematics, Kiev, Ukraine.

Parsiad Azimzadeh, Peter A. Forsyth, and Kenneth R. Vetzal is an associate professor with the School of Accounting and Finance at the University of Waterloo. He holds a PhD in finance and an MA in economics from the University of Toronto. His research interests are in the areas of derivative securities, risk management, the term structure of interest rates and asset pricing. He is an associate editor of the *Journal of Computational Finance.*

Craig A. Wilson and Robert J. Elliott is an associate professor of finance at the University of Saskatchewan, Canada. He researches in the areas of fixed income securities, asset pricing and financial investments, as well as issues dealing with corporate finance and governance. In particular, he combines methods from the theory of stochastic processes and econometrics to model and analyse financial problems.

He received his BSc in mathematics, his BComm in finance and his PhD in finance from the University of Alberta, Canada.

Shu Wu is an associate professor in Economics Department at the University of Kansas. His research interests are in the areas of macroeconomics, financial economics and applied time series econometrics. He has published in *Journal of Monetary Economics, Journal of Money, Credit and Banking, Macroeconomic Dynamics, and International Journal of Theoretical and Applied Finance*, among others. He has coedited the monograph *State Space Models: Applications to Economics and Finance*, a Springer book volume published in 2013. He has held visiting positions at the City University of Hong Kong and Federal Reserve Bank of Kansas City. He obtained his PhD in economics from Stanford University in 2000.

Xiaojing Xi is currently with the Bank of Montréal in Ottawa. She received her PhD in applied mathematics from the University of Western Ontario, Canada. Her research specialisations are on the financial applications of higher-order Markov models, recursive filtering techniques and dynamic model parameter estimation. She has published in *Economic Modelling, Computational Economics, and Journal of Mathematical Modelling and Algorithms*. Her research works also appeared in peer-reviewed monographs comprising the first volume of the Springer's Series in Statistics and Finance and also the first volume of World Scientific's Advances in Statistics, Probability and Actuarial Science. She obtained her MSc and BA in mathematics from Wilfrid Laurier University in Waterloo.

George Yin joined the Department of Mathematics, Wayne State University, after receiving her PhD in applied mathematics from Brown University in 1987. He became a professor in 1996. His research interests include stochastic systems, applied stochastic processes and applications. He severed on many technical committees. He is an associate editor of *SIAM Journal on Control and Optimization* and is on the editorial board of a number of other journals and book series; he was an associate editor of *Automatica* and *IEEE Transactions on Automatic Control*. He is a fellow of IEEE.

Rudi Zagst is professor of mathematical finance, director of the Center of Mathematics and head of the Chair of Mathematical Finance at Technische Universität München (TUM), Germany. He is also president of risklab GmbH, a German-based consulting company offering advanced asset management solutions. He is a consultant and a professional trainer to a number of leading institutions. His current research interests are in financial engineering, risk and asset management.

Yong Zeng is a professor in the Department of Mathematics and Statistics at University of Missouri at Kansas City. His main research interests include mathematical finance, financial econometrics, stochastic nonlinear filtering and Bayesian statistical analysis. Notably, he has developed the statistical analysis via filtering for financial ultra-high-frequency data. He has published in *Mathematical Finance, International Journal of Theoretical and Applied Finance, Applied Mathematical Finance, Applied Mathematics and Optimization, IEEE Transactions on Automatic Control*, and *Statistical Inference for Stochastic Processes*, among others. He has coedited the monograph *State Space Models: Applications to Economics and Finance*, a Springer book volume published in 2013. He has held visiting associate professorships at Princeton University and the University of Tennessee. He received his BS from Fudan University in 1990, MS from University of Georgia in 1994 and PhD from University of Wisconsin at Madison in 1999. All degrees were in statistics.

Qing Zhang is a professor of mathematics at University of Georgia. He received his PhD from Brown University in 1988. His recent research interests include stochastic control, filtering and applications in finance. He was an associate editor of *IEEE Transactions on Automatic Control* and *SIAM Journal on Control and Optimization*. He is currently an associate editor of *Automatica* and the corresponding editor of *SIAM Journal on Control and Optimization*.

Chapter 1
Robustification of an On-line EM Algorithm for Modelling Asset Prices Within an HMM

Christina Erlwein-Sayer and Peter Ruckdeschel

Abstract In this paper, we establish a robustification of Elliott's on-line EM algorithm for modelling asset prices within a hidden Markov model (HMM). In this HMM framework, parameters of the model are guided by a Markov chain in discrete time, parameters of the asset returns are therefore able to switch between different regimes. The parameters are estimated through an on-line algorithm, which utilizes incoming information from the market and leads to adaptive optimal estimates. We robustify this algorithm step by step against additive outliers appearing in the observed asset prices with the rationale to better handle possible peaks or missings in asset returns.

1.1 Introduction

Realistic modelling of financial time series from various markets (stocks, commodities, interest rates, etc.) is often achieved through hidden Markov or regime-switching models. One major advantage of regime-switching models is their flexibility to capture switching market conditions or switching behavioural aspects of market participants resulting in a switch in the volatility or mean value.

Regime-switching models were first applied to issues in financial markets by Hamilton [14], where he established a Markov switching AR-model to model the GNP of the U.S. His results show promising effects of including possible regime-switches into the characterization of a financial time series. Further regime-switching models for financial time series followed, e.g., switching ARCH or switching GARCH models (see for example Cai [2] and Gray [12]), amongst many other applications.

C. Erlwein-Sayer (✉) • P. Ruckdeschel
Department of Financial Mathematics, Fraunhofer ITWM, Fraunhofer-Platz 1,
D-67663 Kaiserslautern, Germany
e-mail: Christina.Erlwein@itwm.fraunhofer.de; Peter.Ruckdeschel@itwm.fraunhofer.de

Various algorithms and methods for statistical inference are applied within these model setups, including as famous ones as the Baum-Welch algorithm and Viterbi's algorithm for an estimation of the optimal state sequence. HMMs in finance, both in continuous and in discrete time often utilize a filtering technique which was developed by Elliott [3]. Adaptive filters are derived for processes of the Markov chain (jump process, occupation time process and auxiliary processes) which are in turn used for recursive optimal parameter estimates of the model parameters. This filter-based Expectation-Maximization (EM) algorithm leads to an on-line estimation of model parameters. Our model setup is based on Elliott's filtering framework.

This HMM can be applied to questions, which arise in asset allocation problems. An investor typically has to decide, how much of his wealth shall be invested into which asset or asset class and when to optimally restructure a portfolio. Asset allocation problems were examined in a regime-switching setting by Ang and Beckaert [1], where high volatility and high correlation regimes of asset returns were discovered. Guidolin and Timmermann [13] presented an asset allocation problem within a regime-switching model and found four different possible underlying market regimes. A paper by Sass and Haussmann [31] derives optimal trading strategies and filtering techniques in a continuous-time regime-switching model setup. Optimal portfolio choices were also discussed in Elliott and van der Hoek [5] and Elliott and Hinz [4] amongst others. Here, Markowitz's famous mean-variance approach (see Markowitz [23]) is transferred into an HMM and optimal weights are derived. A similar Markowitz based approach within an HMM was developed in Erlwein et al. [7], where optimal trading strategies for portfolio decisions with two asset classes are derived. Trading strategies are developed herein to find optimal portfolio decisions for an investment in either growth or value stocks. Elliott's filtering technique is utilized to predict asset returns.

However, most of the optimal parameter estimation techniques for HMMs in the literature only lead to reasonable results, when the market data set does not contain significant outliers. The handling of outliers is an important issue in many financial models, since market data might be unreliable at times or high peaks in asset returns, which might occur in the market from time to time shall be considered separately and shall not negatively influence the parameter estimation method. In general, higher returns in financial time series might belong to a separate regime within an HMM. This flexibility is already included in the model setup. However, single outliers, which are not typical for any of the regimes considered, shall be handled with care, a separate regime would not reflect the abnormal data point. In this paper, we will develop a robustification of Elliot's filter-based EM-algorithm. In Sect. 1.2 we will set the HMM framework, which is applied (either in a one- or multi-dimensional setting) to model asset or index returns. The general filtering technique is described in Sect. 1.3. The asset allocation problem which clarifies the effect outliers can have on the stability of the filters is developed in Sect. 1.4. Section 1.5 then states the derivation of a robustification for various steps in the filter equations. The robustification of a reference probability measure is derived as well as a robust version of the filter-based EM-algorithm. An application of the robust filters is shown in Sects. 1.6 and 1.7 finishes our work with some conclusions and possible future applications.

1.2 Hidden Markov Model Framework for Asset Returns

For our problem setting we first review a filtering approach for a hidden Markov model in discrete time which was developed in [3]. The logarithmic returns of an asset follow the dynamics of the observation process y_k, which can be interpreted as a discretized version of the Geometric Brownian motion, which is a standard process to model stock returns. The underlying hidden Markov chain $\mathbf{X}_k$ cannot be directly observed. The parameters of the observation process are governed by the Markov chain and are therefore able to switch between regimes over time.

We work under a probability space $(\Omega, \mathscr{F}, P)$ under which $\mathbf{X}_k$ is a homogeneous Markov chain with finite state space $I = \{1, \ldots, N\}$ in discrete time ($k = 0, 1, 2 \ldots$). Let the state space of $\mathbf{X}_k$ be associated with the canonical basis $\{\mathbf{e}_1, \mathbf{e}_2, \ldots, \mathbf{e}_N\} \in \mathbb{R}^N$ with $\mathbf{e}_i = (0, \ldots, 0, 1, 0, \ldots, 0)^\top \in \mathbb{R}^N$. The initial distribution of $\mathbf{X}_0$ is known and $\mathbf{\Pi} = (\pi_{ji})$ is the transition probability matrix with $\pi_{ji} = P(\mathbf{X}_{k+1} = e_j | \mathbf{X}_k = e_i)$. Let $\mathscr{F}_k^{\mathbf{X}_0} = \sigma\{\mathbf{X}_0, \ldots, \mathbf{X}_k\}$ be the σ-field generated by $\mathbf{X}_0, \ldots, \mathbf{X}_k$ and let $\mathscr{F}_k^{\mathbf{x}}$ be the complete filtration generated by $\mathscr{F}_k^{\mathbf{X}_0}$. Under the real world probability measure P, the Markov chain $\mathbf{x}$ has the dynamics

$$\mathbf{X}_{k+1} = \mathbf{\Pi}\mathbf{X}_k + \mathbf{v}_{k+1}, \qquad (1.1)$$

where $\mathbf{v}_{k+1} := \mathbf{X}_{k+1} - \mathbf{\Pi}\mathbf{X}_k$ is a martingale increment (see Lemma 2.3 in [3]).

The Markov chain $\mathbf{X}_k$ is "hidden" in the log returns y_{k+1} of the stock price S_k. Our observation process is given by

$$y_{k+1} = \ln \frac{S_{k+1}}{S_k} = f(\mathbf{X}_k) + \sigma(\mathbf{X}_k)w_{k+1}, \qquad (1.2)$$

where $\mathbf{X}_k$ has finite state space and w_k's constitute a sequence of i.i.d. random variables independent of $\mathbf{X}$. The real-valued process y can be re-written as

$$y_{k+1} = \langle \boldsymbol{f}, \mathbf{X}_k \rangle + \langle \boldsymbol{\sigma}, \mathbf{X}_k \rangle w_{k+1}. \qquad (1.3)$$

Note that $\mathbf{f} = (f_1, f_2, \ldots, f_N)^\top$ and $\boldsymbol{\sigma} = (\sigma_1, \sigma_2, \ldots, \sigma_N)^\top$ are vectors, furthermore $f(\mathbf{X}_k) = \langle \mathbf{f}, \mathbf{X}_k \rangle$ and $\sigma(\mathbf{X}_k) = \langle \boldsymbol{\sigma}, \mathbf{X}_k \rangle$, where $\langle \mathbf{b}, \mathbf{c} \rangle$ denotes the Euclidean scalar product in $\mathbb{R}^N$ of the vectors $\mathbf{b}$ and $\mathbf{c}$. We assume $\sigma_i \neq 0$. Let $\mathscr{F}_k^y$ be the filtration generated by the $\sigma(y_1, y_2, \ldots, y_k)$ and $\mathscr{F}_k = \mathscr{F}_k^{\mathbf{x}} \vee \mathscr{F}_k^y$ is the global filtration.

1.3 Essential Steps in Elliott's Algorithm

1.3.1 Change of Measure

A widely used concept in filtering applications, going back to Zakai [32] for stochastic filtering, is a change of probability measure technique. A measure change to a reference measure $\bar{P}$ is applied here, under which filters for the Markov chain and related processes are derived. Under $\bar{P}$, the underlying Markov chain still has the dynamics $\mathbf{X}_{k+1} = \Pi\mathbf{X}_k + \mathbf{v}_{k+1}$ but is independent of the observation process and the observations y_k are $\mathcal{N}(0,1)$ i.i.d. random variables.

Following the change of measure technique which was outlined in Elliott et al. [6], the adaptive filters for the Markov chain and related processes are derived under this "idealized" measure $\bar{P}$. Changing back to the real world is done by constructing P from $\bar{P}$ through the Radon-Nikodým derivative $\left.\frac{dP}{d\bar{P}}\right|_{\mathscr{F}_k} = \Lambda_k$. To construct Λ_k we define the process λ_l

$$\lambda_l := \frac{\phi\left[\sigma(\mathbf{X}_{l-1})^{-1}(y_l - f(\mathbf{X}_{l-1}))\right]}{\sigma(\mathbf{X}_{l-1})\phi(y_l)}, \qquad (1.4)$$

where $\phi(z)$ is the probability density function of a standard normal random variable Z and set $\Lambda_k := \prod_{l=1}^k \lambda_l$, $k \geq 1$, $\Lambda_0 = 1$. Under P the sequence of variables $w_1, w_2, \ldots$, is a sequence of i.i.d. standard normals, where we have $w_{k+1} = \sigma(\mathbf{X}_k)^{-1}(y_{k+1} - f(\mathbf{X}_k))$.

1.3.2 Filtering for General Adapted Processes

The general filtering techniques and the filter equations which were established in [3] for Markov chains observed in Gaussian noise are stated in this subsection. This filter-based EM-algorithm is adaptive, which enables fast calculations and filter updates. Our robustification partly keeps this adaptive structure of the algorithm, although the recursivity cannot be kept completely.

In general, filters for four types of processes related to the Markov chain, namely the state space process, the jump process, the occupation time process and auxiliary processes including terms of the observation process are derived. Information on these processes can be filtered out from our observation process and can in turn be used to find optimal parameter estimates.

To determine the expectation of any $\mathscr{F}$-adapted stochastic process H given the filtration $\mathscr{F}_k^y$, consider the reference probability measure $\bar{P}$ defined as $P(A) = \int_A \Lambda \, d\bar{P}$. From Bayes' theorem a filter for any adapted process H is given by $\mathrm{E}[H_k \mid \mathscr{F}_k^y] = \overline{\mathrm{E}}[H_k \Lambda_k \mid \mathscr{F}_k^y] \big/ \overline{\mathrm{E}}[\Lambda_k \mid \mathscr{F}_k^y]$. The characterization of the conditional

distribution of $\mathbf{X}_k$ under P given $\mathscr{F}_k^y$ can be written as $\hat{p}_k = \mathrm{E}\left[\mathbf{X}_k \mid \mathscr{F}_k^y\right]$
$= \frac{\overline{E}\left[\mathbf{X}_k \Lambda_k \mid \mathscr{F}_k^y\right]}{\overline{E}\left[\Lambda_k \mid \mathscr{F}_k^y\right]}$. We know that $\sum_{i=1}^N \langle \mathbf{X}_k, \mathbf{e}_i \rangle = 1$. With $\sum_{i=1}^N \langle \overline{E}\left[\mathbf{X}_k \Lambda_k \mid \mathscr{F}_k^y\right], \mathbf{e}_i \rangle =$
$\overline{E}\left[\Lambda_k \sum_{i=1}^N \langle \mathbf{X}_k, \mathbf{e}_i \rangle \mid \mathscr{F}_k^y\right] = \overline{E}\left[\Lambda_k \mid \mathscr{F}_k^y\right]$ we have

$$\hat{p}_k = \frac{\overline{E}\left[\mathbf{X}_k \Lambda_k \mid \mathscr{F}_k^y\right]}{\sum_{i=1}^N \langle \overline{E}\left[\mathbf{X}_k \Lambda_k \mid \mathscr{F}_k^y\right], \mathbf{e}_i \rangle}.$$

Now, we define $\eta_k(H_k) := \overline{\mathrm{E}}\left[H_k \Lambda_k \mid \mathscr{F}_k^y\right]$, so that $\mathrm{E}\left[H_k \mid \mathscr{F}_k^y\right] = \eta_k(H_k) / \eta_k(1)$. A recursive relationship between $\eta_k(H_k)$ and $\eta_{k-1}(H_{k-1})$ has to be derived, where $\eta_0(H_0) = E[H_0]$. However, a recursive formula for the term $\eta_{k-1}(H_{k-1} \mathbf{X}_{k-1})$ is found. To relate $\eta_k(H_k)$ and $\eta_k(H_k \mathbf{X}_k)$ we note that with $\langle \mathbf{1}, \mathbf{X}_k \rangle = 1$

$$\langle \mathbf{1}, \eta_k(H_k \mathbf{X}_k) \rangle = \eta_k(H_k \langle \mathbf{1}, \mathbf{X}_k \rangle) = \eta_k(H_k). \tag{1.5}$$

Therefore

$$\mathrm{E}\left[H_k \mid \mathscr{F}_k^y\right] = \frac{\langle \mathbf{1}, \eta_k(H_k \mathbf{X}_k) \rangle}{\langle \mathbf{1}, \eta_k(\mathbf{X}_k) \rangle}. \tag{1.6}$$

A general recursive filter for adapted processes was derived in [3]. Suppose H_l is a scalar $\mathscr{F} = \sigma((\mathbf{X}_t, Y_t)_t)$–adapted process, H_0 is $\mathscr{F}_0^{\mathbf{X}}$ measurable and $H_l = H_{l-1} + a_l + \langle b_l, \mathbf{v}_l \rangle + g_l f(y_l)$, where a, b and g are $\mathscr{F}$-predictable, f is a scalar-valued function and $\mathbf{v}_l = \mathbf{X}_l - \Pi \mathbf{X}_{l-1}$. A recursive relation for $\eta_k(H_k \mathbf{X}_k)$ is given by

$$\begin{aligned}\eta_k(H_k \mathbf{X}_k) = \sum_{i=1}^N \Gamma^i(y_k) \big[&\langle \mathbf{e}_i, \eta_{k-1}(H_{k-1} \mathbf{X}_{k-1}) \rangle \Pi \mathbf{e}_i \\ &+ \langle \mathbf{e}_i, \eta_{k-1}(a_k \mathbf{X}_{k-1}) \rangle \Pi \mathbf{e}_i \\ &+ (\mathrm{diag}(\Pi \mathbf{e}_i) - (\Pi \mathbf{e}_i)(\Pi \mathbf{e}_i)') \eta_{k-1}(b_k \langle \mathbf{e}_i, \mathbf{X}_{k-1} \rangle) \\ &+ \eta_{k-1}(g_k \langle \mathbf{e}_i, \mathbf{X}_{k-1} \rangle) f(y_k) \Pi \mathbf{e}_i \big]. \end{aligned} \tag{1.7}$$

Here, for any column vectors $\mathbf{z}$ and $\mathbf{y}$, $\mathbf{zy}'$ denotes the rank-one (if $\mathbf{z} \neq \mathbf{0}$ and $\mathbf{y} \neq \mathbf{0}$) matrix $\mathbf{zy}^\top$. The term $\Gamma^i(y_k)$ denotes the component-wise Radon-Nikodŷm derivative λ_k^i,

$$\Gamma^i(y_k) = \phi\big(\frac{y_k - f_i}{\sigma_i}\big) / \sigma_i \phi(y_k).$$

Now, filters for the state of the Markov chain as well as for three related processes: the jump process, the occupation time process and auxiliary processes of the Markov chain are derived. These processes can be characterized as special cases of the general process H_l.

The estimator for the state $\mathbf{X}_k$ is derived from $\eta_k(H_k \mathbf{X}_k)$ by setting $H_k = H_0 = 1$, $a_k = 0$, $b_k = 0$ and $g_k = 0$. This implies that

$$\eta_k(\mathbf{X}_k) = \sum_{i=1}^{N} \Gamma^i(y_k)\langle \mathbf{e}_i, \eta_{k-1}(\mathbf{X}_{k-1})\rangle \Pi \mathbf{e}_i \ . \tag{1.8}$$

The initialization for $k=1$ includes Π_0, which is the initial distribution of $\mathbf{X}_0$. Therefore we have $\eta_1(\mathbf{X}_1) = \sum_{i=1}^{N} \Gamma^i(y_1)\langle \mathbf{e}_i, \Pi_0\rangle \Pi \mathbf{e}_i$.

The first related process is the number of jumps of the Markov chain $\mathbf{X}_k$ from state $\mathbf{e}_r$ to state $\mathbf{e}_s$ in time k, $J_k^{sr} = \sum_{l=1}^{k}\langle \mathbf{X}_{l-1}, \mathbf{e}_r\rangle\langle \mathbf{X}_l, \mathbf{e}_s\rangle$.

Setting $H_k = J_k^{sr}$, $H_0 = 0$, $a_k = \langle \mathbf{X}_{k-1}, \mathbf{e}_r\rangle \pi_{sr}$, $b_k = \langle \mathbf{X}_{k-1}, \mathbf{e}_r\rangle \mathbf{e}_s'$ and $g_k = 0$ in Eq. (1.7) we get

$$\eta_k(J_k^{sr}\mathbf{X}_k) = \sum_{i=1}^{N} \Gamma^i(y_k)\langle \eta_{k-1}(J_{k-1}^{sr}\mathbf{X}_{k-1}), \mathbf{e}_i\rangle \Pi \mathbf{e}_i$$
$$+ \Gamma^r(y_k)\eta_{k-1}(\langle \mathbf{X}_{k-1}, \mathbf{e}_r\rangle)\pi_{sr}\mathbf{e}_s \ . \tag{1.9}$$

The second process O_k^r denotes the occupation time of the Markov process $\mathbf{X}$, which is the length of time $\mathbf{X}$ spent in state r up to time k. Here, $O_k^r = \sum_{l=1}^{k}\langle \mathbf{X}_{l-1}, \mathbf{e}_r\rangle = O_{k-1}^r + \langle \mathbf{X}_{k-1}, \mathbf{e}_r\rangle$. We set $H_k = O_k^r$, $H_0 = 0$, $a_k = \langle \mathbf{X}_{k-1}, \mathbf{e}_r\rangle$, $b_k = 0$ and $g_k = 0$ in Eq. (1.7) to obtain

$$\eta_k(O_k^r\mathbf{X}_k) = \sum_{i=1}^{N} \Gamma^i(y_k)\langle \eta_{k-1}(O_{k-1}^r\mathbf{X}_{k-1}), \mathbf{e}_i\rangle \Pi \mathbf{e}_i$$
$$+ \Gamma^r(y_k)\langle \eta_{k-1}(\mathbf{X}_{k-1}), \mathbf{e}_r\rangle \Pi \mathbf{e}_r \ . \tag{1.10}$$

Finally, consider the auxiliary process $T_k^r(g)$, which occur in the maximum likelihood estimation of model parameters. Specifically, $T_k^r(g) = \sum_{l=1}^{k}\langle \mathbf{X}_{l-1}, \mathbf{e}_r\rangle g(y_l)$, where g is a function of the form $g(y) = y$ or $g(y) = y^2$. We apply formula (1.7) and get

$$\eta_k(T_k^r(g)\mathbf{X}_k) = \sum_{i=1}^{N} \Gamma^i(y_k)\langle \eta_{k-1}(T_{k-1}^r(g(y_{k-1}))\mathbf{X}_{k-1}), \mathbf{e}_i\rangle \Pi \mathbf{e}_i$$
$$+ \Gamma^r(y_k)\langle \eta_{k-1}(\mathbf{X}_{k-1}), \mathbf{e}_r\rangle g(y_k)\Pi \mathbf{e}_r \ . \tag{1.11}$$

We choose to set the initial values of the processes $\eta_k(J_k^{sr}\mathbf{X}_k)$, $\eta_k(O_k^r\mathbf{X}_k)$, $\eta_k(T_k^r(g)\mathbf{X}_k)$ to zero. In the first step of the algorithm we therefore get:

$$\eta_1(J_1^{sr}\mathbf{X}_1) = \Gamma^r(y_1)\langle \eta_0(\mathbf{X}_0), \mathbf{e}_r\rangle \pi_{sr}\mathbf{e}_s$$
$$\eta_1(O_1^r\mathbf{X}_1) = \Gamma^r(y_1)\langle \eta_0(\mathbf{X}_0), \mathbf{e}_r\rangle \Pi \mathbf{e}_r$$
$$\eta_1(T_1^r(g)\mathbf{X}_1) = \Gamma^r(y_1)\langle \eta_0(\mathbf{X}_0), \mathbf{e}_r\rangle g(y_1)\Pi \mathbf{e}_r. \tag{1.12}$$

The recursive optimal estimates of J, O and T can be calculated using Eq. (1.5).

1.3.3 Filter-Based EM Algorithm

The derived adapted filters for processes of the Markov chain can now be utilized to derive optimal parameter estimates through a filter-based EM-algorithm. The set of parameters ρ, which determines the regime-switching model is

$$\rho = \{\pi_{ji}, 1 \leq i,j \leq N, f_i, \sigma_i, 1 \leq i \leq N\}. \tag{1.13}$$

Initial values for the EM algorithm are assumed to be given. Starting from these values updated parameter estimates are derived which maximize the conditional expectation of the log-likelihoods. The M-step of the algorithm deals with maximizing the following likelihoods:

- The likelihood in the global $\mathscr{F}$-model is given by

$$\log \Lambda_t(\sigma,f;(\mathbf{X}_s,y_s)_{s\leq t}) = -\frac{1}{2}\sum_{s=1}^{t}\left(\log\langle\sigma,\mathbf{X}_{s-1}\rangle + \frac{(y_s - \langle f,\mathbf{X}_{s-1}\rangle)^2}{\langle\sigma,\mathbf{X}_s\rangle}\right).$$

- In the $\mathscr{F}^y$-model, where the Markov chain is not observed, we obtain

$$L_t(\sigma,f;(y_s)_{s\leq t}) = \mathrm{E}[\log\Lambda_t(\sigma,f;(\mathbf{X}_s,y_s)_{s\leq t}) \mid \mathscr{F}_t^y] =$$
$$= -\frac{1}{2}\sum_{k=1}^{N}\left(\log\sigma_k\hat{O}_t^k + (\hat{T}_t^k(y^2) - 2\hat{T}_t^k(y)f_k + \hat{O}_t^k f^2)/\sigma_k^2\right). \tag{1.14}$$

The maximum likelihood estimates of the model parameters can be expressed through the adapted filters. Whenever new information is available on the market, the filters are updated and, respectively, updated parameter estimates can be obtained.

Theorem 1.1 (Optimal parameter estimates). *Write $\hat{H}_k = E[H_k|\mathscr{F}_k^y]$ for any $\mathscr{F}$-adapted process H. With $\hat{J}$, $\hat{O}$ and $\hat{T}$ denoting the best estimates for the processes J, O and T, respectively, the optimal parameter estimates $\hat{\pi}_{ji}, \hat{f}_i$ and $\hat{\sigma}_i$ are given by*

$$\hat{\pi}_{ji} = \frac{\hat{J}_k^{ji}}{\hat{O}_k^i} = \frac{\eta_k(J_k^{ji})}{\eta_k(O_k^i)} \tag{1.15}$$

$$\hat{f}_i = \frac{\hat{T}_k^{(i)}}{\hat{O}_k^{(i)}} = \frac{\eta(T^{(i)}(y))_k}{\eta(O^{(i)})_k} \tag{1.16}$$

$$\hat{\sigma}_i = \sqrt{\frac{\hat{T}_k^{(i)}(y^2) - 2\hat{f}_i\hat{T}_k^{(i)}(y) + \hat{f}_i^2\hat{O}_k^{(i)}}{\hat{O}_k^{(i)}}}. \tag{1.17}$$

Proof. The derivation of the optimal parameter estimates can be found in [6]. □

1.3.4 Summary of Algorithm

The filter-based EM-algorithm runs in batches of n data points (n typically equals a minimum of 10 up to a maximum of 50) over the given time series. The parameters are updated at the end of each batch. The batches are not overlapping. The algorithm comprises the following steps:

(0)   Find suitable starting values for Π and f, σ.
(RN) Determine the RN-derivative for the measure change to $\bar{P}$.
(E)   Recursively, compute filters $\hat{J}_t^{ji}$, $\hat{O}_t^i$, and $\hat{T}_t^i(g)$.
(M1) Obtain ML-estimators $\hat{f} = (\hat{f}_1, \ldots, \hat{f}_N)$ and $\hat{\sigma} = (\hat{\sigma}_1, \ldots, \hat{\sigma}_N)$.
(M2) Obtain ML-estimators $\hat{\Pi}$.
(Rec) Go to (RN) to compute the next batch.

1.4 Outliers in Asset Allocation Problem

1.4.1 Outliers in General

In the following sections we derive a robustification of the Algorithm 1.3.4 to stabilize it in the presence of outliers in the observation process. To this end let us discuss what makes an observation an outlier. First of all, outliers are exceptional events, occurring **rarely**, say with probability 5–10 %. Rather than captured by usual randomness, i.e., by some distributional model, they belong to what Knight [21] refers to *uncertainty*: They are **u**ncontrollable, of **u**nknown distribution, **u**npredictable, their distribution may change from observation to observation, so they are non-recurrent and do not form an additional state. They cannot be used to enhance predictive power, and, what makes their treatment difficult, they often cannot be told with certainty from ideal observations.

Still, the majority of the observations in a realistic sample should resemble an ideal (distributional) setting closely, otherwise the modeling would be questionable. Here we understand closeness as in a distributional sense, as captured, e.g., by goodness-of-fit distances like Kolmogorov, total variation or Hellinger distance. More precisely, ideally, this closeness should be compatible to the usual convergence mode of the Central Limit Theorem, i.e., with weak topology. In particular, closeness in moments is incompatible with this idea.

Topologically speaking, one would most naturally use balls around a certain element, i.e., the set of all distributions with a suitable distance no larger than some given radius $\varepsilon > 0$ to the distribution assumed in the ideal model.

Conceptually, the most tractable neighborhoods are given by the so-called *Gross Error Model*, defining a neighborhood $\mathscr{U}$ about a distribution F as the set of all distributions given by

$$\mathscr{U}_c(F, \varepsilon) = \{G \,|\, \exists H : G = (1-\varepsilon)F + \varepsilon H\}. \tag{1.18}$$

They can also be thought of as the set of all distributions of (*re*alistic) random variables X^{re} constructed as

$$X^{\mathrm{re}} = (1-U)X^{\mathrm{id}} + UX^{\mathrm{di}}, \tag{1.19}$$

where X^{id} is a random variable distributed according to the *id*eal distribution. U is an independent $\mathrm{Bin}(1,\varepsilon)$ switching variable. In most cases it lets you see X^{id} but in some cases replaces it by some contaminating or *di*storting variable X^{di} which has nothing to do with the original situation.

1.4.2 Time-Dependent Context: Exogenous and Endogenous Outliers

In our time dependent setup in addition to the i.i.d. situation, we have to distinguish whether the impact of an outlier is propagated to subsequent observations or not. Historically there is a common terminology due to Fox [9], who distinguishes *innovation outliers* (or IO's) and *additive outliers* (or AO's). Non-propagating AO's are added at random to single observations, while IO's denote gross errors affecting the innovations. For consistency with literature, we use the same terms, but use them in a wider sense. IO's stand for general endogenous outliers entering the state layer (or the Markov chain in the present context), hence with propagated distortion. As in our Markov chain, the state space is finite, IO's are much less threatening as they are in general.

Correspondingly, wide-sense AO's denote general exogenous outliers which do not propagate, hence also comprise substitutive outliers or SO's as defined in a simple generalization of (1.19) to the state space context in Eqs. (1.20)–(1.23).

$$Y^{\mathrm{re}} = (1-U)Y^{\mathrm{id}} + UY^{\mathrm{di}}, \qquad U \sim \mathrm{Bin}(1,r), \tag{1.20}$$

for U independent of $(X, Y^{\mathrm{id}}, Y^{\mathrm{di}})$ and some arbitrary distorting random variable Y^{di} for which we assume

$$Y^{\mathrm{di}}, X \quad \text{independent} \tag{1.21}$$

and the law of which is arbitrary, unknown and uncontrollable. As a first step consider the set $\partial \mathscr{U}^{\mathrm{SO}}(r)$ defined as

$$\partial \mathscr{U}^{\mathrm{SO}}(r) = \left\{ \mathscr{L}(X, Y^{\mathrm{re}}) \,|\, Y^{\mathrm{re}} \text{ acc. to (1.20) and (1.21)} \right\}, \tag{1.22}$$

where $\mathscr{L}(Z)$ denotes the law of a random variable Z. Because of condition (1.21), in the sequel we refer to the random variables Y^{re} and Y^{di} instead of their respective (marginal) distributions only. In the common gross error model, reference to the respective distributions would suffice. Condition (1.21) also entails that in general, contrary to the gross error model, $\mathscr{L}(X, Y^{\mathrm{id}})$ is not element of $\partial \mathscr{U}^{\mathrm{SO}}(r)$.

As corresponding (convex) neighborhood we define

$$\mathscr{U}^{SO}(r) = \bigcup_{0 \leq s \leq r} \partial \mathscr{U}^{SO}(s). \tag{1.23}$$

Of course, $\mathscr{U}^{SO}(r)$ contains $\mathscr{L}(X, Y^{id})$. In the sequel where clear from the context we drop the superscript SO and the argument r.

Due to their different nature, as a rule, IO's and AO's require different policies: As AO's are exogenous, we would like to damp their effect, while when there are IO's, something has happened in the system, so the usual goal will be to track these changes as fast as possible.

1.4.3 Evidence for Robustness Issue in Asset Allocation

In this section we examine the robustness of the filter and parameter estimation technique. The filter technique is implemented and applied to monthly returns of MSCI index between 1994 and 2009. The MSCI World Index is one of the leading indices on the stock markets and a common benchmark for global stocks. The algorithm is implemented with batches of ten data points, therefore the adaptive filters are updated whenever ten new data points are available on the market. The recursive parameter estimates, which utilize this new information, are updated as well, the algorithm is self-tuning. Care has to be taken when choosing the initial values for the algorithm, since the EM-algorithm in its general form converges to a local maximum. In this implementation we choose the initial values with regard to mean and variance of the first ten data points. Figure 1.1 shows the original return series, optimal parameter estimates for the index returns as well as the one-step ahead forecast of the index.

To highlight the sensitivity of the filter technique towards exogenous outliers we plant unusual high returns within the time series. Considerable SO outliers are included at time steps $t = 40, 80, 130, 140$. The optimal parameter estimation through the filter-based EM-algorithm of this data set with outliers can be seen in Fig. 1.2. The filter still finds optimal parameter estimates, although the estimates are visibly affected by the outliers.

In a third step, severe outliers are planted into the observation sequence. Data points $t = 40, 80, 130, 140$ now show severe SO outliers as can be seen from the first panel in Fig. 1.3. The filters cannot run through any longer, optimal parameter estimates cannot be established in a setting with severe outliers.

In practice, asset or index return time series can certainly include outliers from time to time. This might be due to wrong prices in the system, but also due to very unlikely market turbulence for a short period of time. It has to be noted, that the type of outliers which we consider in this study does not characterize an additional state of the Markov chain. In the following, we develop robust filter equations, which can handle exogenous outliers.

1 Robustification of an On-line EM Algorithm for Modelling Asset Prices 11

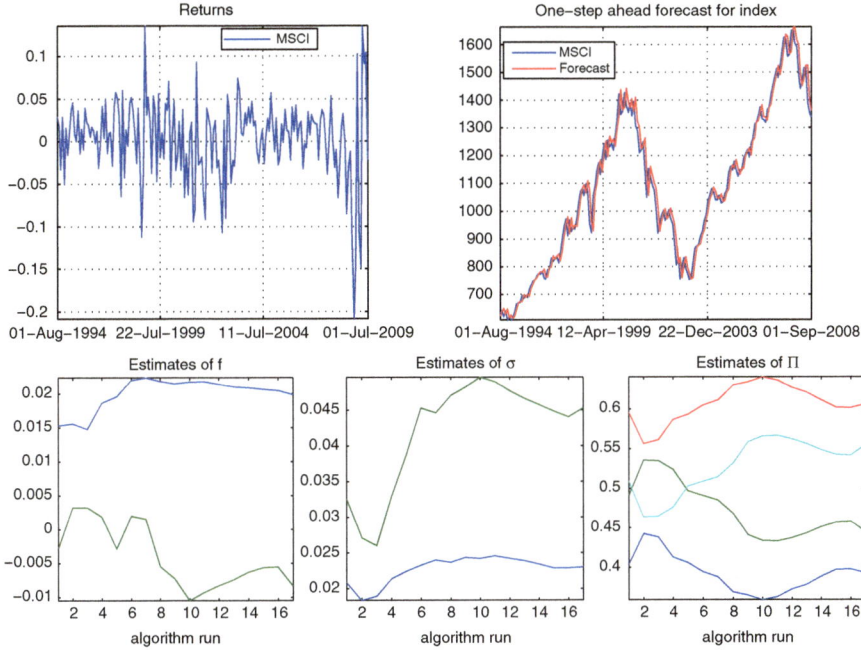

Fig. 1.1 Optimal parameter estimates for monthly MSCI returns between 1994 and 2009

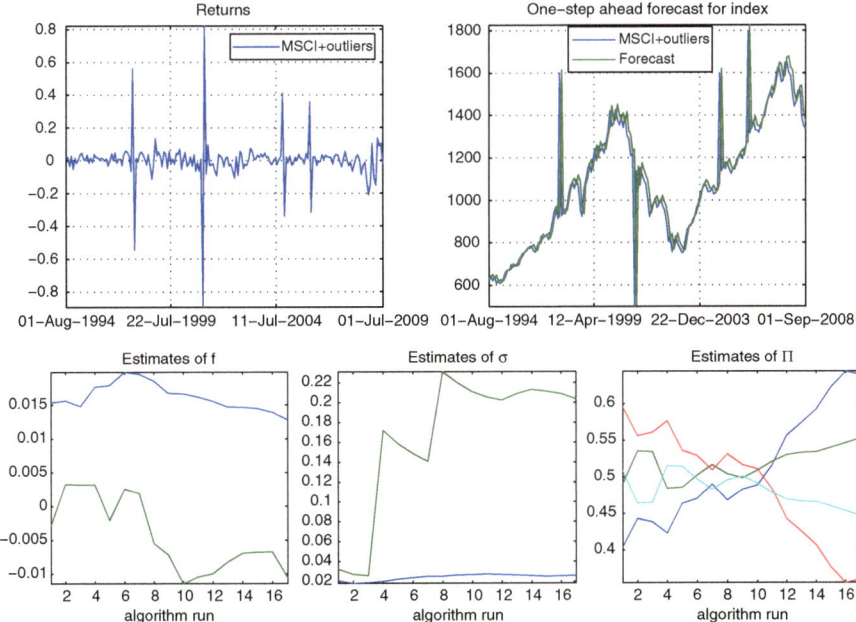

Fig. 1.2 Optimal parameter estimates for monthly MSCI returns with planted outliers

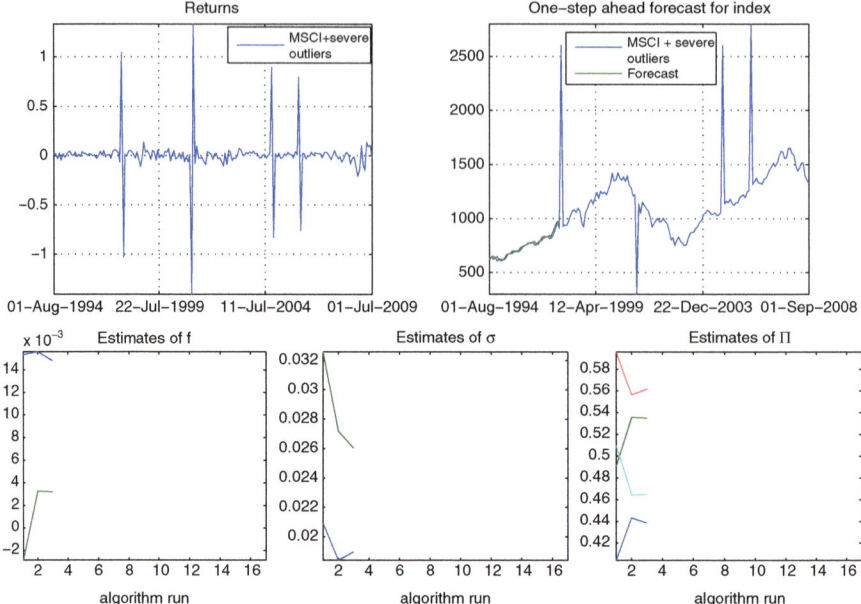

Fig. 1.3 Filter-based EM algorithm for observation sequence with severe outliers; from obs. 40 onwards (corresponding to algorithm run 3 onwards), the algorithm breaks down and no longer returns any value

1.5 Robust Statistics

To overcome effects like in Fig. 1.3 we need more stable variants of the Elliott type filters discussed so far. This is what robust statistics is concerned with. Excellent monographs on this topic are e.g., Huber [17], Hampel et al. [16], Rieder [25], Maronna et al. [24]. This section provides necessary concepts and results from robust statistics needed to obtain the optimally-robust estimators used in this article.

1.5.1 Concepts of Robust Statistics

The central mathematical concepts of continuity, differentiability, or closeness to singularities may in fact serve to operationalize stability quite well already. To make these available in our context, it helps to consider a statistical procedure, for example an estimator, a predictor, a filter, or a test as a function of the underlying distribution. In a parametric context, this amounts to considering functionals T mapping distributions to the parameter set Θ. An estimator will then simply be T applied to the empirical distribution $\hat{F}_n$. For filtering or prediction, the range of such a functional will rather be the state space but otherwise the arguments run in parallel.

To formulate continuity in the case of outliers, we use corresponding neighborhoods which essentially capture convergence in distribution. With these neighborhoods, we now may easily translate the notions of continuity, differentiability and closest singularity to this context, compare [16, Sects. 2.1, 2.2]. In the sequel, we focus on the robustness concepts of *influence function* and *breakdown point*.

The influence function (IF) is the derivative[1] of such a functional and reflects the infinitesimal influence of a single observation on the estimator. If it is bounded, the functional is called *locally robust*. Under additional assumptions, many of the asymptotic properties of an estimator are expressions in the IF ψ. For instance, the asymptotic variance of the estimator in the ideal model is the second moment of ψ. Infinitesimally, i.e., for contamination rate $\varepsilon \to 0$, the maximal bias on a neighborhood $\mathscr{U}$ as in Eq. (1.18) is just $\sup|\psi|$, where $|\cdot|$ denotes Euclidean norm. $\sup|\psi|$ is then also called *gross error sensitivity* (GES). Hence, seeking robust optimality amounts to finding optimal IFs.

The *breakdown point* of an estimator is a global robustness concept: it is the maximal contamination rate ε the estimator can cope with without producing an arbitrary large bias, see [16, Sect. 2.2 Definitions 1,2] for formal definitions.

Usually, the classically optimal estimators (MLE in many circumstances) are non-robust, both locally and globally. Robust estimators on the other hand pay a certain price for this stability as expressed by an *asymptotic relative efficiency* (ARE) strictly lower than 1 in the ideal model, where ARE is the ratio of the two asymptotic (co)variances of the classically optimal estimator and its robust alternative.

To rank various robust procedures among themselves, other quality criteria are needed, though, summarizing the behavior of the procedure on a whole neighborhood $\mathscr{U}$ around the ideal model, as in (1.18). A natural candidate for such a criterion is maximal MSE (maxMSE) on $\mathscr{U}$. Other popular choices consider the maximal bias (maxBias) on $\mathscr{U}$, or (the trace of) the variance in the ideal model subject to a bias bound on $\mathscr{U}$. This is the choice proposed in the famous Lemma 5 of Hampel [15]. The respective optimal estimators for these criteria are called OMSE (Optimal MSE estimator), MBRE (Most Bias Robust Estimator), and OBRE (Optimally Bias Robust Estimator).[2] In the (M) step we will use the MBRE.

In our context we encounter two different situations where we want to apply robust ideas: recursive filtering in the (E)-step and estimation in the (M)-step. In the (E)-step we only add a single new observation, which precludes asymptotic arguments. In the (M)-step, however, the preceding application of Girsanov's theorem turns our situation into an i.i.d. setup, where each observation becomes (uniformly) asymptotically negligible and asymptotics apply in a standard form.

---

[1] In mathematical rigor, the IF, when it exists, is the Gâteaux derivative of T into the direction of the tangent $\delta_x - F$. For details, also on stronger notions like Hadamard or Fréchet differentiability, see, e.g., Fernholz [8] or [25, Chap. 1].

[2] Names OBRE and MBRE are used, e.g., in Hampel et al. [16] while OMSE is used in Ruckdeschel and Horbenko [30].

1.5.2 Our Robustification of the HMM: General Strategy

As a robustification of the whole estimation process in this only partially observed model would lead to computationally intractable terms and would drop the key feature of recursivity. Therefore, we instead propose to robustify each of the steps in Elliott's algorithm separately. Doing so, the whole procedure will be robust, but will loose robust optimality in general, since the Bellmann principle does not hold for optimal robust multi-step procedures. Table 1.1 lists all essential steps in Elliot's algorithm related to our proposed robustification approach.

Table 1.1 Classical algorithm setting and robustified version of each step

Classical setting	Robust version

Initialization: find suitable starting values for Π, f, σ and $\mathbf{X}_0$

Build N clusters on first batches	Build $N+1$ clusters on first batches, distribute points in outlier cluster randomly on other clusters
Use first and second moment of each cluster as initial values for f and σ	Use median and MAD of clusters for f and σ
Choose Π and $\mathbf{X}_0$ according to cluster probabilities	Choose Π and $\mathbf{X}_0$ according to cluster probabilities

E-step: determine RN-derivative and calculate E-Step

Find RN-derivative Λ	Robustified version of Λ through suitable clipped version of λ_k
Estimate recursive filters J_k^{ji}, O_k^i and $T_k^i(g)$	No further robustification needed; take over J_k^{ji}, O_k^i unchanged and skip $T_k^i(g)$

M-step 1: obtain estimates for f and σ

MLE-estimates for f and σ through recursive filters O_k^i and $T_k^i(g)$	Likelihoods re-stated; they are expressed as weighted sums of the observations y_k
Recursive filters are substituted into likelihood	Robustified version of MLE through asymptotic linear estimators
Estimates updated after each batch	Estimates updated after each batch, recursivity cannot be preserved completely

M-step 2: obtain ML-estimators Π

MLE-estimation, recursive filters J_k^{ji} and O_k^i are substituted into likelihood	Robustification through robust version of filters J_k^{ji} and O_k^i, no further observation y_k has to be considered

Rec: algorithm runs on next batch

Go to (RN) to compute the next batch	Go to (RN) to compute the next batch

1.5.3 Robustification of Step (0)

So far, little has been said about the initialization of the algorithm. Basically, all we have to do is to make sure that the EM algorithm converges. In prior applications of this algorithms (Mamon et al. [22] and Erlwein et al. [7] amongst others), one approach was to fill Π with entries $1/N$, that is with uniform (and hence non-informative) distribution over all states, independent from state X_0. As to f_i and σ_i, an ad hoc approach would estimate the global mean and variance over all states and then, again in a non-informative way jitter the state-individual moments, adding independent noise to it. In our preliminary experiments, it turned out that this naïve approach could drastically fail in the presence of outliers. Therefore we instead propose a more sophisticated approach which can also be applied in a classical non-robust setting: In a first step we ignore the time dynamics and interpret our observations as realizations of a Gaussian Mixture Model, for which we use R package mclust (Fraley and Raftery [10] and Raftery et al. [11]) to identify the mixture components. For each of these, we individually determine the moments f_i and σ_i. As to Π, we again assume independence of X_0, but fill the columns according to the estimated frequencies of the mixture components. In case of the non-robust setting we would use N mixture components and for each of them determine f_i and σ_i by their ML estimators assuming independent observations. For a robust approach, we use $N+1$ mixture components. One of them – the one with the lowest frequency – is a pure noise component capturing outliers. For each non-noise component we retain the ML estimates for f_i and σ_i. The noise component is then randomly distributed amongst the remaining components, respecting their relative frequencies prior to this redistribution.

We are aware of the fact that reassigning the putative outliers at random could be misleading in ideal situations without outliers. Since we assume a given number of states, one of the estimated cluster could needlessly be split into two. The smaller offspring of the cluster would in part be reassigned to wrong other clusters. On the other hand, our choice of building initial clusters works reasonably well in most cases, more sophisticated strategies are questions for model selection and are deferred to further work.

Based on the f_i and σ_i, for each observation j and each state i, we get weights $0 \leq w_{i,j} \leq 1$, $\sum_i w_{i,j} = 1$ for each j, representing the likelihood that observation j is in state i. For each i, again we determine robustified moment estimators f'_i, σ'_i as weighted medians and scaled weighted MADs (medians of absolute deviations). For a suitable definition of weighted medians and scaled weigthed MADs see the Appendix.

1.5.4 Robustification of the E-Step

As indicated, in this step we cannot recur to asymptotics, but rather have to appeal to a theory suited for this recursive setting. In particular, the SO-neighborhoods introduced in (1.20) turn out to be helpful here.

1.5.4.1 Crucial Optimality Theorem

Consider the following optimization problem of reconstructing the ideal observation Y^{id} by means of the realistic/possibly contaminated Y^{re} on an SO-neighborhood.

Minimax-SO problem

Minimize the maximal MSE on an SO-neighborhood, i.e., find a Y^{re}-measurable reconstruction f_0 for Y^{id} which minimizes

$$\max_{\mathcal{U}} \mathrm{E}_{re} |Y^{id} - f(Y^{re})|^2. \quad (1.24)$$

The solution is given by

Theorem 1.2 (Minimax-SO). *In this situation, there is a saddle-point $(f_0, P_0^{Y^{di}})$ for Problem* (1.24)

$$f_0(y) := EY^{id} + H_\rho(D(y)), \qquad H_b(z) = z \min\{1, b/|z|\}, \quad (1.25)$$

$$P_0^{Y^{di}}(dy) := \frac{1-r}{r}(|D(y)|/\rho - 1)_+ P^{Y^{id}}(dy), \quad (1.26)$$

where $\rho > 0$ ensures that $\int P_0^{Y^{di}}(dy) = 1$ and

$$D(y) = y - EY^{id}. \quad (1.27)$$

The value of the minimax risk of Problem (1.24) *is*

$$\mathrm{tr}\,\mathrm{Cov}(Y^{id}) - (1-r)E_{id}\big[\min\{|D(Y^{id})|, \rho\}^2\big]. \quad (1.28)$$

Proof. See the Appendix. □

The optimal procedure in Eq. (1.25) has an appealing interpretation. It is a compromise between the (unobservable) situations where (a) one observes the ideal Y^{id} and (b) one observes Y^{di}, something completely unrelated to Y^{id}. In the first case one would use it unchanged, whereas in case (b) one would use the best prediction for Y^{id} in MSE-sense without any information, hence the unconditional expectation EY^{id}. The decision on how much to tend to case (a) and how much to case (b) is taken according to the length of the discrepancy $D(Y^{re})$ between observed signal Y^{re} and EY^{id}. If this length is smaller than ρ, we keep Y^{re} unchanged, otherwise we modify EY^{id} by adding a clipped version of $D(Y^{re})$.

1.5.4.2 Robustification of Steps (RN), (E)

In the change of measure step we recall that the corresponding likelihood ratio here is just

1 Robustification of an On-line EM Algorithm for Modelling Asset Prices

$$\lambda_s := \frac{\varphi\Big((y_s - f(\mathbf{X}_{s-1}))/\sigma(\mathbf{X}_{s-1})\Big)/\sigma(\mathbf{X}_{s-1})}{\varphi(y_s)}. \tag{1.29}$$

Apparently λ_s can take both, values close to 0 and very large values, in particular for small values of $\sigma(\mathbf{X}_{s-1})$. So bounding the λ_s is crucial to avoid effects like in Fig. 1.3.

A first (non-robust) remedy uses a data driven reference measure. Instead of $\mathcal{N}(0,1)$, we use $\mathcal{N}(0,\bar{\sigma}^2)$ where $\bar{\sigma}$ is a global scale measure taken over all observations, ignoring time dependence and state-varying σ's. A robust proposal would take $\bar{\sigma}$ to be the MAD of all observations (tuned for consistency at the normal distribution). This leads to

$$\tilde{\lambda}_s := \frac{\varphi\Big((y_s - f(\mathbf{X}_{s-1}))/\sigma(\mathbf{X}_{s-1})\Big)/\sigma(\mathbf{X}_{s-1})}{\varphi(y_s/\bar{\sigma})/\bar{\sigma}}. \tag{1.30}$$

Eventually, in both estimation and filtering/prediction, $\bar{\sigma}$ cancels out as a common factor in numerator and denominator, so it is irrelevant in the subsequent steps. Its mere purpose is to stabilize the terms in numeric aspects.

To take into account time dynamics in our robustification, we want to use Theorem 1.2. To this end, we need second moments, which for λ_s need not exist. So instead, we apply the theorem to $Y^{\text{id}} = \sqrt{\tilde{\lambda}_s}$, which means that $\tilde{\lambda}_s = (Y^{\text{id}})^2$ is robustified by

$$\bar{\lambda}_s = \Big(\mathrm{E}_{\text{id}}\sqrt{\tilde{\lambda}_s} + H_b(\sqrt{\tilde{\lambda}_s} - \mathrm{E}_{\text{id}}\sqrt{\tilde{\lambda}_s})\Big)^2 \tag{1.31}$$

for $H_b(x) = x\min\{1,b/|x|\}$. Clipping height b is chosen such that $\mathrm{E}\bar{\lambda}_s = \alpha$, $\alpha = 0.95$ for instance. As in the ideal situation $\mathrm{E}\lambda_s = 1$. In a last step with a consistency factor c'_s determined similarly to c_i in the initialization step for the weighted MADs, we pass over to $\bar{\lambda}^0_s = c_s\bar{\lambda}_s$ such that $\mathrm{E}\bar{\lambda}^0_s = 1$.

Similarly, in the remaining parts of the E-step, for each of the filtered processes generically denoted by G and the filtered one by $\hat{G}$, we could replace $\hat{G}$ by

$$\bar{G} = \mathrm{E}_{\text{id}}\hat{G} + H_b(\hat{G} - \mathrm{E}_{\text{id}}\hat{G}) \tag{1.32}$$

for G any of J_k^{ji}, O_k^i, and $T_k^i(f)$ and again suitably chosen b.

It turns out though, that it is preferable to pursue another route. The aggregates $T_k^i(f)$ are used in the M-step in (1.14). But for a robustification of this step, it is crucial to be able to attribute individual influence to each of the observations. So, instead of using the aggregates processes $T_k^i(f)$, we split up the terms of the filtered neg-loglikelihood into summands

$$w_{i,j}/\hat{O}_k^i[(y_j - f_i)^2/\sigma_i^2 + \log \sigma_i]$$

for $j = 1,\ldots,k$. Hence, we may skip a robustification of $T_k^i(f)$. Similarly, as J_t^{ji}, O_t^i are filtered observations of multinomial-like variables, and any contribution of a single observation to these variables can at most be of absolute value 1. A robustification is of limited use, since the processes are bounded anyway. Hence in the E-step, we only robustify λ_s. By splitting up the aggregates $T_k^i(f)$ into summands we lose strict recursivity. For k observations in one batch, we now have to store the values $w_{i,j}/\hat{O}_k^i$ for $j = 1,\ldots,k$. Building up from $j = 1$, at observation time $j = j_0$ within the batch, we construct $w_{i,j;j_0}$, $j = 1,\ldots,j_0$ from the values $w_{i,j;j_0-1}$, $j = 1,\ldots,j_0-1$, so we have a growing triangle of weight values. This would lead to increasing memory requirements, if we had not chosen to work in batches of fixed length k, which puts an upper bound onto memory needs.

1.5.5 Robustification of the (M)-Step

As mentioned before, contrast to the (E)-step, in this estimation step, we may work with classical gross error neighborhoods (1.18) and with the standard independent setting.

1.5.5.1 Shrinking Neighborhood Approach

By the Bienaymé formula, variance usually is $\mathcal{O}(1/n)$ for sample size n, while for robust estimators, the maximal bias is proportional to the neighborhood radius ε. Hence unless ε is appropriately scaled in n, bias will dominate eventually for growing n. This is avoided in the shrinking neighborhood approach by setting $\varepsilon = \varepsilon_n = r/\sqrt{n}$ for some initial radius $r \in [0,\infty)$, compare Rieder [25], Kohl et al. [20]. We use this approach for the (M) step.

One could see this shrinking as indicating that with growing n, diligence is increasing so the rate of outliers is decreasing. This is perhaps overly optimistic; instead, we prefer the interpretation that the severeness of the robustness problem with 10% outliers at sample size 100 should not be compared with the one with 10% outliers at sample size 10,000 but rather with the one with 1% outliers at this sample size.

In this setting, with mathematical rigor, optimization of the robust criteria from Sect. 1.5.1 can be deferred to the respective IFs, we construct an estimator to a given (optimally-robust) IF. More specifically, in the (M)-step, we are heading for the MBRE, so we will determine the IF of the MBRE in our context. We will construct a respective estimator to this IF with a one-step construction.

This is achieved by the concept of *asymptotically linear estimators* (ALEs), see [25, Sect. 4.2.3], as it arises canonically in most proofs of asymptotic normality.

Definition 1.1 (ALE). Let $\mathscr{P} = \{P_\theta, \ \theta \in \Theta\}$ be a smooth (L_2-differentiable) parametric model with parameter domain $\Theta \subset \mathbb{R}^d$ for i.i.d observations $X_i \sim P_\theta$. Then based on the scores[3] Λ_θ and its finite Fisher information $\mathscr{I}_\theta = \mathrm{E}_\theta \Lambda_\theta \Lambda_\theta^\tau$, we define the set $\Psi_2(\theta)$ of influence functions as the subset of the square integrable functions ψ_θ with d coordinates $L_2^d(P_\theta)$ for which $\mathrm{E}_\theta \psi_\theta = 0$ and $\mathrm{E}_\theta \psi_\theta \Lambda_\theta^\tau = \mathbb{I}_d$ where $\mathbb{I}_d$ is the d-dimensional unit matrix. Then a sequence of estimators $S_n = S_n(X_1, \ldots, X_n)$ is called an ALE if there is some influence function $\psi_\theta \in \Psi_2(\theta)$ s.t.

$$S_n = \theta + \frac{1}{n}\sum_{i=1}^{n} \psi_\theta(X_i) + \mathrm{o}_{P_\theta^n}(n^{-1/2}). \tag{1.33}$$

Usually the MLE is an ALE with influence function $\psi_\theta^{\mathrm{MLE}} = \mathscr{I}_\theta^{-1} \Lambda_\theta$. Most other common estimators also have a representation (1.33) with a different ψ.

For given IF ψ we get an ALE $\hat{\theta}_n$ by a one-step construction, often called one-step-reweighting. With given suitably consistent starting estimator θ_n^0 (such that $R_n^0 = \theta_n^0 - \theta = \mathrm{o}_{P_\theta^n}(n^{-1/4+0})$) we define

$$\hat{\theta}_n = \theta_n^0 + \frac{1}{n}\sum_{j=1}^{n} \psi_{\theta_n^0}(X_j). \tag{1.34}$$

Then indeed $\hat{\theta}_n = \theta + \frac{1}{n}\sum_{j=1}^n \psi_\theta(X_j) + R_n$ and $R_n = \mathrm{o}_{P_\theta^n}(n^{-1/2})$. So, $\hat{\theta}_n$ forgets about θ_n^0 as to its asymptotic variance and GES. However its breakdown point is inherited from θ_n^0 once Θ is unbounded and ψ is bounded. Hence for the starting estimator, we seek for θ_n^0 with high breakdown point. For a more detailed account on this approach, see [25].

In the sequel we fix the true $\theta \in \Theta$ and suppress it from notation where clear from context.

1.5.5.2 Shrinking Neighborhood Approach with Weighted Observations

Coming from the E-step, observations y_j are not equally likely to contribute to state i. Hence, we are in a situation with weighted observations, where we may pass over to standardized weights $w_{i,j}^0 = w_{i,j}/\sum_{j'} w_{i,j'}$ summing up to 1.

In the (M1)-step, we now want to set up an ALE with the IF of the MBRE in our context. To this end, we suppress state index i from notation here. With the weights w_j^0 given above, we use the vector of weighted median and scaled weighted MAD for state i as starting estimator θ_n^0. The IF ψ of the MBRE in the one-dimensional Gaussian location and scale model at $\mathcal{N}(0,1)$ is given by

$$\psi(y) = bY(y)/|Y(y)|, \qquad Y(y) = (y, A(y^2-1)-a), \tag{1.35}$$

with numerical values for A, a, b up to four digits taken from R package RobLox, Kohl [18]. The values are

---

[3] Usually Λ_θ is the logarithmic derivative of the density w.r.t. the parameter, i.e., $\Lambda_\theta(x) = \partial/\partial\theta \log p_\theta(x)$.

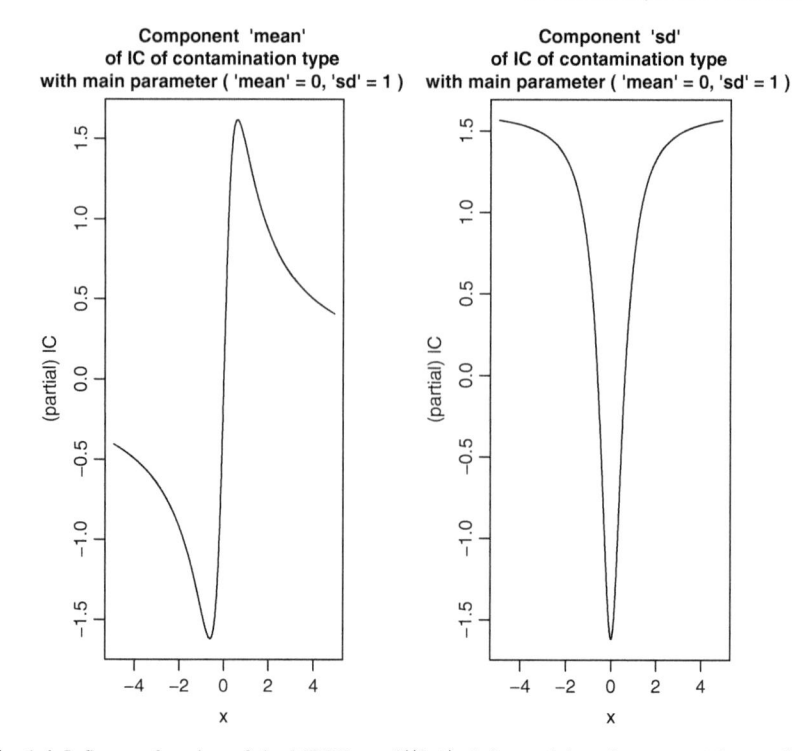

Fig. 1.4 Influence function of the MBRE at $\mathcal{N}(0,1)$; *left panel*: location part; *right panel*: scale part

$$A = 0.7917, \qquad a = -0.4970, \qquad b = 1.8546, \tag{1.36}$$

ψ is illustrated in Fig. 1.4. Setting up a (weighted) one-step construction as in Eq. (1.34), for $\psi_\theta(y) = \sigma \psi((y-f)/\sigma)$, we get our MBRE as

$$\hat{\theta}_n = \theta_n^0 + \sum_{j=1}^n w_j^0 \psi_{\theta_n^0}(y_j). \tag{1.37}$$

We warrant positivity of σ and want to maintain a high breakdown point even in the presence of inliers, which might drive σ towards 0. Therefore, instead of (1.37), we use the asymptotically equivalent form for the scale component

$$\hat{\sigma}_n = \sigma_n^0 \exp\left[\sum_{j=1}^n w_j^0 \psi_{\text{scale}}\left((y_j - \mu_n^0)/\sigma_n^0\right)\right]. \tag{1.38}$$

Using the MBRE on first glance could be seen as overly cautious. Detailed simulation studies, compare e.g., Kohl and Deigner [19], show that for our typical batch lengths of 10–20, the MBRE also is near to optimal in the sense of Rieder et al. [26] in the situation where nothing is known about the true outlier rate (including, of course the situation where no outliers at all occur).

1.5.5.3 Robustification of Steps (M1) and (M2)

Now, we derive robust estimators of the model parameters f_i and σ_i and justify the transition to weighted ALEs as in (1.37). In particular we specify the weights $w_j^0 = w_{i,j}^0$ therein. Recall, that the M1-step in the classical algorithm gives the optimal parameter estimates stated in Theorem 1.1. We now build ALEs, which can be achieved, when the MLEs of the parameters f_i and σ_i are stated as weighted sums of the observations y_k.

Theorem 1.3. *With $w_{i,l}^0 = \langle \hat{\mathbf{X}}_{l-1}, \mathbf{e}_i \rangle / \eta(O^{(i)})_k$, the optimal parameter estimates $\hat{f}_i$ and $\hat{\sigma}_i$ are given by*

$$\hat{f}_i = \sum_{l=1}^{k} w_{i,l}^0 y_l \text{ and} \tag{1.39}$$

$$\hat{\sigma}_i^2 = \sum_{l=1}^{k} w_{i,l}^0 (y_l - f_i)^2. \tag{1.40}$$

Proof. To find the optimal estimate for f consider

$$\Lambda_k^* := \prod_{l=1}^{k} \lambda_l^* \text{ with } \lambda_l^* := \exp\Big(\frac{(y_l - \langle f, \mathbf{X}_{l-1} \rangle)^2 - (y_l - \langle \hat{f}, \mathbf{X}_{l-1} \rangle)^2}{2 \langle \sigma, \mathbf{X}_{l-1} \rangle^2}\Big).$$

Up to constants irrelevant for optimization, the filtered log-likelihood is then

$$\mathrm{E}[\ln(\Lambda_k^*) \mid \mathscr{F}_k^y] = \sum_{i=1}^{N} \sum_{l=1}^{k} w_{i,l}^0 (y_l - \hat{f}_i)^2. \tag{1.41}$$

Maximising the log-likelihood $\mathrm{E}[\ln(\Lambda_k^*) \mid \mathscr{F}_k^Y]$ in $\hat{f}_i$ hence leads to the optimal parameter estimate

$$\sum_{l=1}^{k} \langle \mathbf{X}_{l-1}, \mathbf{e}_i \rangle (2y_l \hat{f}_i - \hat{f}_i^2) = 0 \quad \Longrightarrow \quad \hat{f}_i(k) = \sum_{l}^{k} w_{i,l}^0 y_l.$$

In an analogue way, for σ_i we define

$$\Lambda_k^+ := \prod_{l=1}^{k} \lambda_l^+ \text{ with } \lambda_l^+ := \frac{\langle \sigma, \mathbf{X}_{l-1} \rangle}{\langle \hat{\sigma}, \mathbf{X}_{l-1} \rangle} \exp\Big(\frac{(y_l - \langle f, \mathbf{X}_{l-1} \rangle)^2}{2 \langle \sigma, \mathbf{X}_{l-1} \rangle^2} - \frac{(y_l - \langle f, \mathbf{X}_{l-1} \rangle)^2}{2 \langle \hat{\sigma}, \mathbf{X}_{l-1} \rangle^2}\Big)$$

and hence, again up to irrelevant terms

$$E[\ln(\Lambda_k^+) \mid \mathscr{F}_k^y] = \sum_{i=1}^{N}\sum_{l=1}^{k} \langle \hat{\mathbf{X}}_{l-1}, \mathbf{e}_i \rangle \Big[\frac{(y_l - \hat{f}_i)^2}{2\sigma_i^2} + \log(\sigma_i)\Big]. \qquad (1.42)$$

From this term, which has to be minimized for $\hat{\sigma}_i$ we get

$$\sum_{l=1}^{k} \langle \mathbf{X}_{l-1}, \mathbf{e}_i \rangle \Big[-\frac{1}{2\hat{\sigma}^3}(y_l - \hat{f}_i)^2 + \frac{1}{\hat{\sigma}_i}\Big] = 0 \quad \Longrightarrow \quad \hat{\sigma}_i^2 = \sum_{l=1}^{k} w_{i,l}^0 (y_l - \hat{f}_i)^2.$$

Note that (1.42) takes its minimum in σ_i at the same place as

$$\sum_{i=1}^{N}\sum_{l=1}^{k} w_{i,l}^0 [(y_l - \hat{f}_i)^2 - \sigma_i^2]^2. \qquad (1.43)$$

□

For the robustification of the parameter estimation (step M1) we now distinguish two approaches. The first robustification is utilized in the first run over the first batch of data and is therefore called the *initialization* step M1. The robust estimates of the parameters from the second batch onwards are then achieved through a weighted ALE.

Theorem 1.4. *The robust parameter estimates for the model parameters f_i and σ_i in the (1) initialization and (2) all following batches are given by*

1. *Replacing, for initialization, the squares by absolute values in (1.41) and in (1.43), $\hat{f}_i$ and $\hat{\sigma}_i$ are the weighted median and scaled weighted MAD, respectively, of the y_l, $l = 1, \ldots, k$ with weights $(w_{i,l}^0)_l$.*
2. *For further batches, the weighted MBRE is obtained as a one-step construction with the parameter estimate (f_i^0, σ_i^0) from the previous batch as starting estimator and with IF $\psi = (\psi_{\text{loc}}, \psi_{\text{scale}})$ from (1.35) and (1.36). Therefore we have*

$$\hat{f}_i = f_i^0 + \sigma_i^0 \sum_{l=1}^{k} w_{i,l}^0 \psi_{\text{loc}}\big((y_l - f_i^0)/\sigma_i^0\big) \quad \text{and} \qquad (1.44)$$

$$\hat{\sigma}_i = \sigma_i^0 \exp\Big(\sum_{l=1}^{k} w_{i,l}^0 \psi_{\text{scale}}\big((y_l - f_i^0)/\sigma_i^0\big)\Big). \qquad (1.45)$$

Proof. 1. Initialization: With absolute values instead of squares, (1.41) becomes $\hat{f}_i = \arg\min_{f_i} \sum_{l=1}^{k} w_{i,l}^0 |y_l - f_i|$. Now if $w_{i,l}^0$ is constant in l, this leads to the empirical median as unique minimizer justifying the name. For the scaled weighted MAD, the argument parallels the previous one, leading to consistency factor $c_i = \Phi(3/4)$ for Φ the cdf of $\mathcal{N}(0,1)$.
2. M1 in further batches: Apparently, by definition, $(\hat{f}_i, \hat{\sigma}_i)$ is an ALE, once we show that ψ is square integrable, $E(\psi) = 0$, $E(\psi\Lambda') = \mathbb{I}_2$. The latter two properties can be checked numerically, while by boundedness square integrability is

obvious. In addition it has the necessary form of an MBRE in the i.i.d. setting as given in [25, Theorem 5.5.1]. To show that this also gives the MBRE in the context of weighted observations, we would need to develop the theory of ALEs for triangular schemes similar to the one in the Lindeberg Feller Theorem. This has been done, to some extent, in Ruckdeschel [29, Sect. 9]. In particular, for each state i, we have to assume a Noether condition excluding observations overly influential for parameter estimation in this particular state, i.e.,

$$\lim_{k\to\infty}\max_{l=1,\ldots,k}(w^0_{i,l;k})^2/\sum_{j=1}^{k}(w^0_{i,l;k})^2=0. \qquad (1.46)$$

We do not work this out in detail here, though. □

Consider again our filtering algorithm and recall, that the filter runs over the data set in batches of roughly 10–50 data points. To determine the ALE for our parameters, we have to calculate the weights $w^0_{i,l} = \langle \hat{\mathbf{X}}_{l-1}, \mathbf{e}_i \rangle / \hat{O}^i_k$. Therefore, our algorithm has to know all values of $\hat{\mathbf{X}}_l$ from 1 to k in each batch. With this, our robustification of the algorithm cannot obtain the same recursiveness as the classical algorithm. However, since we only have to determine and save the estimates of $\mathbf{X}_l$ in each batch, the algorithm still is numerically efficient, the additional costs are low. In general, the ALEs are fastly computed robust estimators, which lead in our case to a fast and, over batches, recursive algorithm.

The additional computational burden to store all the weights $w^0_{i,l}$ arising in the robustification of the M1-step is more than paid off by the additional benefits they offer for diagnostic purposes beyond the mere EM-algorithm. They tell us which of the observations, due to their likelihood to be in state i carry more information on the respective parameters f_i and σ_i than others. The same goes for the terms $\psi_\theta(y_l)$ which capture the individual information of observation y_l for the respective parameters. Even more though, the coordinates of $\psi_\theta(y_j)/|\psi_\theta(y_l)|$ tell us how much of the information in observation y_l is used for estimating f_i and how much for σ_i. In addition the function $y \mapsto w^0_{i,l}\psi_\theta(y)$ can be used for sensitivity analysis, telling us what happens to the parameter estimates for small changes in observation y. Finally, using the unclipped, classically optimal IF of the MLE, but evaluated at the robustly estimated parameters, we may identify outliers not fitting to the "usual" states.

1.6 Implementation and Simulation

The classical algorithm as well as the robust version are implemented in R; we plan to release the code in form of a contributed package on CRAN at a later stage. The implementation builds up on, respectively uses contributed packages RobLox and mclust. At the time of writing we are preparing a thorough simulation study to explore our procedure in detail and in a quantitative way. For the moment, we restrict ourselves to assess the procedure in a qualitative way, illustrating how it can cope with a situation like in Fig. 1.3.

In Fig. 1.5, we see the paths of the robust parameter estimates for f, σ, and Π; due to the new initialization procedure, the estimates – in particular those for Π – differ a little from those of Fig. 1.1. Still, all the estimators behave very reasonable and are not too far from the classical ones.

In the outlier situations from Figs. 1.2 and 1.3, illustrated in Fig. 1.6, the estimates for f and σ remain stable at large as desired. The estimates for Π however do get irritated, essentially flagging out one state as outlier state. Some more work remains to be done to better understand this. Aside from this, our algorithm achieves its goals: our procedure never breaks down – contrary to the classical one, we are able to find parameter estimates even in the case of large outliers in the observation process.

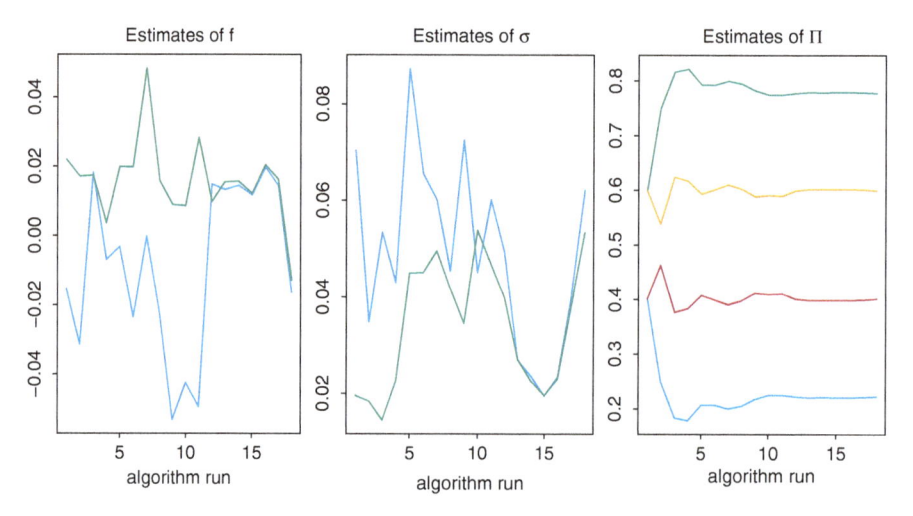

Fig. 1.5 Robust parameter estimates for monthly MSCI returns between 1994 and 2009 – analogue to Fig. 1.1

1.7 Conclusion

In financial applications, we often have to consider the case of outliers in our data set, which can occur from time to time due to either wrong values in the financial database or unusual peaks or lows in volatile markets. Conventional parameter estimation methods cannot handle these specific data characteristics well.

1.7.1 Contribution of This Paper

Our contribution to this issue is two fold: First, we analyse step by step the general filter-based EM-algorithm for HMMs by Elliott [3] and highlight, which problems

1 Robustification of an On-line EM Algorithm for Modelling Asset Prices

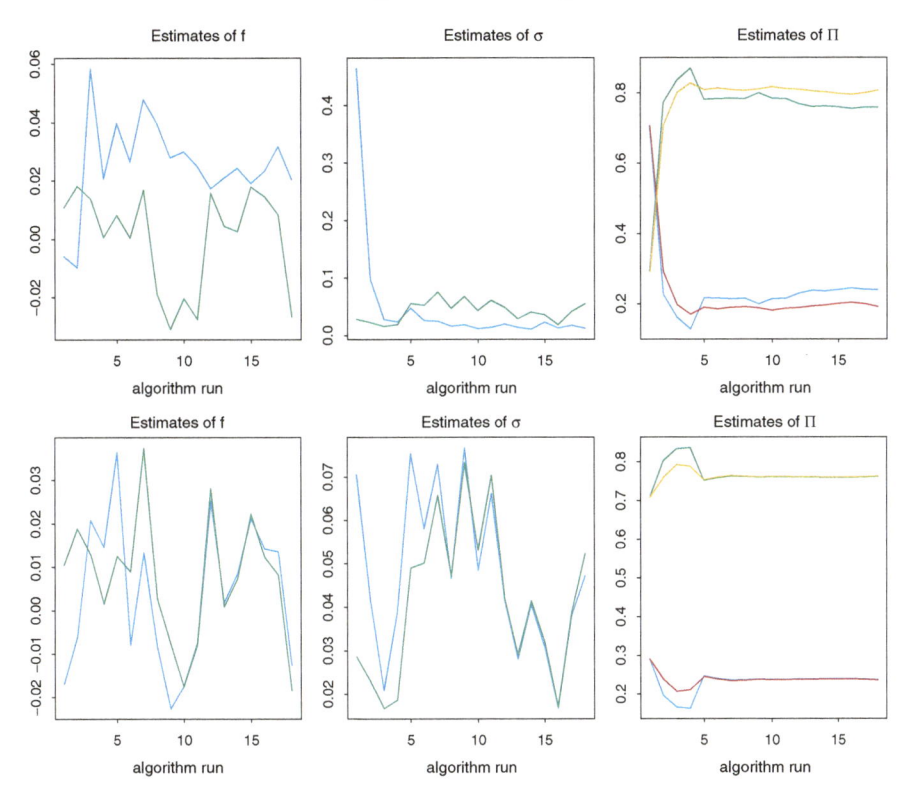

Fig. 1.6 Robust parameter estimates for monthly MSCI with planted outliers – analogue to Figs. 1.2 and 1.3

can occur in case of extreme values. We extend the classical algorithm by a new technique to find initial values, taking into account the N-state setting of the HMM. In addition, for numerical reasons we use a data-driven reference measure instead of the standard normal distribution.

Second, we have proposed a full robustification of the classical EM-algorithm. Our robustified algorithm is stable w.r.t. outliers in the observation process and is still able to estimate processes of the Markov chain as well as optimal parameter estimates with acceptable accuracy. The robustification builds up on concepts from robust statistics like SO-optimal filtering and asymptotic linear estimators. Due to the non-i.i.d. nature of the observations as apparent from the non-uniform weights $w_{i,l}^0$ attributed to the observations, these concepts had to be generalized for this situation, leading to weighted medians, weighted MADs and weighted ALEs. Similarly, the SO-optimal filtering (with focus on state reconstruction) is not directly applicable for robustifying the Radon-Nikodym terms λ_s, where we (a) had to "clean" the observations themselves and (b) had to pass over to $\sqrt{\lambda_s}$ for integrability reasons.

Our robust algorithm is computationally efficient. Although complete recursivity cannot be obtained, the algorithm runs over batches and keeps its recursivity there additionally storing the filtered values of the Markov chain. This additional burden is outweighed though by the benefits of these weights and influence function terms for diagnostic purposes.

As in the original algorithm, the model parameters, which are guided by the state of the Markov chain, are updated after each batch, using a robust ALE however. The robustification therefore keeps the characteristic of the algorithm, that new information, which arises in the observation process, is included in the recent parameter update – there is no forward-backward loop. The forecasts of asset prices, which are obtained through the robustified parameter estimates, can be utilized to make investment decisions in asset allocation problems. To sum up, our forecasts are robust against additive outliers in the observation process and able to handle switching regimes occurring in financial markets.

1.7.2 Outlook

It is pretty obvious how to generalize our robustification to a multivariate setting: The E-step is not affected by multivariate observations, and the initialization technique using Gaussian Mixture Models ideas already is available in multivariate settings. Respective robust multivariate scale and location estimators for weighted situations still have to be implemented, though, a candidate being a weighted variant of the (fast) MCD-estimator, compare Rousseeuw and Leroy [27], Rousseeuw and van Driessen [28].

In addition, an automatic selection criterion for the number of states would be desirable. This is a question of model selection, where criteria like BIC still have to be adopted for robustness.

Future work will hence translate our robustification to a multivariate setting to directly apply the algorithm to asset allocation problems for portfolio optimization. Furthermore, investment strategies shall be examined within this robust HMM setting to enable investors a view on their portfolio, which includes possible outliers or extreme events. The implementation of the algorithms shall be part of an R package, including a thorough simulation study of the robustified algorithm and its application in portfolio optimization.

Acknowledgements We thank an anonymous referee and the editor for valuable comments. Financial support for C. Erlwein-Sayer from Deutsche Forschungsgemeinschaft (DFG) within the project "Regimeswitching in zeitstetigen Finanzmarktmodellen: Statistik und problemspezifische Modellwahl" (RU-893/4-1) is gratefully acknowledged.

Appendix

Definition: Weighted Medians and MADs

For weights $w_j \geq 0$ and observations y_j, the weighted median $m = m(y,w)$ is defined as
$$m = \mathrm{argmin}_f \sum_j w_j |y_j - f|.$$
With $y'_j = |y_j - m|$, the scaled weighted MAD $s = s(y,w)$ is defined as
$$s = c^{-1} \mathrm{argmin}_t \sum_j w_j |y'_j - t|,$$
where c is a consistency factor. It warrants consistent estimation of σ in case of Gaussian observations, i.e., $c = \mathrm{E}\,\mathrm{argmin}_t \sum_j w_j \big||\tilde{y}_j| - t\big|$ for $\tilde{y}_j \overset{\mathrm{i.i.d.}}{\sim} \mathcal{N}(0,1)$. c can be obtained empirically for a sufficiently large sample size M, e.g., $M = 10{,}000$, setting $c = \frac{1}{M} \sum_{k=1}^M c_k$, $c_k = \mathrm{argmin}_t \sum_j w_j \big||y''_{j,k}| - t\big|$, $y''_{j,k} \overset{\mathrm{i.i.d.}}{\sim} \mathcal{N}(0,1)$.

As to the (finite sample) breakdown point $FSBP$ of the weighted median (and at the same time for the scaled weighted MAD), we define $w_j^0 = w_{i,j}/\sum_{j'} w_{j'}$, and for each i define the ordered weights $w_{(j)}^0$ such that $w_{(1)}^0 \geq w_{(2)}^0 \geq \ldots \geq w_{(k)}^0$. Then the FSBP in both cases is $k^{-1} \min\{j_0 = 1, \ldots, k \mid \sum_{j=1}^{j_0} w_{(j)}^0 \geq k/2\}$ which (for equivariant estimators) can be shown to be the largest possible value. So using weighted medians and MADs, we achieve a decent degree of robustness against outliers. E.g., assume we have 10 observations with weights $5 \times 0.05; 3 \times 0.1; 0.2; 0.25$. Then we need at least three outliers (placed at weights $0.1, 0.2, 0.25$, respectively) to produce a breakdown.

Proof of Theorem 1.2

Proof. Let us solve $\max_{\partial \mathcal{U}} \min_f [\ldots]$ first, which amounts to $\min_{\partial \mathcal{U}} \mathrm{E}_{\mathrm{re}}\big[|\mathrm{E}_{\mathrm{re}}[Y^{\mathrm{id}}|Y^{\mathrm{re}}]|^2\big]$. For fixed element $P^{Y^{\mathrm{di}}}$ assume a dominating σ-finite measure μ, i.e., $\mu \gg P^{Y^{\mathrm{di}}}$, $\mu \gg P^{Y^{\mathrm{id}}}$; this gives us a μ-density $q(y)$ of $P^{Y^{\mathrm{di}}}$. Determining the joint (real) law $P^{Y^{\mathrm{id}}, Y^{\mathrm{re}}}(d\tilde{y}, dy)$ as

$$P(Y^{\mathrm{id}} \in A, Y^{\mathrm{re}} \in B) = \int I_A(\tilde{y}) I_B(y) [(1-r) I(\tilde{y} = y) + r q(y)] p^{Y^{\mathrm{id}}}(\tilde{y}) \mu(d\tilde{y}) \mu(dy), \tag{1.47}$$

we deduce that $\mu(dy)$-a.e.

$$\mathrm{E}_{\mathrm{re}}[Y^{\mathrm{id}}|Y^{\mathrm{re}}=y] = \frac{rq(y)\mathrm{E}Y^{\mathrm{id}}+(1-r)yp^{Y^{\mathrm{id}}}(y)}{rq(y)+(1-r)p^{Y^{\mathrm{id}}}(y)} =: \frac{a_1q(y)+a_2(y)}{a_3q(y)+a_4(y)}. \tag{1.48}$$

Hence we have to minimize

$$F(q) := \int \frac{|a_1q(y)+a_2(y)|^2}{a_3q(y)+a_4(y)} \mu(dy)$$

in $M_0 = \{q \in L_1(\mu) \mid q \geq 0, \int q\,d\mu = 1\}$. To this end, we note that F is convex on the non-void, convex cone $M = \{q \in L_1(\mu) \mid q \geq 0\}$ so, for some $\tilde{\rho} \geq 0$, we may consider the Lagrangian

$$L_{\tilde{\rho}}(q) := F(q) + \tilde{\rho} \int q\,d\mu$$

for some positive Lagrange multiplier $\tilde{\rho}$. Pointwise minimization in y of $L_{\tilde{\rho}}(q)$ gives

$$q_s(y) = \frac{1-r}{r}(|D(y)|/s - 1)_+ p^Y(y)$$

for some constant $s = s(\tilde{\rho}) = (|\mathrm{E}Y^{\mathrm{id}}|^2 + \tilde{\rho}/r)^{1/2}$, Pointwise in y, $\hat{q}_s$ is antitone and continuous in $s \geq 0$ and $\lim_{s \to 0[\infty]} q_s(y) = \infty[0]$, hence by monotone convergence,

$$H(s) = \int \hat{q}_s(y)\mu(dy)$$

too, is antitone and continuous and $\lim_{s \to 0[\infty]} H(s) = \infty[0]$. So by continuity, there is some $\rho \in (0,\infty)$ with $H(\rho) = 1$. On M_0, $\int q\,d\mu = 1$, but $\hat{q}_\rho = q_{s=\rho} \in M_0$ and is optimal on $M \supset M_0$ hence it also minimizes F on M_0. In particular, we get representation (1.26) and note that, independently from the choice of μ, the least favorable $P_0^{Y^{\mathrm{di}}}$ is dominated according to $P_0^{Y^{\mathrm{di}}} \ll P^{Y^{\mathrm{id}}}$, i.e.; non-dominated $P^{Y^{\mathrm{di}}}$ are even easier to deal with.

As next step we show that

$$\max_{\partial \mathcal{U}} \min_f [\ldots] = \min_f \max_{\partial \mathcal{U}} [\ldots] \tag{1.49}$$

To this end we first verify (1.25) determining $f_0(y)$ as $f_0(y) = \mathrm{E}_{\mathrm{re};\hat{P}}[X|Y^{\mathrm{re}} = y]$. Writing a sub/superscript "re; P" for evaluation under the situation generated by $P = P^{Y^{\mathrm{di}}}$ and $\hat{P}$ for $P_0^{Y^{\mathrm{di}}}$, we obtain the risk for general P as

$$\mathrm{MSE}_{\mathrm{re};P}[f_0(Y^{\mathrm{re},P})] = (1-r)\mathrm{E}_{\mathrm{id}}|Y^{\mathrm{id}} - f_0(Y^{\mathrm{id}})|^2 + r\mathrm{tr}\mathrm{Cov}Y^{\mathrm{id}} + \\ + r\mathrm{E}_P \min(|D(Y^{\mathrm{di};q})|^2, \rho^2). \tag{1.50}$$

This is maximal for any P that is concentrated on the set $\{|D(Y^{\text{di};q})|>\rho\}$, which is true for $\hat{P}$. Hence (1.49) follows, as for any contaminating P

$$\text{MSE}_{\text{re};P}[f_0(Y^{\text{re};P})] \leq \text{MSE}_{\text{re};\hat{P}}[f_0(Y^{\text{re};\hat{P}})].$$

Finally, we pass over from $\partial \mathscr{U}$ to $\mathscr{U}$: Let f_r, $\hat{P}_r$ denote the components of the saddle-point for $\partial \mathscr{U}(r)$, as well as $\rho(r)$ the corresponding Lagrange multiplier and w_r the corresponding weight, i.e., $w_r = w_r(y) = \min(1, \rho(r)/|D(y)|)$. Let $R(f, P, r)$ be the MSE of procedure f at the SO model $\partial \mathscr{U}(r)$ with contaminating $P^{Y^{\text{di}}} = P$. As can be seen from (1.26), $\rho(r)$ is antitone in r; in particular, as $\hat{P}_r$ is concentrated on $\{|D(Y)| \geq \rho(r)\}$ which for $r \leq s$ is a subset of $\{|D(Y)| \geq \rho(s)\}$, we obtain

$$R(f_s, \hat{P}_s, s) = R(f_s, \hat{P}_r, s) \qquad \text{for } r \leq s.$$

Note that $R(f_s, P, 0) = R(f_s, Q, 0)$ for all P, Q – hence passage to $\tilde{R}(f_s, P, r) = R(f_s, P, r) - R(f_s, P, 0)$ is helpful – and that

$$\text{tr}\text{Cov}Y^{\text{id}} = \text{E}_{\text{id}}\left[\text{tr}\text{Cov}_{\text{id}}[Y^{\text{id}}|Y^{\text{id}}] + |D(Y^{\text{id}})|^2\right]. \tag{1.51}$$

Abbreviate $\bar{w}_s(Y^{\text{id}}) = 1 - (1 - w_s(Y^{\text{id}}))^2 \geq 0$ to see that

$$\tilde{R}(f_s, P, r) = r\left\{\text{E}_{\text{id}}\left[|D(Y^{\text{id}})|^2 \bar{w}_s(Y^{\text{id}})\right] + \text{E}_P \min(|D(Y^{\text{id}})|, \rho(s))^2\right\} \leq$$
$$\leq r\left\{\text{E}_{\text{id}}\left[|D(Y^{\text{id}})|^2 \bar{w}_s(Y^{\text{id}})\right] + \rho(s)^2\right\} = \tilde{R}(f_s, \hat{P}_r, r) < \tilde{R}(f_s, \hat{P}_s, s).$$

Hence the saddle-point extends to $\mathscr{U}(r)$; in particular the maximal risk is never attained in the interior $\mathscr{U}(r) \setminus \partial \mathscr{U}(r)$. (1.28) follows by plugging in the results. □

References

1. Ang, A., Bekaert, G.: International asset allocation with regime shifts. Rev. Financ. Stud. **15**, 1137–1187 (2002)
2. Cai, J.: A Markov model of switching-regime ARCH. J. Bus. Econ. Stat. **12**, 309–316 (1994)
3. Elliott, R.J.: Exact adaptive filters for Markov chains observed in Gaussian noise. Automatica **30**, 1399–1408 (1994)
4. Elliott, R.J., Hinz, J.: A method for portfolio choice. Appl. Stoch. Models Bus. Ind. **19**, 1–11 (2003)
5. Elliott, R.J., van der Hoek, J.: An application of hidden Markov models to asset allocation problems. Financ. Stoch. **1**, 229–238 (1997)
6. Elliott, R.J., Aggoun, L., Moore, J.B.: Hidden Markov Models: Estimation and Control. Applications of Mathematics, vol. 29. Springer, New York (1995)

7. Erlwein, C., Mamon, R., Davison, M.: An examination of HMM-based investment strategies for asset allocation. Appl. Stoch. Model. Bus. Ind. **27**(3), 204–221 (2009)
8. Fernholz, L.T.: Von Mises Calculus for Statistical Functionals. Lecture Notes in Statistics, vol. 19. Spinger, New York (1983)
9. Fox, A.J.: Outliers in time series. J. R. Stat. Soc. Ser. B **34**, 350–363 (1972)
10. Fraley, C., Raftery, A.E.: Model-based clustering, discriminant analysis and density estimation. J. Am. Stat. Assoc. **97**, 611–631 (2002)
11. Fraley, C., Raftery, A.E., Murphy, T.B., Scrucca, L.: mclust version 4 for R: normal mixture modeling for model-based clustering, classification, and density estimation. Technical report No. 597, Department of Statistics, University of Washington (2012)
12. Gray, S.F.: Modeling the conditional distribution of interest rates as a regime-switching process. J. Financ. Econ. **42**, 27–62 (1996)
13. Guidolin, M., Timmermann, A.: Asset allocation under multivariate regime switching. J. Econ. Dyn. Control **31**, 3503–3544 (2007)
14. Hamilton, J.D.: A new approach to the economic analysis of nonstationary time series and the business cycle. Econom. J. Econom. Soc. **57**, 357–384 (1989)
15. Hampel, F.R.: Contributions to the theory of robust estimation. Dissertation, University of California, Berkely (1968)
16. Hampel, F.R., Ronchetti, E.M., Rousseeuw, P.J. Stahel, W.A.: Robust Statistics: The Approach Based on Influence Functions. Wiley, New York (1986)
17. Huber, P.J.: Robust Statistics. Wiley, New York (1981)
18. Kohl, M.: RobLox: optimally robust influence curves and estimators for location and scale. R package version 0.8.2. http://robast.r-forge.r-project.org/ (2012)
19. Kohl, M., Deigner, H.P.: Preprocessing of gene expression data by optimally robust estimator. BMC Bioinform. **11**, 583 (2010)
20. Kohl, M., Rieder, H., Ruckdeschel, P.: Infinitesimally robust estimation in general smoothly parametrized models. Stat. Methods Appl. **19**, 333–354 (2010)
21. Knight, F.H.: Risk, Uncertainty, and Profit. Houghton Mifflin, Boston (1921)
22. Mamon, R., Erlwein, C., Gopaluni, B.: Adaptive signal processing of asset price dynamics with predictability analysis. Inform. Sci. **178**, 203–219 (2008)
23. Markowitz, H.: Portfolio selection. J. Financ. **7**(1), 77–91 (1952)
24. Maronna, R.A., Martin, R.D. Yohai, V.J.: Robust Statistics: Theory and Methods. Wiley, Chichester (2006)
25. Rieder, H.: Robust Asymptotic Statistics. Springer, New York (1994)
26. Rieder, H., Kohl, M., Ruckdeschel, P.: The cost of not knowing the radius. Stat. Methods Appl. **17**(1), 13–40 (2008)
27. Rousseeuw, P.J., Leroy, A.M.: Robust Regression and Outlier Detection. Wiley, New York (1987)
28. Rousseeuw, P.J., van Driessen, K.: A fast algorithm for the minimum covariance determinant estimator. Technometrics **41**, 212–223 (1999)

29. Ruckdeschel, P.: Ansätze zur Robustifizierung des Kalman Filters. Bayreuther Mathematische Schriften, vol. 64. Mathematisches Inst. der Univ. Bayreuth, Bayreuth (2001)
30. Ruckdeschel, P., Horbenko, N.: Robustness properties of estimators in generalized pareto models. Technical report ITWM N^o182. http://www.itwm.fraunhofer.de/fileadmin/ITWM-Media/Zentral/Pdf/Berichte_ITWM/2010/bericht_182.pdf (2010)
31. Sass, J., Haussmann, U.G.: Optimizing the terminal wealth under partial information: the drift process as a continuous time Markov chain. Financ. Stoch. **8**, 553–577 (2004)
32. Zakai, M.: On the optimal filtering of diffusion processes. Zeitschrift für Wahrscheinlichkeitstheorie und verwandte Gebiete **11**, 230–243 (1969)

Chapter 2
Stochastic Volatility or Stochastic Central Tendency: Evidence from a Hidden Markov Model of the Short-Term Interest Rate

Craig A. Wilson and Robert J. Elliott

Abstract We develop a two-factor model for the short-term interest rate that incorporates additional randomness in both the drift and diffusion components. In particular, the model nests stochastic volatility and stochastic central tendency, and therefore provides a medium for testing the overall importance of both factors. The randomness in the drift and diffusion terms is governed by a hidden Markov chain. The likelihood function is determined through an iterative procedure and maximum likelihood estimates are obtained via numerical maximization. This process allows likelihood ratio testing of nested restrictions. These tests show that stochastic volatility is more important than stochastic central tendency for describing the short rate dynamics.

2.1 Introduction

The risk-free interest rate is one of the most vital inputs in financial and economic theory. There is still much debate about the relationship between rates for differing time horizons. Evidence for and against the expectations hypothesis waxes and wanes as additional complexities are incorporated into interest rate models and as more robust empirical analysis is applied to the various models. Most models of the short-term interest rate combine a mean-reverting drift component with a diffusion

---

C.A. Wilson (✉)
Edwards School of Business, University of Saskatchewan, Saskatoon, SK S7N 5A7, Canada
e-mail: wilson@edwards.usask.ca

R.J. Elliott
Haskayne School of Business, Scurfield Hall, University of Calgary, 2500 University Drive NW, Calgary, AB T2N 1N4, Canada

School of Mathematical Sciences, North Terrace Campus, University of Adelaide, Adelaide, SA 5005, Australia
e-mail: relliott@ucalgary.ca

component. The simplest models assume that the drift is a linear function of the short-term interest rate with constant parameters, and that the diffusion is governed by a constant volatility parameter. This model provides a mathematical framework for describing a central bank trying to control interest rates by pushing rates slowly toward a target rate with a force that is proportional to how far the current rate is from the target, but misjudging the effect of its policy independently of (with additive noise models) or proportionally to (with multiplicative noise models) the current rate in a consistent way (with constant volatility). Guthrie and Wright [11] develop an alternative model of central bank behavior that leads to similar observed interest rate behavior.

The main extensions to these simple models involve allowing for non-linearity in the drift function, allowing for randomness in the drift function, and allowing for randomness in the volatility function. Each of these extensions has been found to be empirically significant, when studied individually; however the purpose of this study is to determine which of these extensions is most important for explaining historical short-term interest rates, and we find that stochastic volatility is by far the more important feature.

Current theory about the dynamics of the short-term, default-free interest rate suggests two alternative methods of modeling: an equilibrium approach, and a no arbitrage approach. The later takes the current term structure as an input so as to force an exact fit to longer-term bond prices and other interest rate derivatives. Examples of this approach include the models of Ho and Lee [14], Hull and White [15], and Heath et al. [13]. On the other hand, equilibrium models such as those of Vasicek [22], and Cox et al. [7] generally do not predict values that exactly match current term structures. In this sense, such models are not arbitrage free. However, this shortcoming is often made up for by the model's applicability to future time periods, since they usually lead to a stationary sequence of interest rates. Also, because of the limitations of financial data, the term structures used as inputs for no arbitrage models are finite, so in practice, arbitrage free predictions can often be achieved by equilibrium type models with a sufficiently large number of parameters.

In the particular case of the equilibrium type model used by Chan et al. [5], the interest rate is supposed to follow a mean reverting process described by dynamics of the form

$$dr_t = \alpha(\bar{r} - r_t)dt + \sigma r_t^\gamma dW_t. \tag{2.1}$$

In unconstrained estimation by Chan et al. [5] it is found that the variance elasticity, γ, is approximately 1.5, (using GMM estimation on U.S. interest rates based on monthly observations between June 1964 and December 1989), which causes the previous SDE to have a non-stationary solution, (i.e. the variance increases without bound as t gets large), which is undesirable for estimation and testing purposes. The above interest rate model has two important features: the drift term is a linear function of the interest rate and the volatility term is deterministic.

Relaxing one or both of these properties could resolve the problem. For instance letting the drift term be non-linear so that it increased the mean-reverting force as the interest rate became large, could resolve the non-stationarity problem [1]. Or if the volatility were allowed to be stochastic, as in Longstaff and Schwartz [18], the need for randomness implicit in the r_t^γ term could be reduced, requiring a smaller elasticity parameter, γ. These issues are addressed by Sun [21], where he finds that stochastic volatility was significant and non-linear drift was not significant. This finding is consistent with Chapman and Pearson [6], who conclude that non-linear drift is not an essential property of the short-term interest rate.

Although Sun [21] describes a model that nests both stochastic volatility and non-linear drift, he is pitting a two-factor model up against a one-factor model when he tests his restrictions. This approach might bias his study toward finding that the stochastic volatility framework dominates the non-linear drift, since such a framework is able to explain some observed term structure phenomena such as yield curve twists that typically cannot be explained in a one-factor model. It turns out that this concern is unfounded, as we also find that stochastic volatility is most important, even when both full and restricted models have two factors of randomness.

Balduzzi et al. [3] develop a model in which the mean-reverting level, or what they call the central tendency of the short rate process provides a second factor of randomness. In this case the second factor enters through the drift, but they still model the drift as being linear in the short rate. If one compares stochastic central tendency against stochastic volatility, both restricted models have two factors of randomness, and they can be compared on a level playing field, which is the approach we take.

A natural model to nest these two phenomena is the three factor model of Balduzzi et al. [2]. However, this model requires the central tendency factor to be independent of the other factors, so testing restrictions of this model pits a two-factor model against a three-factor model, which may again bias toward rejecting the restrictions of constant central tendency or volatility. One solution to this problem would be to implement a special case of this model's extension by Dai and Singleton [8], where both central tendency and volatility are governed by the same Brownian motion.

Unfortunately, implementing this approach in the Chan et al. [5] framework is difficult because it requires relaxing the affine term structure model assumption. We examine estimation techniques on the Chan et al. [5] model where the central tendency level and volatility parameters themselves are prone to switch in accordance with the same Markov chain. In this way, even the full model with both stochastic drift and stochastic volatility has only two factors of randomness: the Markov chain and an independent Brownian motion. Even so, this framework allows an arbitrary correlation between drift and volatility. Such a feature was found to be important by Dai and Singleton [8].

This regime switching framework has been found helpful in explaining interest rate and term structure characteristics in a number of studies including Hamilton [12], Naik and Lee [19], Bansal and Zhou [4], Smith [20], and Kalimipalli and Susmel [16], to name a few. We assume that the state of the Markov chain cannot be observed directly, and must be estimated through filtering observations of the short-term interest rate. In this way we consider the interest rate to be governed as a hidden Markov model.

We consider the case of a discrete time autoregressive stochastic process. An extension of Hamilton's [12] algorithm provides an iterative construction of the likelihood function. Evaluating this likelihood function at the maximum likelihood estimates obtained for full and restricted models allows the testing of various restrictions using likelihood ratio tests. We do this for observations of Canadian and U.S. nominal interest rates. The null hypothesis of constant volatility can be strongly rejected, whereas that of constant mean reverting level or central tendency cannot.

The remainder of this paper is organized as follows. Section 2.2 discusses the general model, and Sects. 2.3 and 2.4 discus techniques to estimate and test this model in the maximum likelihood framework. Section 2.5 specializes the model to the case of short-term interest rates. Section 2.6 discusses the data, Sect. 2.7 analyzes the results, and Sect. 2.8 concludes.

2.2 The Model

The first process that we consider is a finite state (N-dimensional), discrete time, homogeneous Markov chain, $X = \{X_t; t \in \mathbf{N} = \{0, 1, 2, \ldots\}\}$, that takes values in the set of unit (column) vectors, $S = \{\mathbf{e}_1, \ldots, \mathbf{e}_N\}$, which is the canonical basis of $\mathbf{R}^N$, (i.e. $X_t = (0, \ldots, 0, 1, 0, \ldots, 0)^\top$). Denote by $\mathscr{F}^X = \{\mathscr{F}_t^X\}$ the filtration generated by the Markov chain X, and $\mathbf{P}$ its transition matrix, where $\mathbf{P}_{ij} = \Pr\{X_{t+1} = \mathbf{e}_j | X_t = \mathbf{e}_i, \ldots, X_0 = \mathbf{e}_k\}$ is the probability of going from state i to state j. It follows that $E[X_{t+1} | \mathscr{F}_t^X] = \mathbf{P}^\top X_t$, where the conditional expectation gives the vector of conditional probabilities and the right hand side picks out the appropriate row of $\mathbf{P}$. (Note that the entries of $\mathbf{P}$ must be non-negative and that the rows must sum to 1.) This convenient notation motivates our choice of state space and stochastic matrix notation, which was done without loss of generality.

We presume that this Markov chain is hidden, (i.e. it is not directly observable), so that we do not have access to the information $\mathscr{F}^X$. However, we do observe a stochastic process $\{Y_t; t \in \mathbf{N}\}$, which has the form

$$Y_{t+1} = \mu(X_t) + \zeta(X_t)\varepsilon_{t+1}, \tag{2.2}$$

where $\{\varepsilon_t\}$ is a sequence of i.i.d. standard normal random variables (although other distributions could be used), independent with the Markov chain, X. (For now, we consider $\{Y_t\}$ to be a general observation process, but later we will specialize it to observations of short-term interest rates.) It is clear that μ and ζ are Markov chains. Also notice that the function $\mu(X_t)$ has the representation $\mu^\top X_t$, where the vector μ has typical entry $\mu_i = \mu(\mathbf{e}_i)$, and similarly for ζ. In general, the drift and volatility terms μ and ζ could be functions of other independent and observable variables.

Denote by $\mathscr{F}^Y = \{\mathscr{F}_t^Y\}$ the filtration generated by the observed process, Y, $\mathscr{F}^\varepsilon = \{\mathscr{F}_t^\varepsilon\}$ the filtration generated by the noise, ε, and $\mathscr{G} = \{\mathscr{G}_t\} = \{\mathscr{F}_t^X \vee \mathscr{F}_t^Y\} = \{\mathscr{F}_t^X \vee \mathscr{F}_t^\varepsilon\}$ is the joint (or global) filtration. The filtering problem will therefore involve the optimal use of the available information. We wish to make inferences about $\mathscr{G}$-adapted processes by conditioning on the filtration $\mathscr{F}^Y$. This procedure gives a best, (in mean square error sense), estimate of the unobservable processes, based on information obtained from observing the process Y [10].

2.3 Maximum Likelihood Estimation

The model requires estimates for the transition probabilities, $\mathbf{P}_{ij}$ and the entries of the vectors μ and ζ. The class of maximum likelihood estimators, (MLE's), has several desirable properties such as consistency, efficiency, and robustness [17]. We therefore attempt to find MLE's for the various parameters. The problem is that MLE's can often be difficult to calculate directly or explicitly. We form the likelihood function iteratively and solve it numerically via the EM algorithm. The procedure we use is a modification of Elliott's [10] filter.

Since the Markov chain is unobservable, we have particular difficulty in estimating the probability matrix. This estimation can be done by using a change of probability technique. Based on the time series of observations and arbitrary starting parameter values, we filter information about the Markov chain's state, which is used to obtain optimal (in the sense of expectation maximization) parameter estimates. The EM algorithm continues by finding new filtered processes using the previous optimal parameter estimates and using the new processes to find new optimal parameter estimates. A fixed point in the parameter space corresponds to maximum likelihood parameter estimates.

The previous algorithm gives maximum likelihood estimates, but not the likelihood function. To perform likelihood ratio (LR) tests, we need the likelihood function evaluated at the optimal parameters for various restrictions. We use a modification of Hamilton's [12] algorithm to obtain the likelihood function and evaluate it at the values found by the EM algorithm. This approach involves manipulating conditional probability mass and density functions at each time and adding their logarithms to get the log-likelihood function. Unfortunately this function cannot be obtained in closed form, which is why we use the EM algorithm to maximize it.

2.4 The Likelihood Function

This section describes an algorithm similar to Hamilton's [12] algorithm. We have a hidden Markov chain $\{X_t\}$ and a sequence of observations $\{Y_t\}$, which are presumed to depend upon the previous state of the Markov chain, and on noise that is independent with the Markov chain. Notice that we are considering probability mass functions and probability density functions in this section and we denote them f and g respectively. (These functions can be thought of as Radon-Nikodym derivatives with respect to counting measure or Lebesgue measure.) Functions of more than one variable refer to joint probability mass or density functions. For ease of notation, the dependence on the parameters is suppressed. It is implied that all of the mass and density functions that follow depend on a common parameterization.

The goal is to obtain the current filter, which is the probability mass function $f(x_t;y_t,\ldots,y_0) = \Pr\{X_t = x_t|Y_t = y_t,\ldots,Y_0 = y_0\}$ when starting with the previous filter $f(x_{t-1};y_{t-1},\ldots,y_0) = \Pr\{X_{t-1} = x_{t-1}|Y_{t-1} = y_{t-1},\ldots,Y_0 = y_0\}$. This function provides a filter for the state of the Markov chain. We assume that the conditional density function $g(y_t|x_{t-1},y_{t-1},\ldots,y_0)$ is known. In particular, for our general model we use the normal density

$$g(y_t|x_{t-1},y_{t-1},\ldots,y_0) = \frac{1}{\sqrt{2\pi}\zeta^T x_{t-1}} \exp\left\{-\frac{(y_t - \mu^T x_{t-1})^2}{2(\zeta^T x_{t-1})^2}\right\}; \qquad (2.3)$$

however, other densities could be used. A consequence of the algorithm is that the conditional density function $g(y_t|y_{t-1},\ldots,y_0)$ is obtained, which can be used to construct the likelihood function $L(\theta;y) = g(y_T,\ldots,y_1|y_0) = \prod_{t=1}^{T} g(y_t|y_{t-1},\ldots,y_0)$. Hamilton advocates maximizing the log-likelihood function numerically to obtain maximum likelihood estimates for the parameters. Knowing the likelihood function explicitly allows likelihood ratio tests to be applied to test equality constraints on the parameters in a straight forward manner: With r distinct equality restrictions, the logarithm of the square of the ratio of the likelihood function evaluated at the unrestricted MLE to that evaluated at the restricted MLE has a central χ^2 distribution with r degrees of freedom (under certain regularity conditions see Lehmann [17] for example).

We outline the algorithm as follows: Assume we know $f(x_0|y_0)$, we have iterated through the algorithm to the tth observation to get $f(x_t|y_t,\ldots,y_0)$, and $g(y_{t+1}|x_t,y_t,\ldots,y_0)$ is given as in Eq. 2.3. Then

1. $g(y_{t+1},x_t|y_t,\ldots,y_0) = g(y_{t+1}|x_t,y_t,\ldots,y_0)f(x_t|y_t,\ldots,y_0)$
2. $g(y_{t+1}|y_t,\ldots,y_0) = \sum_i g(y_{t+1},\mathbf{e}_i|y_t,\ldots,y_0)$
3. $f(x_t|y_{t+1},\ldots,y_0) = \frac{g(y_{t+1},x_t|y_t,\ldots,y_0)}{g(y_{t+1}|y_t,\ldots,y_0)}$
4. $f(x_{t+1},x_t|y_{t+1},\ldots,y_0) = f(x_{t+1}|x_t,y_{t+1},\ldots,y_0)f(x_t|y_{t+1},\ldots,y_0)$
5. $f(x_{t+1}|y_{t+1},\ldots,y_0) = \sum_i f(x_{t+1},\mathbf{e}_i|y_{t+1},\ldots,y_0)$

Each step follows from the definition of conditional probability or a straightforward application of Bayes' theorem. In Steps 2 and 5, the term $\mathbf{e}_i$ in the joint density or mass function refers to the case when $x_t = \mathbf{e}_i$. In Step 4, the conditional mass is $f(x_{t+1}|x_t,y_{t+1},\ldots,y_0) = f(x_{t+1}|x_t)$, (by the Markov property and independence between the noise and the Markov chain), which is simply the transition probability. We obtain the likelihood function as the product of conditional density functions found in Step 2.

2.5 The Interest Rate Model

We assume that the interest rate follows a discrete version of the continuous time stochastic process defined by the SDE,

$$dr_t = \alpha_t(\bar{r}_t - r_t)dt + \sigma_t r_t^\gamma dW_t, \tag{2.4}$$

where $\alpha_t = \alpha$, $\bar{r}_t = \bar{r}(X_t)$, $\sigma_t = \sigma(X_t)$, and W is a standard Brownian motion independent with the continuous time Markov chain, X. The continuous time Markov chain is characterized by its transition rate matrix, $\mathbf{Q}$, which is related to the probability transition matrix through the forward and backward Kolmogorov equations. In particular, for a homogeneous Markov chain whose rate matrix doesn't depend on time, we have $\mathbf{P} = e^{\mathbf{Q}\Delta t}$, obtained by the matrix exponential. A well-behaved Markov chain has a rate matrix that is a so-called conservative Q-matrix, which means that $\mathbf{Q}$ has non-negative off-diagonal entries and its rows sum to zero, so the probability transition matrix does turn out to be a stochastic matrix with entries $\mathbf{P}_{ij}$ representing the probability of going from state i at time s to state j at time $s + \Delta t$.

If Δt is small, then an Euler approximation of the above SDE provides the following discrete representation of the interest rate:

$$\Delta r_{t+1} = \alpha\{\bar{r}(X_t) - r_t\}\Delta t + \sigma(X_t)r_t^\gamma\sqrt{\Delta t}\varepsilon_{t+1}, \tag{2.5}$$

where $\{X_t\}$ is a discrete time Markov chain with transition matrix $\mathbf{P} = e^{\mathbf{Q}\Delta t}$ and $\{\varepsilon_t\}$ are i.i.d. standard normal. Here α is the rate of mean reversion, $\bar{r}$ is the mean reverting level or central tendency, σ is the volatility of the interest rate process, and γ is the variance elasticity. We estimate the following equation

$$\Delta r_{t+1} = \beta_0(X_t) + \beta_1 r_t + \varsigma(X_t)r_t^\gamma \varepsilon_{t+1}, \tag{2.6}$$

and then transform the coefficients to the more meaningful term structure coefficients.

We now turn our attention to what this model implies about the behavior of interest rates. First of all the short-term rates, following this model will be positively auto-correlated through time. The auto-correlation is $\rho = 1 - \alpha\Delta t$, which is less than 1 whenever the mean reversion rate, α, is positive, where α measures the rate at which r is expected to approach the mean reverting level, $\bar{r}$. If $\rho > 1$, then the process will drift away from the mean. If $\rho = 1$, then μ must equal zero and thus r follows a random walk.

Allowing the parameters to depend on a Markov chain means that the central tendency and interest rate volatility will change from time to time. When it does change, the interest rate will begin to converge toward the new central tendency level, when it changes back, the interest rate will turn around and begin to converge back. It seems intuitive that such a data generating process would describe a cyclical pattern, but with a random cycle length. In particular, such a data generating process would be able to create large cycles with a relatively small volatility parameter, as is typically seen in a series of interest rates.

Since short-term interest rates are essentially controlled by the central bank, it might seem unreasonable that they switch so violently. However, whatever the underlying variable that the bank is primarily controlling through its choice of interest rates, be it inflation, exchange rates, or unemployment, etc, it might not be unreasonable to model these, exogenously, as having the impact of a Markov chain switching the converging level. Furthermore, the continuous time rate process has continuous sample paths, which provides a certain smoothness in interest changes that is often desired by central banks. This intuition suggests that randomness in the central tendency or drift term of the short rate process will be the more natural. Surprisingly, it is randomness in the volatility that seems to be more important.

2.6 Data

For the Canadian data we use monthly and weekly observations of the short term interest rate implied by Government of Canada 3-month Treasury bills. The monthly rates were obtained from the Bank of Canada website, excerpted from *Selected Canadian and International Interest Rates Including Bond Yields and Interest Arbitrage*. The data set includes bills from March 1934, when the first public tender occurred, until December 2004. The rates quoted in this data set are measured in units of percent and quoted as a discount style of interest rate. To be consistent with our modeling, we convert these rates to unitless annual continuously compounded values. Monthly rates are based on the last Wednesday of the month. Data with weekly observations starts Wednesday January 3, 1962. The weekly Canadian data was taken from the CANSIM website (series V121778).

For the U.S. data, we also use monthly and weekly observations of the U.S. 3-month Treasury bill returns provided by the St. Louis Federal Reserve website. Monthly returns are provided from January 1934 to December 2004 and weekly returns are provided from January 1954; however, to be consistent with our Canadian data, we restrict attention to the post 1962 period. These returns are based on averages over the week or month, so we use daily data and choose data from each Wednesday or the last Wednesday of the month for our observations. For those Wednesdays that land on a holiday, we use data from the following Thursday. The returns in these data sets are discrete discount returns, so we convert them into continuously compounded annual returns.

Interest rates obtained from 3-month T-bills are used because these products have much longer time series of data available than 1-month T-bills. The interest rate model in the previous section is a discrete time approximation of a continuous time model of the infinitesimally short term risk-free rate. As such it would be better to use a shorter term product with more frequent observations; however, the institutional features of the interest rate market and data availability leave us with the current compromise.[1]

---

[1] Moving to daily observations may also introduce too much serial correlation, which may lead to inconsistent estimators.

Before we proceed to the general case, it is informative to first consider the simpler case where the parameters are constant and do not depend on the Markov chain. First we rewrite Eq. 2.6 as $\Delta \mathbf{r} = \mathbf{X}\beta + \mathbf{u}$, where $\mathbf{u} \sim N(\mathbf{0}, \Omega)$, $\Omega = \varsigma^2 \text{diag}[r_0^{2\gamma}, \ldots, r_{T-1}^{2\gamma}]$, $\mathbf{X} = [\mathbf{1} \ \ \ell_1(\mathbf{r})]$, $\mathbf{1}$ is a column vector of 1's, $\ell_1(\mathbf{r}) = (r_0, \ldots, r_{T-1})^\mathsf{T}$, and $\beta = (\alpha \bar{r} \Delta t, 1 - \alpha \Delta t)^\mathsf{T}$. This form makes it clear that for any known γ, β can be estimated by generalized (or weighted) least squares, $\hat{\beta}(\gamma) = (\mathbf{X}^\mathsf{T} \Omega^{-1} \mathbf{X})^{-1} \mathbf{X}^\mathsf{T} \Omega^{-1} \Delta \mathbf{r}$, which is the output of the OLS regression $\eta \Delta \mathbf{r} = \eta \mathbf{X} \beta + \eta \mathbf{u}$, for $\eta = \text{diag}[r_0^{-\gamma}, \ldots, r_{T-1}^{-\gamma}]$.

On the other hand, if β is known, then the log-likelihood function is

$$\ell(\mathbf{r}; \beta, \gamma, \hat{\sigma}) = \{-T\ln(2\pi) - T - T\ln(\hat{\sigma}^2) - 2\gamma \sum_{t=1}^{T} \ln(r_{t-1})\}/2, \qquad (2.7)$$

where $\hat{\sigma}^2 = \frac{1}{T} \sum \left(\frac{\Delta r_t - \mathbf{X}\beta}{r_{t-1}^\gamma} \right)^2$. Given β, this equation can be maximized numerically over the one unknown variable γ.

With normally distributed errors, $\hat{\beta}(\gamma)$ is a maximum likelihood estimate conditional on γ. Since γ is independent of β terms, we can iterate back and forth maximizing conditional on γ, then conditional on β, etc. In fact, thinking of $\hat{\beta}(\gamma)$ as a function of γ allows us to maximize over γ in one step. Details of this approach can be found in Davidson and MacKinnon [9].

The output of this estimation is provided in Table 2.1. All of the parameters have the expected signs; however, none of them are significantly different from zero. Furthermore, none of the series are expected to differ from zero as is seen by the F statistics. This observation foreshadows our more general finding that once the proper diffusion parameters are employed, the drift parameters are not very important. None of the four series differs significantly from a unit root as tested by the Dickey-Fuller τ statistics, all four series have residuals that exhibit significant serial correlation as seen by the Durbin-Watson statistics, and the residuals of all four series have a significantly non-normal distribution according to the Jarque-Bera statistics. These negative results, together with the very low R^2 statistics demonstrate that this type of linear drift one-factor model does not adequately describe 3-month T-bill rates in either country, at least not for such a long time period.[2] Particularly troubling is the combination of serial correlation in the residuals and regressing on lagged dependent variables, which can cause the maximum likelihood estimators to be inconsistent. Combining this observation with the non-normality of the residuals implies that the parameter estimates may not be very accurate.

Nevertheless, to compare the parameters and to allow them to be more easily understood, it is helpful to convert them to the form in Eq. 2.5. This conversion is done in the second panel of Table 2.1. The parameters should be approximately

---

[2] For robustness we repeated the analysis for the shorter time period from May 1990 to December 2004 using 3-month T-bill rates, 3-month LIBOR rates, and 1-month LIBOR rates, and although the parameters differed substantially from the longer period, the findings were generally similar. The only noteworthy differences were that β_1 was significantly less than zero, ruling out a unit root in each case, and for the Canadian short rates, serial correlation in the residuals was no longer present. Data for the LIBOR rates was obtained at the British Bankers' Association website.

CEV	US-T-3-m	US-T-3-w	Can-T-3-m	Can-T-3-w
β_0	0.000123	2.59×10^{-5}	0.000552	6.81×10^{-5}
t stat	1.137823	0.83111	1.498601	1.065669
β_1	−0.00037	−0.00013	−0.00784	−0.00098
DF τ stat	−0.09936	−0.13147	−1.15416	−0.82891
std err	0.012858	0.002313	0.001547	0.000208
F stat	1.542211	0.668327	1.183339	0.594016
γ	1.156092	1.134993	0.770751	0.772217
r^2	0.009546	0.000596	0.004592	0.00053
DW stat	1.621623	1.831274	1.584126	1.564956
JB stat	241.351	4030.834	2596.504	88514.14
# obs	515	2,243	515	2,243
Δt	0.083333	0.019231	0.083333	0.019231
$\bar{r}$	0.330697	0.200768	0.070411	0.06966
α	0.004479	0.006698	0.094047	0.050839
σ	0.392037	0.34664	0.135987	0.103896
γ	1.156092	1.134993	0.770751	0.772217

Table 2.1 Quasi-maximum likelihood estimation of the regression equation $\Delta r_{t+1} = \beta_0 + \beta_1 r_t + \varsigma r_t^\gamma \varepsilon_{t+1}$, where ε_{t+1} are independent standard normal random variables. For a given γ this equation can be estimated independently of the distribution of ε by GLS. The normal distribution is used to obtain the maximum likelihood estimate for γ. Here $\Delta r_{t+1} = r_{t+1} - r_t$, and r_t is the unitless, annual, continuously-compounded yield to maturity at date t on a US or Canadian T-bill maturing 3 months later. These interest rate observations occur monthly or weekly from January 1962 to December 2004, with weekly observations occurring each Wednesday (or the next available date), and monthly observations occurring on the last Wednesday of each month (or the last weekly observation of the month). The column notation is used to indicate the country, interest rate type (T-bill v. LIBOR), maturity (3 month v. 1 month), and observation frequency (monthly v. weekly) of the short rate process. (Although only 3-month T-bill data was available from 1962, other short rates became available from May 1990 and were considered for robustness.) The regression coefficients have Student t statistics reported, which in the case of β_1 is actually a Dickey-Fuller τ statistic used to test against a unit root. The F statistic tests against the fully restricted model with both coefficients being zero. The Durbin-Watson statistic is used to test for serial correlation in the residuals, and the Jarque-Bera statistic is used to test whether the residuals are normally distributed. The values for Δt are $1/12$ for monthly and $1/52$ for weekly observations, which are used to convert the regression coefficient estimates into interest rate model coefficients for the equation $\Delta r_{t+1} = \alpha(\bar{r} - r_t)\Delta t + \sigma r_t^\gamma \sqrt{\Delta t} \varepsilon_{t+1}$

equal when comparing the monthly to the weekly observed time series, which is the case for all parameters except the rate of mean reversion parameter α, being difficult to estimate anyway [5, 2, 21].

Comparing parameter estimates for both countries yields a further contrast. The Canadian rates have a central tendency ($\bar{r}$) around 0.07 and a rate of mean reversion (α) between 0.05 and 0.09, whereas the U.S. rates have a much higher central tendency (greater than 0.2) and a much slower rate of mean reversion between 0.0045 and 0.0067. The data suggests that U.S. rates tend toward a fairly high interest rate at a fairly slow rate and Canadian rates tend more quickly toward a modest interest rate. This high US-low Canadian central tendency is even more puzzling when

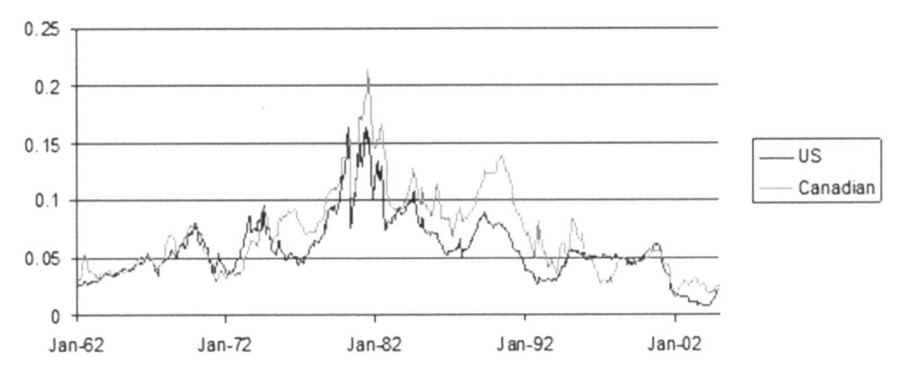

Fig. 2.1 The graph depicts US and Canadian 3-month T-bill rates from January 1962 to December 2004. The rates are in unitless continuously compounded form and they are based on monthly observations on the last Wednesday in each month

considering Fig. 2.1, which shows that US rates never once reach 20%, and Canadian rates peak higher and typically are higher for most of this period.

Finally, we notice, like Chan et al. [5], that the elasticity parameter for this period is greater than 1 for the U.S. data (between 1.135 and 1.156), which provides a good opportunity to see if incorporating stochastic volatility will reduce the elasticity to an acceptable level as suggested by Sun [21]. It turns out that we find that elasticity is not affected in a predictable way by incorporating stochastic volatility. (In particular, we find that elasticity is decreased in the Canadian case, which was already sufficiently low, but it is not changed significantly in US rates, which could be considered too high for purposes of term structure modeling.)

Yong [23] provides some conditions on the variance elasticity parameter γ for a continuous time model similar to Eq. 2.4, in which the parameter functions α_t, $\bar{r}_t$, and σ_t are bounded, deterministic functions of time. In that case, there does not exist a continuous associated wealth process bounded in expectation when $\gamma > 1/2$. In particular, the condition $E[\exp(\lambda \int_0^T r_t dt)] < \infty$ does not hold for any $T > 0$ and $\lambda > 0$. Yong [23] also shows that Novikov's condition fails when $\gamma \geq 1/2$, which causes problems for determining the existence of an equivalent martingale measure, when using this interest model in conjunction with a Black-Scholes type model for the risky asset(s). Failure of the existence of an equivalent martingale measure implies that the model permits arbitrage.

In empirical investigation, we find $\gamma > 1/2$ for all of our samples except for weekly observations of Canadian T-bill rates. This indicates that a linear interest rate model of the type described by Eq. 2.4 may not adequately fit with historical interest rate data. One approach to deal with this problem could be to apply a non-linear model, such as that put forward by Aït-Sahalia [1]. However, two main distinctions between our model and Yong's [23] model could help explain these problematic empirical results: First, we actually estimate the discrete model given by Eqs. 2.5 and 2.6, instead of the continuous model given by Eq. 2.4, so the parameters may not exactly coincide with those of a continuous model. Second, we do not permit the parameter functions to vary directly and deterministically with time. By permitting the parameters to vary directly with time, information from the term structure could be used to estimate the parameter functions, which might decrease estimates of the elasticity parameter. However, since our main objective in this paper is to compare the relative importance of stochastic volatility and stochastic central tendency, we leave a detailed investigation of this issue for future research.

Figure 2.2 plots the monthly observed US 3-month T-bill rates and the residuals to the estimation equation used in this section. Note that the residuals seem to cluster into high and low volatility regimes, which provides further motivation for our method.

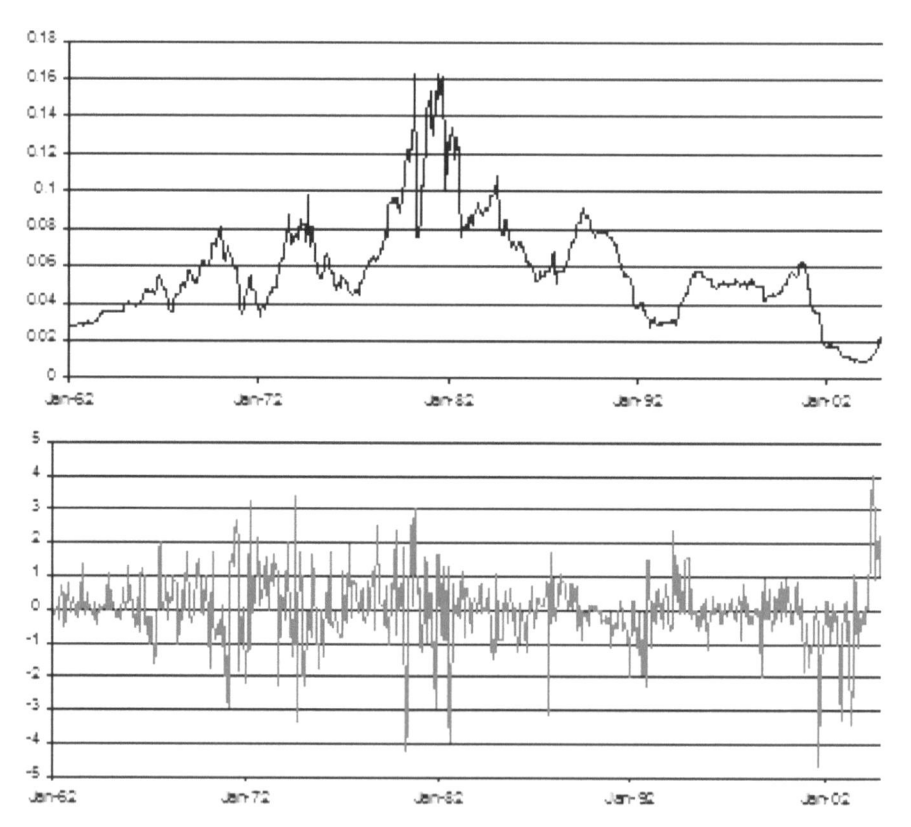

Fig. 2.2 The *upper panel* depicts the US 3-month T-bill rate between January 1962 and December 2004. The *lower panel* plots the standardized residuals from the constant elasticity of volatility model $\Delta r_{t+1} = \beta_0 + \beta_1 r_t + \sigma r_t^\gamma \varepsilon_{t+1}$ estimated via maximum likelihood

2.7 Results

We implement the model for a 2-state Markov chain.[3] Because the columns of the Markov matrix must sum to 1, the matrix is associated with two free parameters, also the central tendency and volatility components of the model are associated with two free parameters each. Together with the mean reversion rate and volatility elasticity parameters the model has a total of eight parameters. The estimates obtained by the algorithm are presented in Tables 2.2–2.5. Tables 2.2 and 2.3 use monthly and weekly observed US data respectively and Tables 2.4 and 2.5 use the equivalent Canadian data.

These tables provide parameter estimates for hidden Markov interest rate models. Each table is associated with a separate time series. The first column reports values for the fully restricted model, in which neither parameter is permitted to switch according to the Markov chain. The remaining columns relax various parameter restrictions. We consider restrictions on the stochastic nature of the mean-reverting level or volatility parameters. These restrictions force the level or volatility to be constant by requiring them to take the same values in each state, (although they allow the constant to be arbitrary). For each restricted model, we report maximum likelihood parameter estimates and the value of the log-likelihood function evaluated at the MLE. We also report likelihood ratio statistics, and Durbin-Watson and Jarque-Bera statistics, which are constructed from each model's residuals. The partially restricted models are associated with only one fewer degrees of freedom, (seven free parameters instead of eight); however, the fully restricted model has four fewer degrees of freedom since the stochastic matrix parameters are no longer relevant in that case.

By looking at the likelihood ratio statistics, we see that for all short rate series, the constant volatility restriction can be rejected and the constant central tendency restriction cannot be rejected. The small improvement in likelihood from relaxing the constant level restriction suggests that the parsimonious model is best. A caveat to this finding is that it depends heavily on the second factor of randomness being present. When comparing the fully restricted model with constant central tendency and constant volatility to a stochastic central tendency model, the likelihood does improve substantially. We don't report statistics in the tables, but the smallest likelihood ratio statistic would be about 40, which is highly significant.

This result shows that on its own, stochastic central tendency seems to be very important, which is consistent with Balduzzi et al. [3]. However, when compared on an equal footing with an alternative 2-factor model having stochastic volatility, it is found to be almost completely unimportant in explaining historical interest rates, which suggests that it is the second factor of randomness in general that is important rather than the stochastic central tendency in particular.

---

[3] For robustness, the complete analysis was repeated for a 3-state Markov chain and the results were qualitatively the same: Stochastic central tendency was important only when compared to the 1-factor model. It was unimportant when compared with a stochastic volatility model.

US-T-3-m	Constant	Constant ς	Constant β_0	Full
$\beta_0(e_1)$	0.000123	0.00184	8.32×10^{-5}	6.68×10^{-5}
$\beta_0(e_2)$		0.00011		0.000173
β_1	−0.000373	−0.029164	-2.12×10^{-8}	-2.1×10^{-8}
$\varsigma(e_1)$	0.113171	0.156445	0.053643	0.053674
$\varsigma(e_2)$			0.167192	0.167746
γ	1.156092	1.286568	1.159406	1.160018
P_{11}		0.990704	0.919711	0.918096
P_{12}		0.009296	0.080289	0.081904
P_{21}		0.044543	0.120808	0.123999
P_{22}		0.955457	0.879192	0.876001
loglikelihood	2161.35	2181.783	2245.361	2245.467
LR-stat	168.2335	127.3688	0.211469	
DW-stat	1.621623	1.723194	1.677849	1.67412
JB-stat	241.351	328.0739	166.5115	183.8003
# obs	515	515	515	515
Δt	0.083333	0.083333	0.083333	0.083333
$\bar{r}(e_1)$	0.330697	0.0631	3930.426	3157.727
$\bar{r}(e_2)$		0.003787		8157.085
α	0.004479	0.349969	2.54×10^{-7}	2.54×10^{-7}
$\sigma(e_1)$	0.392037	0.541941	0.185826	0.185933
$\sigma(e_2)$			0.579172	0.581091
γ	1.156092	1.286568	1.159406	1.160018
Q_{11}		−0.114663	−1.075674	−1.1005
Q_{12}		0.114663	1.075674	1.100502
Q_{21}		0.549448	1.618516	1.666099
Q_{22}		−0.549448	−1.618516	−1.6661

Table 2.2 Quasi-maximum likelihood estimation of the regression equation $\Delta r_{t+1} = \beta_0(X_t) + \beta_1 r_t + \varsigma(X_t) r_t^\gamma \varepsilon_{t+1}$, where ε_{t+1} are independent standard normal random variables and X_t is a hidden Markov chain. Here $\Delta r_{t+1} = r_{t+1} - r_t$, and r_t is the unitless, annual, continuously-compounded yield to maturity at date t on a US T-bill maturing 3 months later based on monthly observations from January 1962 to December 2004 on the last Wednesday of each month. The column notation is used to indicate various model restrictions from the full model with both β_0 and ς allowed to take different values in different states of the Markov chain. The columns "constant ς" "constant β_0" and "constant" imply that ς, β_0 or both respectively take the same value in each state of the Markov chain. The matrix P is the transition probability matrix with P_{ij} being the probability of switching from state i to state j. The loglikelihood is used to calculate the likelihood ratio statistics, which are based on the difference in loglikelihood between the full model and each particular restriction. The Durbin-Watson statistic is used to test for serial correlation in the residuals, and the Jarque-Bera statistic is used to test whether the residuals are normally distributed. The value for Δt is used to convert the regression coefficient estimates into interest rate model coefficients for the equation $\Delta r_{t+1} = \alpha(\bar{r}(X_t) - r_t)\Delta t + \sigma(X_t) r_t^\gamma \sqrt{\Delta t} \varepsilon_{t+1}$. The matrix Q is the transition rate matrix $P = e^{Q\Delta t}$

2 Stochastic Volatility: Evidence from a Hidden Markov Model

US-T-3-w	Constant	Constant ς	Constant β_0	Full
$\beta_0(\mathbf{e}_1)$	2.59×10^{-5}	4.76×10^{-5}	2.01×10^{-5}	1.86×10^{-5}
$\beta_0(\mathbf{e}_2)$		-0.004438		2.65×10^{-5}
β_1	-0.000129	-3.79×10^{-8}	-1.75×10^{-11}	-1.75×10^{-11}
$\varsigma(\mathbf{e}_1)$	0.04807	0.045367	0.026938	0.026941
$\varsigma(\mathbf{e}_2)$			0.077429	0.077445
γ	1.134993	1.137806	1.151832	1.151875
$\mathbf{P}_{11}$		0.993943	0.967915	0.9679
$\mathbf{P}_{12}$		0.006057	0.032085	0.0321
$\mathbf{P}_{21}$		0.343071	0.060586	0.060636
$\mathbf{P}_{22}$		0.656929	0.939414	0.939364
loglikelihood	11195.18	11269.19	11612.03	11612.04
LR-stat	833.7179	685.7047	0.028266	
DW-stat	1.831274	0.858285	1.853169	1.853177
JB-stat	4030.834	12186.02	5194.324	5194.194
# obs	2,243	2,243	2,243	2,243
Δt	0.019231	0.019231	0.019231	0.019231
$\bar{r}(\mathbf{e}_1)$	0.200768	1255.337	1,150,858	1,067,508
$\bar{r}(\mathbf{e}_2)$		-116999.2		1,514,390
α	0.006698	1.97×10^{-6}	9.08×10^{-10}	9.08×10^{-10}
$\sigma(\mathbf{e}_1)$	0.34664	0.327146	0.194254	0.194272
$\sigma(\mathbf{e}_2)$			0.558349	0.558466
γ	1.134993	1.137806	1.151832	1.151875
Q_{11}		-0.387444	-1.750856	-1.751751
Q_{12}		0.387444	1.750856	1.751751
Q_{21}		21.94357	3.306166	3.308974
Q_{22}		-21.94357	-3.306166	-3.308974

Table 2.3 Quasi-maximum likelihood estimation of the regression equation $\Delta r_{t+1} = \beta_0(X_t) + \beta_1 r_t + \varsigma(X_t) r_t^\gamma \varepsilon_{t+1}$, where ε_{t+1} are independent standard normal random variables and X_t is a hidden Markov chain. Here $\Delta r_{t+1} = r_{t+1} - r_t$, and r_t is the unitless, annual, continuously-compounded yield to maturity at date t on a US T-bill maturing 3 months later based on weekly observations from January 1962 to December 2004 on the last Wednesday of each month. The column notation is used to indicate various model restrictions from the full model with both β_0 and ς allowed to take different values in different states of the Markov chain. The columns "constant ς" "constant β_0" and "constant" imply that ς, β_0 or both respectively take the same value in each state of the Markov chain. The matrix **P** is the transition probability matrix with $\mathbf{P}_{ij}$ being the probability of switching from state i to state j. The loglikelihood is used to calculate the likelihood ratio statistics, which are based on the difference in loglikelihood between the full model and each particular restriction. The Durbin-Watson statistic is used to test for serial correlation in the residuals, and the Jarque-Bera statistic is used to test whether the residuals are normally distributed. The value for Δt is used to convert the regression coefficient estimates into interest rate model coefficients for the equation $\Delta r_{t+1} = \alpha(\bar{r}(X_t) - r_t)\Delta t + \sigma(X_t) r_t^\gamma \sqrt{\Delta t} \varepsilon_{t+1}$. The matrix **Q** is the transition rate matrix $\mathbf{P} = e^{\mathbf{Q}\Delta t}$

Can-T-3-m	Constant	Constant ς	Constant β_0	Full
$\beta_0(\mathbf{e}_1)$	0.000552	0.014384	0.000277	0.000273
$\beta_0(\mathbf{e}_2)$		0.000433		−0.000328
β_1	−0.007837	−0.012064	−0.001139	-6.23×10^{-5}
$\varsigma(\mathbf{e}_1)$	0.039256	0.034997	0.011153	0.010552
$\varsigma(\mathbf{e}_2)$			0.038387	0.036238
γ	0.770751	0.788342	0.553748	0.539309
$\mathbf{P}_{11}$		0.269987	0.932768	0.933597
$\mathbf{P}_{12}$		0.730013	0.067232	0.066403
$\mathbf{P}_{21}$		0.02281	0.191737	0.177576
$\mathbf{P}_{22}$		0.97719	0.808263	0.822424
loglikelihood	2038.84	2073.68	2143.932	2144.174
LR-stat	210.6678	140.9884	0.485521	
DW-stat	1.584126	0.647096	1.528304	1.528396
JB-stat	2596.504	31575.5	1117.866	1076.651
# obs	515	515	515	515
Δt	0.083333	0.083333	0.083333	0.083333
$\bar{r}(\mathbf{e}_1)$	0.070411	1.19231	0.243123	4.38532
$\bar{r}(\mathbf{e}_2)$		0.035868		−5.27452
α	0.094047	0.144767	0.013662	0.000747
$\sigma(\mathbf{e}_1)$	0.135987	0.121232	0.038637	0.036552
$\sigma(\mathbf{e}_2)$			0.132975	0.125534
γ	0.770751	0.788342	0.553748	0.539309
$\mathbf{Q}_{11}$		−16.26365	−0.933712	−0.913454
$\mathbf{Q}_{12}$		16.26365	0.933712	0.913454
$\mathbf{Q}_{21}$		0.508179	2.662839	2.442781
$\mathbf{Q}_{22}$		−0.508179	−2.662839	−2.442781

Table 2.4 Quasi-maximum likelihood estimation of the regression equation $\Delta r_{t+1} = \beta_0(X_t) + \beta_1 r_t + \varsigma(X_t) r_t^\gamma \varepsilon_{t+1}$, where ε_{t+1} are independent standard normal random variables and X_t is a hidden Markov chain. Here $\Delta r_{t+1} = r_{t+1} - r_t$, and r_t is the unitless, annual, continuously-compounded yield to maturity at date t on a Canadian T-bill maturing 3 months later based on monthly observations from January 1962 to December 2004 on the last Wednesday of each month. The column notation is used to indicate various model restrictions from the full model with both β_0 and ς allowed to take different values in different states of the Markov chain. The columns "constant ς" "constant β_0" and "constant" imply that ς, β_0 or both respectively take the same value in each state of the Markov chain. The matrix $\mathbf{P}$ is the transition probability matrix with $\mathbf{P}_{ij}$ being the probability of switching from state i to state j. The loglikelihood is used to calculate the likelihood ratio statistics, which are based on the difference in loglikelihood between the full model and each particular restriction. The Durbin-Watson statistic is used to test for serial correlation in the residuals, and the Jarque-Bera statistic is used to test whether the residuals are normally distributed. The value for Δt is used to convert the regression coefficient estimates into interest rate model coefficients for the equation $\Delta r_{t+1} = \alpha(\bar{r}(X_t) - r_t)\Delta t + \sigma(X_t) r_t^\gamma \sqrt{\Delta t} \varepsilon_{t+1}$. The matrix $\mathbf{Q}$ is the transition rate matrix $\mathbf{P} = e^{\mathbf{Q}\Delta t}$

Can-T-3-w	Constant	Constant ς	Constant β_0	Full
$\beta_0(\mathbf{e}_1)$	6.81×10^{-5}	0.008376	2.45×10^{-5}	2.54×10^{-5}
$\beta_0(\mathbf{e}_2)$		7.37×10^{-5}		-1.17×10^{-5}
β_1	-0.000978	-0.00254	-7.84×10^{-5}	-6.09×10^{-5}
$\varsigma(\mathbf{e}_1)$	0.014408	0.012304	0.002357	0.002356
$\varsigma(\mathbf{e}_2)$			0.010228	0.010221
γ	0.772217	0.771539	0.431682	0.431884
$\mathbf{P}_{11}$		0.296443	0.933413	0.933518
$\mathbf{P}_{12}$		0.703557	0.066587	0.066482
$\mathbf{P}_{21}$		0.008998	0.172207	0.170958
$\mathbf{P}_{22}$		0.991002	0.827793	0.829042
loglikelihood	11140.16	11373.37	11825.4	11825.44
LR-stat	1370.562	904.1384	0.074158	
DW-stat	1.564956	0.322833	1.59396	1.594426
JB-stat	88514.14	358305.9	12302.88	12244.49
# obs	2,243	2,243	2,243	2,243
Δt	0.019231	0.019231	0.019231	0.019231
$\bar{r}(\mathbf{e}_1)$	0.06966	3.297269	0.312126	0.416996
$\bar{r}(\mathbf{e}_2)$		0.029008		-0.192659
α	0.050839	0.13209	0.004077	0.003166
$\sigma(\mathbf{e}_1)$	0.103896	0.088725	0.016999	0.016991
$\sigma(\mathbf{e}_2)$			0.073757	0.073705
γ	0.772217	0.771539	0.431682	0.431884
Q_{11}		-64.01087	-3.95636	-3.946761
Q_{12}		64.01087	3.95636	3.946761
Q_{21}		0.818628	10.23188	10.14906
Q_{22}		-0.818628	-10.23188	-10.14906

Table 2.5 Quasi-maximum likelihood estimation of the regression equation $\Delta r_{t+1} = \beta_0(X_t) + \beta_1 r_t + \varsigma(X_t) r_t^{\gamma} \varepsilon_{t+1}$, where ε_{t+1} are independent standard normal random variables and X_t is a hidden Markov chain. Here $\Delta r_{t+1} = r_{t+1} - r_t$, and r_t is the unitless, annual, continuously-compounded yield to maturity at date t on a Canadian T-bill maturing 3 months later based on weekly observations from January 1962 to December 2004 on the last Wednesday of each month. The column notation is used to indicate various model restrictions from the full model with both β_0 and ς allowed to take different values in different states of the Markov chain. The columns "constant ς" "constant β_0" and "constant" imply that ς, β_0 or both respectively take the same value in each state of the Markov chain. The matrix $\mathbf{P}$ is the transition probability matrix with $\mathbf{P}_{ij}$ being the probability of switching from state i to state j. The loglikelihood is used to calculate the likelihood ratio statistics, which are based on the difference in loglikelihood between the full model and each particular restriction. The Durbin-Watson statistic is used to test for serial correlation in the residuals, and the Jarque-Bera statistic is used to test whether the residuals are normally distributed. The value for Δt is used to convert the regression coefficient estimates into interest rate model coefficients for the equation $\Delta r_{t+1} = \alpha(\bar{r}(X_t) - r_t)\Delta t + \sigma(X_t) r_t^{\gamma} \sqrt{\Delta t} \varepsilon_{t+1}$. The matrix $\mathbf{Q}$ is the transition rate matrix $\mathbf{P} = e^{\mathbf{Q}\Delta t}$

The remaining rows report estimates for the equivalent parameterization of Eq. 2.5. These estimates are more meaningful and they are comparable with each other among the different observation frequencies. The entries of the matrix $\mathbf{Q}$ represent the rate that a continuous Markov chain switches between states. They are obtained by solving $\mathbf{P} = e^{\mathbf{Q}\Delta t}$. The values of these observation frequency-robust estimates generally do seem to be quite close to each other; however the LR statistics indicate a significant difference (e.g. comparing monthly to weekly observations for US 3-month T-bills yields a LR statistic of 7.28).

As in Table 2.1, most models predict that β_1 is close to 0, which is associated with a rate of mean reversion α close to 0. The exception is the case with stochastic drift and constant volatility. In that case the mean reversion rate is quite high, which is not too surprising since, for the central tendency to be important, the mean reversion rate can't be too close to zero in our models.

One problem that occurs in some cases is a very high central tendency, (around 151 million percent for the US weekly observed rate). On the other hand this very high central tendency was also associated with a very slow rate of mean reversion in all but the Canadian stochastic central tendency and constant volatility cases, which had high values for P_{12} suggesting that the process switches out of state 1 soon after it enters, but the potential for unreasonably high interest rates to be generated by this process does still exist.[4]

Finally, for the Canadian data, the elasticity parameter does seem to decrease with the introduction of stochastic volatility. On the other hand it increases slightly for the U.S. data. There doesn't seem to be a conclusive empirical finding on this result.

Furthermore, it seems that the potential for our more general models to alleviate the observed serial correlation and non-normality of the residuals for the constant models is not achieved. While the stochastic volatility models tend to increase the Durbin-Watson statistics and reduce the Jarque-Bera statistics, these statistics are nowhere near their optimal values of 2 and 0 respectively. A caveat to this finding is that maximum likelihood estimation does not necessarily minimize the sum of squared residuals, so the reported statistics are not as meaningful as they are for regression models.

Another quantity of interest is the filter for the Markov chain. Figure 2.3 plots the conditional probability of being in the high-volatility state over time as estimated by the full model on monthly observations US 3-month T-bill rates. Several features to notice are that the state probability seems to traverse quickly between high and low values of approximately 0.88 and 0.17. Furthermore, it seems to switch out of each state quite frequently, (and more frequently down than up, consistent with the transition probability estimates of 0.082 for switching up and 0.124 for switching down). Also, the volatility seems to be more likely in the high state between the late 1960s and 1982 and more likely in the low volatility state in the post 1982 period. However, the frequent switching of the Markov chain suggests that it is picking up

---

[4] For robustness, we further restricted the stochastic volatility models by requiring all drift parameters to be 0. Such restrictions had little effect on the log-likelihoods and the LR statistics were all less than 0.4.

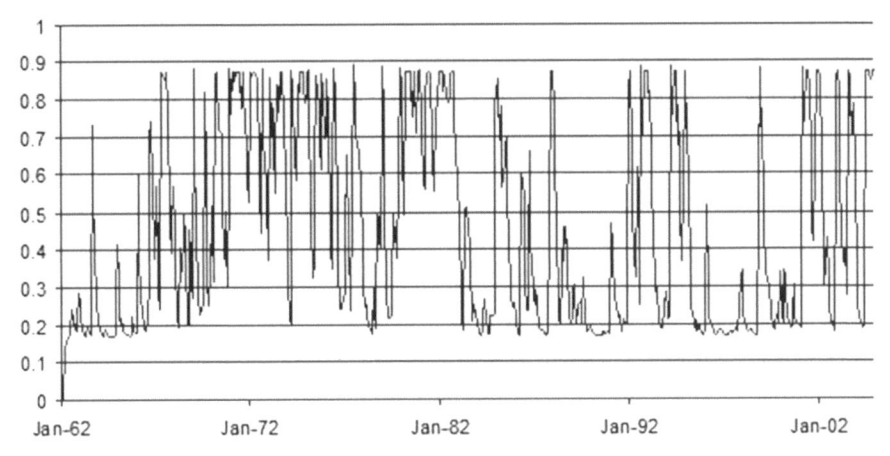

Fig. 2.3 The filter for the Markov chain. The conditional probability (given the observations up to date) of the high volatility (and high central tendency) state as estimated from the hidden Markov model $\Delta r_{t+1} = \beta_0(X_t) + \beta_1 r_t + \sigma(X_t) r_t^\gamma \varepsilon_{t+1}$ via maximum likelihood using monthly observations of the US 3-month T-bill rate, where $\{X_t\}$ is an unobserved 2-state Markov chain

more than just a structural break at 1982. It is more likely that the split is due to the rising and falling long-term trends apparent in Fig. 2.2 together with the fact that the high volatility state is also the high central tendency state in this case.

2.8 Conclusion

We develop a 2-factor model for the short-term interest rate where the second factor enters through both the volatility and the drift, (and in particular, the central tendency). We develop a method for estimating and testing the model. Since the decrease in the likelihood is much greater for the constant volatility restriction, it seems that a stochastic volatility is very important, and far more important than stochastic drift components for explaining nominal interest rate movements. This conclusion is true for both Canadian and U.S. interest rates using both monthly and weekly observations.

A potential weakness of this finding is the limitation of a linear stochastic drift. The combination of non-linear drift and stochastic central tendency may affect our conclusions. This issue is particularly relevant in light of our estimates of the volatility elasticity parameter γ being greater than 1/2 for most samples, which is inappropriate for term structure modeling. We leave this problem to future research.

References

1. Aï t-Sahalia, Y.: Testing continuous-time models of the spot interest rate. Rev. Financ. Stud. **9**, 385–426 (1996)
2. Balduzzi, P., Das, S.R., Foresi, S., Sundaram, R.: A simple approach to three-factor term structure models. J. Fixed Income **6**, 43–53 (1996)
3. Balduzzi, P., Das, S.R., Foresi, S.: The central tendency: a second factor in bond yields. Rev. Econ. Stat. **80**, 62–72 (1998)
4. Bansal, R., Zhou, H.: Term structure of interest rates with regime shifts. J. Finance **57**, 1997–2043 (2002)
5. Chan, K., Karolyi, G., Longstaff, F., Sanders, A.: An empirical comparison of alternative models of the short-term interest rate. J. Finance **47**, 1209–1227 (1992)
6. Chapman, D.A., Pearson, N.D.: Is the short rate drift actually nonlinear? J. Finance **55**, 355–388 (2000)
7. Cox, J.C., Ingersoll, J.E., Ross, S.A.: A theory of the term structure of interest rates. Econometrica **53**, 385–408 (1985)
8. Dai, Q., Singleton, K.J.: Specification analysis of affine term structure models. J. Finance **55**, 1943–1978 (2000)
9. Davidson, R., MacKinnon, J.G.: Estimation and Inference in Econometrics. Oxford University Press, New York (1993)
10. Elliott, R.J.: Exact adaptive filters for Markov chains observed in Gaussian noise. Automatica **30**, 1399–1408 (1994)
11. Guthrie, G., Wright, J.: The optimal design of interest rate target changes. J. Money Credit Bank. **36**(1), 115–137 (2004)
12. Hamilton, J.D.: Rational expectations economic analysis of changes in regime. J. Econ. Dyn. Control **12**, 385–423 (1988)
13. Heath, D., Jarrow, R., Morton, A.: Bond pricing and the term structure of interest rates: a new methodology for contingent claims valuation. Econometrica **60**, 77–105 (1992)
14. Ho, T.S.Y., Lee, S.B.: Term structure movements and pricing interest rate contingent claims. J. Finance **41**, 1011–1029 (1986)
15. Hull, J., White, A.: Pricing interest-rate derivative securities. Rev. Financ. Stud. **3**, 573–592 (1990)
16. Kalimipalli, M., Susmel, R.: Regime-switching stochastic volatility and short-term interest rates. J. Empir. Finance **11**, 309–329 (2004)
17. Lehmann, E.L.: Theory of Point Estimation. Wiley, New York (1983)
18. Longstaff, F., Schwartz, E.: Interest rate volatility and the term structure: a two-factor general equilibrium model. J. Finance **47**, 1259–1282 (1992)
19. Naik, V., Lee, M.H.: Yield curve dynamics with discrete shifts in economic regimes: theory and estimation. Working paper, University of British Columbia (1997)
20. Smith, D.R.: Markov switching and stochastic volatility diffusion models of short-term interest rates. J. Bus. Econ. Stat. **20**(2), 183–197 (2002)

21. Sun, L.: Nonlinear drift and stochastic volatility: an empirical investigation of short-term interest rate models. J. Financ. Res. **26**, 389–404 (2003)
22. Vasicek, O.: An equilibrium characterization of the term structure. J. Financ. Econ. **5**, 177–188 (1977)
23. Yong, J.: Some estimates on exponentials of solutions to stochastic differential equations. J. Appl. Math. Stoch. Anal. **4**, 287–316 (2004)

Chapter 3
An Econometric Model of the Term Structure of Interest Rates Under Regime-Switching Risk

Shu Wu and Yong Zeng

Abstract This paper develops and estimates a continuous-time model of the term structure of interests under regime shifts. The model uses an analytically simple representation of Markov regime shifts that elucidates the effects of regime shifts on the yield curve and gives a clear interpretation of regime-switching risk premiums. The model falls within the broad class of essentially affine models with a closed form solution of the yield curve, yet it is flexible enough to accommodate priced regime-switching risk, time-varying transition probabilities, regime-dependent mean reversion coefficients as well as stochastic volatilities within each regime. A two-factor version of the model is implemented using Efficient Method of Moments. Empirical results show that the model can account for many salient features of the yield curve in the U.S.

3.1 Introduction

Economic theories relate asset prices, and interest rates in particular, to either observable or latent variables that summarize the state of the aggregate economy. Since one important characteristic of the aggregate economy is the recurrent shifts between distinct phases of the business cycle, economists have long used models that incorporate Markov regime shifts to describe the stochastic behavior of interest

---

S. Wu (✉)
Department of Economics, University of Kansas, 1450 Jayhawk Blvd, Lawrence, KS 66045, USA
e-mail: shuwu@ku.edu

Y. Zeng
Department of Mathematics and Statistics, University of Missouri at Kansas City, 5100 Rockhill Rd, Kansas City, MO 64110, USA
e-mail: zengy@umkc.edu

rates. Some examples include Hamilton [36], Lewis [43], Cecchetti et al. [11], Sola and Driffill [48], Garcia and Perron [34], Gray [35] and Ang and Bekaert [2] among others. Typically these studies model the short-term interest rate as a stochastic process with time-varying parameters that are driven by a Markov state variable. Long-term interest rates can then related to the short rate through the expectation hypothesis. Results from these studies suggest that regime-switching models in general have better empirical performance than their single-regime counterparts. Regime-switching models are shown to be able to capture the non-linearities in the drift and volatility function of the short rate found in non-parametric models [3].

The success of these empirical studies have motivated a growing literature that examine the impact of regime shifts on the entire yield curve using dynamic term structure models. For example, Boudoukh et al. [10] investigates the implications of a 2-regime model of the business cycle based on GDP, consumption and production data for term premiums and volatilities in the bond market. Bansal and Zhou [5] and Bansal et al. [7] incorporate a Markov-switching state variable into the parameters of an otherwise standard multi-factor Cox-Ingersoll-Ross (CIR) model of the term structure of interest rates. A closed-form solution for the yield curve is obtained under log-linear approximation. They find that the key to the better empirical performance of the regime-switching model is the added flexibility of the market price of risk under multiple regimes, and regimes in the term structure model are intimately related to bond risk premiums and the business cycle. Evans [26] develops and estimates a dynamic term structure model under regime shifts for both nominal and real interest rates in Britain. In a similar study, Ang et al. [4] also develops a non-arbitrage regime-switching model of the term structure of interest rates with both nominal bond yields and inflation data to efficiently identify the term structure of real rates and inflation risk premia. Different from the model in Evans [26], Ang et al. [4] allows inflation and real rates to be driven by two different regime variables. Dai et al. [17] emphasizes that not only regime shifts can affect parameters of the state variables, but also regime-switching risk should be priced in dynamic term structure models. Using monthly data on the U.S. Treasury zero-coupon bond yields, they show that the priced regime-switching risk plays a critical role in capturing the time variations in the expected excess bond returns. These studies all use discrete-time models. More recent examples include Ferland et al. [28], Futami [30] and Xiang and Zhu [51] among others.

In this paper we contribute to this literature by developing and estimating a continuous-time model of the term structure of interests under regime shifts. Most of the existing regime-switching models are specified in a discrete-time framework. Compared to those models, our continuous-time model has several advantages. (1) It uses an analytically simple representation of Markov regime shifts that helps elucidate the effect of regime shifts on the yield curve; (2) It offers a clear economic interpretation of the market price of regime-switching risk; (3) It gives a tractable solution of the term structure of interest rates in the presence of time-varying transition probabilities, regime-dependent mean reversion coefficients, priced regime-switching risk, and stochastic volatilities conditional on each regime without using log-linear approximations; (4) A continuous-time model is also more convenient in the applications of the pricing of interest rate derivatives.

The model presented in this paper falls within the broad class of affine models of Duffie and Kan [20], Dai and Singleton [14], Duffee [19], and more recently Aït-Sahalia and Kimmel [1] and Le et al. [42]. The model implies that bond risk premiums include two components under regime shifts. One is a regime-dependent risk premium due to diffusion risk. The other is a regime-switching risk premium that depends on the covariations between the discrete changes in bond prices and the stochastic discount factor across different regimes. This new component of the term premiums is associated with the systematic risk of recurrent shifts in bond prices (or interest rates) due to regime changes and is an important factor that affects bond returns.

One stylized fact about the term structure of interest rates is that long-term rates do not attenuate in volatility. In the standard affine models, volatility of interest rates depends the factor loadings, which can converge to zero quickly unless the underlying state variables are very persistent (under the risk-neutral probability measure). The model in the present paper shows that regime shifts introduce an additional source of volatility that can equally affect both the short and the long end of the yield curve. Therefore the model is able to generate volatile long-term interest rates even when the underlying state variables are not very persistent.

Other continuous-time term structure models under regime shifts include Landen [40] which uses a similar representation for Markov regime shifts as that in the current paper. Landen [40], however, only solves the term structure of interest rates under the risk-neutral probability measure, and is silent on the market price of risk. Dai and Singleton [16] also proposes a continuous-time model of the term structure of interest rates under regime shifts based on a different representation of Markov regime shifts. Both studies did not implement their models empirically. Wu and Zeng [50] develops a general equilibrium model of the term structure of interest rates under regime shifts similar to that in Cox et al. [13]. The focus of their study is on the general equilibrium interpretation of the regime-switching risk. And unlike the present paper, they obtain the solution of the yield curve under log-linear approximations following Bansal and Zhou [5]. Separately, exponential affine models of bond prices under regime switching are also derived in Elliott and Siu [24] and Siu [46].

The rest of the paper is organized as follows. Section 3.2 presents the theoretical model and examines the effects of regime shifts on the term structure of interest rates. A closed-form solution of the term structure of interest rates is obtained for an essentially affine model. Section 3.3 implements a two-factor version of the model using Efficient Method of Moments and discusses the empirical results. Section 3.4 contains some concluding remarks and possible extensions of the model.

3.2 The Model

3.2.1 A Simple Representation of Markov Regime Shifts

In order to obtain a simple and closed-form solution of the yield curve, we first show that Markov regime shifts can be modeled as a marked point process.[1] The main advantage of this new representation of regime shifts is that it allows us to elucidates the role of regime shifts in determining the term structure of interest rates with a clear interpretation of the regime-switching risk premiums.

We assume that there are N possible regimes and denote $S(t)$ as the regime at time t. Let the mark space $U = \{1, 2, \ldots, N\}$ be all possible regimes with the power σ-algebra. We denote u as a generic point in U and A as a subset of U. A marked point process or a random counting measure, $m(t,A)$, is defined as the total number of times we enter a regime that belongs to A during $(0,t]$. For example, $m(t, \{u\})$ simply counts the total number of times we enter regime u during $(0,t]$. We also define η as the usual counting measure on U with the following two properties: For $A \in U$, $\eta(A) = \int I_A \eta(du)$ (i.e. $\eta(A)$ counts the number of elements in A) and $\int_A f(u)\eta(du) = \sum_{u \in A} f(u)$.

The probability laws of the marked point process defined above, $m(t,\cdot)$, can be uniquely characterized by a stochastic intensity kernel,[2] which is assumed to be

$$\gamma_m(dt, du) = h(u; S(t-), X(t))\eta(du)dt, \tag{3.1}$$

where $X(t)$ is a vector of other continuous state variables to be specified below; $h(u, S(t-), X(t))$ is the conditional regime-shift (from regime $S(t-)$ to u) intensity at time t (we assume $h(u, S(t-), X(t))$ is bounded) that is measurable with respect to u, $S(t-)$ and $X(t)$. The $N \times N$ conditional intensity matrix of regime-switching is $H(X(t)) = \{h(j, i, X(t))\}$ with $h(i, j, X(t)) = 0$ when $i = j$. Heuristically, $\gamma_m(dt, du)$ can be thought of as the (time-varying) conditional probability of shifting from Regime $S(t-)$ to Regime u during $[t-, t+dt)$ given $X(t)$ and $S(t-)$. Note that $\gamma_m(t, A)$, the compensator of $m(t, A)$,[3] can be written as

$$\gamma_m(t, A) = \int_0^t \int_A h(u, S(\tau-), X(\tau))\eta(du)d\tau = \sum_{u \in A} \int_0^t h(u, S(\tau-), X(\tau))d\tau.$$

---

[1] In the context of continuous-time models, Landen [40] also uses a marked point process to represent Markov regime shifts in her model of the term structure of interest rates. However, we use a different construction of the mark space that simplifies the corresponding random measure. Other approaches to regime shifts include Hidden Markov Models (e.g. Elliott et al. [22]) and the Conditional Markov Chain models (e.g. Yin and Zhang [52]). An application of Hidden Markov Models to the term structure of interest rates can be found in Elliott and Mamon [23]. Bielecki and Rutkowski [8, 9] are examples of the application of conditional Markov Chain models to the term structure of interest rates.

[2] See Last and Brandt [41] for detailed discussion of marked point process, stochastic intensity kernel and related results.

[3] This simply means that $m(t, A) - \gamma_m(t, A)$ is a martingale.

With the market point process appropriately defined, we now can represent the regime, $S(t)$, as an integral along $m(\cdot,\cdot)$ as that in He et al. [39],

$$S(t) = S(0) + \int_{[0,t] \times U} (u - S(\tau-))m(d\tau, du). \tag{3.2}$$

Note that $m(d\tau, du)$ is 0 most of time and only becomes 1 at a regime-switching time t_i with $u = S(t_i)$, the new regime at time t_i. In other words, the above expression is equivalent to a telescoping sum: $S(t) = S(0) + \sum_{t_i < t}(S(t_i) - S(t_{i-1}))$.

We can also describe the evolution of regimes $S(t)$ in a differential form

$$dS(t) = \int_U (u - S(t-))m(dt, du). \tag{3.3}$$

To see the above differential equation is valid, assuming there is a regime shift from $S(t-)$ to u at time t, then $S(t) - S(t-) = (u - S(t-))$, implying $S(t) = u$.

Alternatively, we can express $dS(t)$ as

$$dS(t) = \int_U (u - S(t-))\gamma_m(dt, du) + \int_U (u - S(t-))[m(dt, du) - \gamma_m(dt, du)]. \tag{3.4}$$

where $\gamma_m(dt, du)$ is the intensity kernel of $m(dt, du)$. And by construction $m(dt, du) - \gamma_m(dt, du)$ is a matingale error term, and hence can be thought of as a *regime-switching shock* whereas the first term is the conditional expectation of $dS(t)$.[4]

3.2.2 Other State Variables

We assume that, in addition to the Markov regime-switching variable $S(t)$, there are L other *continuous* state variables represented by a $L \times 1$ vector $X(t)$. Without loss of generality we assume that $X(t)$ is given by the following stochastic differential equation

$$dX(t) = \Theta(X(t), S(t-))dt + \Sigma(X(t), S(t-))dW(t) \tag{3.5}$$

where $\Theta(X, S)$ is a $L \times 1$ vector; $\Sigma(X, S)$ is a $L \times L$ matrix; $W(t)$ is a $L \times 1$ vector of standard Brownian motions that is independent of $S(t)$. Note that in this specification, both the drift term $\Theta(\cdot, \cdot)$ and the diffusion term $\Sigma(\cdot, \cdot)$ are regime dependent. This general specification also allows stochastic volatility within each regime. The time-path of $X(t)$, however, is continuous.

---

[4] This is analogous to the representation of Markov regime shifts as an AR(1) process in discrete-time models.

3.2.3 The Term Structure of Interest Rates

Let $M(t)$ denote the pricing kernel.[5] We assume that $M(t)$ is given by

$$\frac{dM(t)}{M(t-)} = -r(t-)dt - \lambda'_D(X(t),S(t-))dW(t)$$
$$- \int_U \lambda_S(u,S(t-),X(t))[m(dt,du) - \gamma_m(dt,du)] \quad (3.6)$$

where $r(t)$ is the instantaneous short-term interest rate, $\lambda_D(X,S)$ is a $L \times 1$ vector of market prices of diffusion risk, and $\lambda_S(u,S,X)$ is the market price of regime-switching (from regime $S(t-)$ to regime u) risk given X_t. The interpretations for λ_D and λ_S will become much clearer from the discussions below.

Note that the explicit solution for $M(t)$ can be obtained by Doleans-Dade exponential formula [44] as follows

$$M(t) = \left(e^{-\int_0^t r_\tau - d\tau}\right) \left(e^{-\int_0^t \lambda'_D(X_\tau,S_{\tau-})dW(\tau) - \frac{1}{2}\int_0^t \lambda'_D(X_\tau,S_{t-})\lambda_D(X_\tau,S_{\tau-})d\tau}\right) \times$$
$$\left(e^{\int_0^t \int_U \lambda_S(u,S_{\tau-},X_\tau)\gamma_m(d\tau,du) + \int_0^t \int_U \log(1-\lambda_S)(u,S_{\tau-},X_\tau)m(d\tau,du)}\right) \quad (3.7)$$

The term structure of interest rates can be obtained by a change of probability measure. We first obtain the following two lemmas. The first lemma characterizes the equivalent martingale measure under which the yield curve is determined. The second lemma obtains the dynamic of the state variables under the equivalent martingale measure.

Lemma 3.1. *For fixed $T > 0$, the equivalent martingale measure $\mathbf{Q}$ can be defined by the Radon-Nikodym derivative below*

$$\frac{dQ}{dP} = \xi(T)/\xi_0$$

where for $t \in [0,T]$

$$\xi(t) = \left(e^{-\int_0^t \lambda'_D(X_\tau,S_{\tau-})dW(\tau) - \frac{1}{2}\int_0^t \lambda'_D(X_\tau,S_{\tau-})\lambda_D(X_\tau,S_{\tau-})d\tau}\right) \times$$
$$\left(e^{\int_0^t \int_U \lambda_S(u,S_{\tau-},X_\tau)\gamma_m(d\tau,du) + \int_0^t \int_U \log(1-\lambda_S)(u,S_{\tau-},X_\tau)m(d\tau,du)}\right) \quad (3.8)$$

provided λ_D satisfies Kazamaki or Novikov's criterion and λ_S and h (the stochastic intensity kernel of $m(t,A)$) are all bounded functions.

Lemma 3.2. *Under the risk-neutral probability measure $\mathbf{Q}$, the dynamics of state variables, $X(t)$ and $S(t)$, are given by the following stochastic differential equations respectively*

---

[5] Absence of arbitrage is sufficient for the existence of the pricing kernel under certain technical conditions, as pointed out by Harrison and Kreps [38].

3 An Econometric Model of The Term Structure of Interest Rates 61

$$dX(t) = \tilde{\Theta}(X(t), S(t-))dt + \Sigma(X(t), S(t-))d\tilde{W}(t) \quad (3.9)$$

$$dS(t) = \int_U (u - S(t-))\tilde{m}(dt, du) \quad (3.10)$$

where $\tilde{\Theta}(X,S) = \Theta(X,S) - \Sigma(X,S)\lambda_D(X,S)$, $\tilde{W}(t)$ is a $L \times 1$ standard Brownian motion and $\tilde{m}(t,A)$ is a marked point process with intensity matrix $\tilde{H}(X) = \{\tilde{h}(j,i,X)\} = \{h(j,i,X)(1 - \lambda_S(j,i,X))\}$, under $\mathbf{Q}$, respectively.

Note that the compensator of $\tilde{m}(t,A)$ under $\mathbf{Q}$ becomes

$$\tilde{\gamma}_m(dt, du) = \tilde{h}(u, S(t-), X(t))\eta(du)dt = (1 - \lambda_S(u, S(t-), X(t)))\gamma_m(dt, du).$$

In the absence of arbitrage, the price at time $t-$ of a default-free pure discount bond that matures at T, $P(t-,T)$, can be obtained as

$$P(t-,T) = E^{\mathbf{Q}}\left[e^{-\int_t^T r_\tau - d\tau}|\mathscr{F}_{t-}\right] = E^{\mathbf{Q}}\left[e^{-\int_t^T r_\tau - d\tau}|X(t), S(t-)\right] \quad (3.11)$$

with the boundary condition $P(T,T) = 1$. The last equality comes from the Markov property of $(X(t), S(t))$. Without loss of generality, let $P(t-,T) = f(t, X(t), S(t-), T)$. The following proposition gives the partial differential equation determining the bond price.

Proposition 3.1. *The price of the default-free pure discount bond $f(t, X, S, T)$ defined in (3.11) satisfies the following partial differential equation*

$$\frac{\partial f}{\partial t} + \frac{\partial f}{\partial X'}\tilde{\Theta} + \frac{1}{2}\mathrm{tr}\left(\frac{\partial^2 f}{\partial X \partial X'}\Sigma\Sigma'\right) + \int_U \Delta_S f\,\tilde{h}(u,S,X)\eta(du) = rf \quad (3.12)$$

with the boundary condition $f(T,X,S,T) = 1$, where $\Delta_S f \equiv f(t,X,u,T) - f(t,X,S,T)$.

3.2.4 Bond Risk Premiums Under Regime Shifts

In general equation (3.12) doesn't admit a closed form solution for the bond price. Nonetheless, the equation allows us to illustrate how regime shifts affect bond risk premiums and give a clear interpretation the market price of regime-switching risk, λ_S.

By Ito's formula, we have

$$df = \left[\frac{\partial f}{\partial t} + \frac{\partial f}{\partial X'}\theta(X,S) + \frac{1}{2}\mathrm{tr}\left(\frac{\partial^2 f}{\partial X \partial X'}\Sigma(X,S)\Sigma'(X,S)\right)\right]dt + \frac{\partial f}{\partial X'}\Sigma(X,S)dW$$
$$+ \int_U \Delta_S f\,\gamma_m(dt,du) + \int_U \Delta_S f\,(m(dt,du) - \gamma_m(dt,du))$$

$$(3.13)$$

Using (3.12), Lemma 3.2 as well as the definition of $\gamma_m(dt,du)$ in Eq. (3.1), we can easily obtain

$$E_{t-}\left(\frac{df}{f}\right) - rdt = \left[\frac{1}{f}\frac{\partial f}{\partial X'}\Sigma(X,S)\lambda_D(X,S)\right]dt \\ + \left[\int_U \frac{\Delta_S f}{f}\lambda_S(u,X,S)h(u,X,S)\eta(du)\right]dt \quad (3.14)$$

The left-hand side of (3.14) gives the (instantaneous) expected excess return on a zero-coupon bond, or risk premium. The equation shows that the bond risk premium includes two components under regime shifts. Conditional on regime S, the first term is the minus of the covariance between the bond return, df/f, and the change in the pricing kernel, dM/M, which can be interpreted as investors' marginal utility growth, due to shocks $dW(t)$ (see Eq. (3.6) for the specification of $M(t)$). We refer to this term in this paper as the diffusion risk premium. This risk premium is in general time-varying due to the presence of $X(t)$ and $S(t)$ in Σ and λ_D.[6] Equation (3.14) shows clearly that, compared to single-regime models in which the diffusion risk premium depends only on $X(t)$, regime shifts introduces an additional source of time variation in the risk premiums as $S(t)$ changes randomly over time. Since researchers often attribute the failure of the expectation theory of the term structure of interest rates to time-varying risk premiums, regime shifts therefore can potentially improve the empirical performance of dynamic term structure models. In fact, Bansal and Zhou [5] argues that the regime-dependence of the diffusion risk premium plays a crucial role in enabling their econometric model to account for the failure of the expectation theory.

Equation (3.14) also makes it clear that, under regime shifts, bond risk premiums in general include a second component. To understand more clearly what the second component is about, recall that $-\lambda_S$ simply gives the impact of a regime-switching shock $m(dt,du) - \gamma_m(dt,du)$ on dM/M in Eq. (3.6), whereas $\frac{\Delta_S f}{f}$ has a similar interpretation in Eq. (3.13). Also recall that $h(u,X,S)$ is the regime-switching intensity (from S to u). Therefore $\int_U \frac{\Delta_S f}{f}\lambda_S(X,S)h(u,X,S)\eta(du)$ is again the minus of the *covariance* between the bond return, df/f, and the change of the pricing kernel, dM/M, or the marginal utility growth, under a regime-switching shock $m(dt,du) - \gamma_m(dt,du)$ given $X(t)$ and $S(t-)$. We refer to this second component as the regime-switching risk premium. This risk premium is present not only because regime shifts have a direct impact on the bond price, $\frac{\Delta_S f}{f}$, but also because regime shifts have a direct impact, $-\lambda_S$, on the pricing kernel or investors' marginal utility. As in the case of the diffusion risk premium, if the regime-switching shocks generate movements in the bond return and the pricing kernel (or marginal utility) in the same direction, the covariance is positive and the risk premium is negative as the bond offers investors a hedge against the risk of regime shifts. On the other hand, If regime shifts generate movements in the bond return and the pricing kernel (or

---

[6] It is possible that $\frac{1}{f}\frac{\partial f}{\partial X'}$ depends on $X(t)$ and $S(t)$ as well. Nonetheless in the broad class of affine models, $\frac{1}{f}\frac{\partial f}{\partial X'}$ is a constant that depends only on the bond's maturity.

marginal utility) in the opposite directions, the covariance is negative and the risk premium will be positive. In this case, regime shifts make the bond risky because they decrease the asset's return when investors' marginal utility is high.

In some regime-switching models such as Bansal and Zhou [5], however, it is assumed that regime-switching risk is not priced by fixing λ_S at zero. The assumption is equivalent to assume that regime-switching is not an aggregate risk, and therefore the regime-switching shock $m(dt,du) - \gamma_m(dt,du)$ doesn't have a any impact on $\frac{dM}{M}$ given X and S (see Eq. (3.6)). Since most empirical regime-switching models are motivated by business cycle fluctuations or shifts in monetary policies, it seems important to treat regime shifts as an aggregate risk. Some empirical results such as those from Dai et al. [17], suggest that λ_S not only is statistically significant, but also economically important.

Finally, as we can see from Eq. (3.14), the regime-switching risk premium is in general time-varying. This is simply because both the market price of regime-switching risk λ_S and the regime-switching intensity h can depend on state variables $X(t)$ and $S(t)$. Moreover, as we will show below, the term $\frac{\Delta_S f}{f}$ is also time-varying even in affine models, unlike the constant term $\frac{1}{f}\frac{\partial f}{\partial X}$ in the diffusion risk premium. This property of regime-switching risk premium adds another flexibility to models with multiple regimes.

3.2.5 An Affine Regime-Switching Model

To further illustrate the effects of regime shifts on the term structure of interest rates, we resort to the tractable specifications of affine models that have been widely used in the empirical studies. Duffie and Kan [20] and Dai and Singleton [14] have detailed discussions of affine term structure models under diffusions. Duffie et al. [21] deals with general asset pricing under affine jump-diffusions. Extensions of the standard affine models are discussed, for example, in Duffee [19] and Duarte [18] which propose a class of essentially affine or semi-affine models. In models with regime shifts, Landen [40], Bansal and Zhou [5], Evans [26], Dai et al. [17] and Ang and Bekaert [4] among others all have similar affine structures. The main advantage of affine models is that they can produce analytical solutions of the term structure of interest rates, and yet at the same time are flexible enough to accommodate time-varying risk premiums and stochastic volatilities. The model we discuss below generalizes the single-regime affine models to include regime-dependent mean reversion coefficients, priced regime-switching risk as well as time-varying regime-switching probabilities.

More specifically, we make the following parametric assumptions:

(1) $\Theta(X(t), S(t-)) = \Theta_0(S(t-)) + \Theta_1(S(t-))X(t)$ where $\Theta_0(S)$ is a $L \times 1$ vector and $\Theta_1(S)$ is $L \times L$ matrix.

(2) $\Sigma(X(t), S(t-))$ is a $L \times L$ diagonal matrix with the ith diagonal element given by $[\Sigma(X_t, S_{t-})]_{ii} = \sqrt{\sigma_{0,i}(S_{t-}) + \sigma'_{1,i} X_t}$ for $i = 1, \cdots, L$ where $\sigma_{1,i}$ is a $L \times 1$ vector. We assume both $\sigma_{0,i}$ and $\sigma_{1,i}$ are positive.

(3) $h(u, X(t), S(t-)) = e^{h_0(u, S_{t-}) + h'_1(u, S_{t-}) X_t}$ where $h_1(u, S_{t-})$ is a $L \times 1$ vector.

(4) $\lambda_D(X(t), S(t-)) = \Sigma(X_t, S_{t-})^{-1} \left(\lambda_{0,D}(S_{t-}) + \lambda_{1,D} X_t + \Theta_1(S_{t-}) X_t \right)$ where $\lambda_{0,D}(S_{t-})$ is a $L \times 1$ vector, and $\lambda_{1,D}$ is a $L \times L$ matrix that is constant across regimes.

(5) $1 - \lambda_S(u, S(t-), X(t)) = e^{\phi(u, S_{t-})} / h(u, S_{t-}, X_t)$ where $h(u, S_{t-}, X_t) \neq 0$.

(6) The instantaneous short-term interest $r(t-)$ is a linear function of the state variables $X(t)$ given $S(t-)$

$$r(t-) = \psi_0(S(t-)) + \psi'_1 X(t) \qquad (3.15)$$

where $\psi_0(S)$ is a regime-dependent constant and ψ_1 is a $L \times 1$ vector of constants that are independent of regimes.[7]

The first three assumptions are about the dynamics of the state variables. For $X(t)$, Assumption (1) and (2) implies that its drift and volatility terms are all affine functions of X_t conditional on regimes. In particular $X(t)$ is given by

$$dX(t) = \left[\Theta_0(S(t-)) + \Theta_1(S(t-)) X(t) \right] dt + \Sigma(X(t), S(t-)) dW(t), \qquad (3.16)$$

where

$$\Sigma(X(t), S(t-)) = \begin{pmatrix} \sqrt{\sigma_{0,1}(S_{t-}) + \sigma'_{1,1} X_t} & & \\ & \ddots & \\ & & \sqrt{\sigma_{0,L}(S_{t-}) + \sigma'_{1,L} X_t} \end{pmatrix}$$

Under this specification, the mean-reversion coefficient and the "steady-state" value of $X(t)$ are given by $-\Theta_1(S)$ and $-\Theta_1(S)^{-1} \Theta_0(S)$, both can shift across regimes. Moreover, the model also has a flexible specification for the volatility of $X(t)$ with regime specific $\sigma_{0,i}(S)$ and stochastic volatility in each regime $\sigma'_{1,i} X(t)$. Some empirical studies have shown that inflation in the U.S. has become less volatile and less persistent in recent years compared to earlier periods, probably due to a combination of the moderation of output volatility and changes in the monetary policy that has provided a better anchor for long-run inflation expectations. If the latent factors in $X(t)$ are to capture the fundamental driving forces in the economy, it is important to allow this kind of regime shifts in an empirical model of the term structure of interest rates.[8]

---

[7] If ψ_1 is regime-dependent, an analytical solution of the yield curve is in general unavailable. Bansal and Zhou [5] and Wu and Zeng [50] assume that ψ_1 depends on regimes and obtain the term structure of interest rates under log-linear approximation.

[8] In order to obtain a closed form solution of the term structure of interest rates, we need to restrict σ_1 to be constant across regimes.

3 An Econometric Model of The Term Structure of Interest Rates 65

Assumption (3) implies that the log intensity of regime shifts is an affine function of the state variable X_t conditional on regimes. This assumption ensures the positivity of the intensity function and also allows the transition probability to be time-varying.

The next two assumptions deal with the market prices of risk. In the standard affine models, the market price of (diffusion) risk is assumed to be proportional to the volatility of the state variable X_t. Such a structure is intuitive: risk compensation goes to zero as risk goes to zero. However, since variances are nonnegative, this specification limits the variation of the compensations that investors anticipate to receive when encountering a risk. More precisely, since the compensation is bounded below by zero, it cannot change sign over time. This restriction, however, is relaxed in the essentially affine models of Duffee [19]. Dai and Singleton [15] argues that this extension is crucial for empirical models to account for the failure of the expectation theory of the term structure of interest rates.

Following this literature, we use a similar specification as that of essentially affine models for the market price of the diffusion risk in Assumption (4), but with an extension to include multiple regimes. Under this assumption, conditional on regimes, the diffusion risk premium of bonds will be proportional to $\lambda_{0,D}(S(t-)) + \lambda_{1,D}X(t) + \Theta_1(S(t-))X(t)$, a linear function of the state variable $X(t)$. Moreover Assumption (4) implies that, from Lemma 3.2, $\lambda_{1,D}$ is the *risk neutral* mean reversion coefficient of $X(t)$, which is assumed to be constant across regimes. It turns out this is one of the crucial conditions that are necessary for obtaining closed form solutions of the term structure of interest rates under regime shifts.[9]

For the market price of regime switching risk λ_S, Assumption (5) postulates that, conditional on regimes, $1 - \lambda_S$ is proportional to the inverse of regime-switching intensity. Under this assumption, λ_S can be time-varying, and the higher the regime-switching intensity, the higher the risk compensation. We restrict λ_S to take this particular form because it implies that the *risk neutral* regime-switching intensity is constant conditional on regimes, $\tilde{h}(u, S_{t-}, X_t) = e^{\phi(u, S_{t-})}$. This is another restriction we need to impose on the model in order to obtain a closed-form solution of the term structure of interest rates.

Proposition 3.2. *Under the assumption (1)–(6), the price at time $t-$ of a default-free pure discount bond with maturity τ is given by $P(t-,\tau) = e^{A(\tau, S_{t-}) + B(\tau)'X_t}$ and the τ-period interest rate is given by $R(t-,\tau) = -\frac{A(\tau, S_{t-})}{\tau} - \frac{B(\tau)'X_t}{\tau}$, where $B(\tau) = (B_1(\tau), \cdots, B_L(\tau))'$, and $A(\tau, S)$ and $B(\tau, S)$ are given by the following ordinary integral-differential equations*

$$-\frac{\partial B(\tau)}{\partial \tau} - \lambda'_{1,D}B(\tau) + \frac{1}{2}\Sigma'_1 B^2(\tau) = \psi_1 \qquad (3.17)$$

---

[9] In the regime switching model of Bansal and Zhou [5], the risk-neutral mean reversion coefficient is allowed to shift across regimes. But the term structure of interest rates can only be solved analytically under log linear approximation.

and

$$-\frac{\partial A(\tau,S)}{\partial \tau} + B(\tau)'[\Theta_0(S) - \lambda_{0,D}(S)] + \frac{1}{2}B(\tau)'\Sigma_0(S)B(\tau)$$
$$+ \int_U \left[e^{A(\tau,u)-A(\tau,S)} - 1\right]e^{\phi(u,S)}\eta(du) = \psi_0(S) \quad (3.18)$$

with boundary conditions $A(0,S) = 0$ and $B(0) = 0$, where $B^2(\tau) = (B_1^2(\tau), \cdots, B_L^2(\tau))'$, and Σ_1 and $\Sigma_0(S)$ are $L \times L$ matrices given by

$$\Sigma_1 = \begin{pmatrix} \sigma'_{1,1} \\ \vdots \\ \sigma'_{1,L} \end{pmatrix}, \text{ and } \Sigma_0(S) = \begin{pmatrix} \sigma_{0,1} & & \\ & \ddots & \\ & & \sigma_{0,L} \end{pmatrix}. \quad (3.19)$$

3.2.6 The Effects of Regime Shifts on the Yield Curve

With the analytical solution in Proposition 3.2, we can now further illustrate the effects of regime shifts on the term structure of interest rates. First, the general result for bond risk premiums in Eq. (3.14) is now simplified as

$$E_t\left(\frac{dP_t}{P_{t-}}\right) - r_t dt = [\lambda'_{0,D}(S_{t-}) + X'_t\lambda'_{1,D} + X'_t\Theta'_1(S_{t-})]B(\tau)dt$$
$$+ \int_U \left(e^{A(\tau,u)-A(\tau,S_{t-})} - 1\right)\left(e^{h_0(u;S_{t-})+h'_1(u;S_{t-})X_t} - e^{\phi(u;S_{t-})}\right)\eta(du)dt. \quad (3.20)$$

As in Eq. (3.14), the first term on the right hand side of Eq. (3.20) is the diffusion risk premium and the second term is the regime-switching risk premium. In the standard affine models without regime shifts, the risk premium is determined by a linear function of the state variable $X(t)$, that is $[\lambda'_{0,D} + X'_t\lambda'_{1,D} + X'_t\Theta'_1]B(\tau)$.[10] Moreover, only the factor loadings in the term structure of interest rates, $B(\tau)$, affect the risk premium. The intercept term, $A(\tau)$, doesn't enter the above equation.

By introducing regime shifts, Bansal and Zhou [5] B essentially makes the risk premium a non-linear function of the state variable $X(t)$ because the intercept term $\lambda_{0,D}$ and the slope coefficient $\lambda_{1,D} + \Theta_1$ are now both regime-dependent. Bansal and Zhou [5] shows that it is mainly this feature of their model that provides improved goodness-of-fit over the existing term structure models. One restriction of Bansal and Zhou [5], however, is that they assume that the regime-switching risk is not priced, $\lambda_S(u;S,X) = 0$. In the context of the above affine model, this is equivalent to assume that $e^{h_0(u;S)+h'_1(u;S)X} = e^{\phi(u;S)}$, that is the risk neutral regime-switching

---

[10] In the more restrictive CIR models, $[\lambda'_{0,D}(S_{t-}) + X'_t\lambda_{1,D} + X'_t\Theta_1(S)]$ is further restricted to be proportional to the variance of the state variables.

probabilities, $e^{\phi(u;S)}$, is the same as the physical regime-switching probabilities $e^{h_0(u;S)+h'_1(u;S)X}$. Therefore the bond risk premium is still a linear function of the state variable *conditional on regimes*.

In Dai et al. [17] and Ang and Bekeart [4], $\lambda_S(u,S,X)$ is not restricted to be zero.[11] Equation (3.20) shows that this extension provides an additional flexibility for the model to account for time-varying risk premiums observed in the data, because the second-term on the right-hand side of Eq. (3.20) is a highly non-linear function of the state variable $X(t)$ even conditional on regimes. The equation also shows that the regime-switching risk, λ_S, directly affects the term structure of interest rates through the intercept term, $A(\tau,S)$,[12] while the diffusion risk, λ_D, affect the yield curve through the factor loading, $B(\tau)$.[13]

One caveat of the affine models such as the one obtained in Proposition 3.2, however, is the tension between the transition probabilities between regimes and the market price of regime-switching risk. To obtain a closed form solution, affine models have to restrict the transition probabilities across regimes under the *risk neutral probability measure* to be constant. In other words, $\tilde{h}(u,S(t-),X(t))$ needs to be independent of $X(t)$. In Eq. (3.20), that term is given by $e^{\phi(u,S_{t-})}$. On the other hand, the transition probabilities under the *the physical measure*, $h(u,S(t-),X(t))$, are given by $e^{h_0(u;S_{t-})+h'_1(u;S_{t-})X_t}$ in Eq. (3.20). If the model allows the transition probabilities under the physical measure to be time varying, i.e. $h_1 \neq 0$, as many regime-switching models do,[14] we would impose that the regime-switching risk is priced. This is because $\lambda_S(u,S,X)h(u,S,X) = h(u,S,X) - \tilde{h}(u,X,S)$, and in affine models $\lambda_S(u,S,X)h(u,S,X) = e^{h_0(u;S_{t-})+h'_1(u;S_{t-})X_t} - e^{\phi(u,S_{t-})}$, which implies that $\lambda_S(u,S,X)$ is not zero and must be time-varying as well. It is possible to loose this restriction with more general models (i.e. with time-varying regime-switching probabilities and zero market price of regime-switching risk), but probably at the cost of not being able to obtain a closed form solution to the term structure of interest rates.

One stylized fact of the term structure of interest rates is that long-term interest rates do not attenuate in volatility. In affine models without regime shifts, interest rates are given by $R(t,\tau) = -\frac{A(\tau)}{\tau} - \frac{B(\tau)'X_t}{\tau}$. Therefore the volatility of interest rates is determined by the factor loading $-\frac{B(\tau)}{\tau}$ alone, where $B(\tau)$ is given by the differential equation (3.17). To illustrate why the volatility of long-term interest rates might pose a challenge to affine models, let's consider the one-factor Gaussian model for example. In this case, the solution to $B(\tau)$ depends on the value of $\lambda_{1,D}$. If $\lambda_{1,D} \gg 0$, $B(\tau)$ will converge very quickly (at the rate of $e^{-\tau \lambda_{1,D}}$) to a constant as τ increases,

---

[11] In Ang and Bekeart [4], however, the market price of regime-switching risk is not explicitly defined. $\lambda_S(u,S,X)$ can be derived from the specification of the pricing kernel.

[12] Of course, $A(\tau,S)$ also depends on the factor loading $B(\tau)$ through the differential equation (3.18).

[13] See Siu [47] for a discussion of the pricing of regime-switching risk in equity market.

[14] Hamilton [37] and Filardo [29] are examples of regime-switching models of business cycles with time-varying transitions probabilities. In regime-switching models of interest rates, time-varying transition probabilities are assumed in Gray [35], Boudoukh et al. [10] and more recently Dai et al. [17] among others.

hence so does the interest rate $R(t,\tau)$. In order to generate volatile long-term interest rates, we need $\lambda_{1,D} \approx 0$, which implies $B(\tau) \approx -\tau$ and $R(t,\tau) \approx -\frac{A(\tau)}{\tau} + X(t)$. Long-term interest rates would be as volatile as short-term interest rates. But note that $\lambda_{1,D}$ is the mean reversion coefficient of the state variable $X(t)$ under the risk-neutral probability measure. The requirement that $\lambda_{1,D} \approx 0$, therefore, is to assume that $X(t)$ is close to a unit root process under the risk-neutral probability measure.

In models with regime shifts, $-\frac{A(\tau,S(t-))}{\tau}$ is stochastic and adds another source of volatility for the interest rate $R(t,\tau)$. Since the volatility of $-\frac{A(\tau,S(t-))}{\tau}$ will not attenuate, this would translate directly into the volatility of long-term interest rates even if $\lambda_{1,D} \gg 0$.

3.3 Empirical Results

3.3.1 Data and Summary Statistics

The data used in this study are monthly interest rates from June 1964 to December 2001 obtained from the Center for Research in Security Prices (CRSP).[15] These are yields on zero-coupon bonds extracted from U.S. Treasury securities. There are eight interest rates with maturities ranging from 1 month to 5 years. Table 3.1 contains their summary statistics. We can see that the yield curve is on average upward-sloping and the large skewness and kurtosis suggest significant departure from Gaussian distribution. As Timmermann [49] shows, Markov switching models can generate such large skewness and kurtosis. Another feature of the data is that long-term interest rates (for example, the 5-year rate) are almost as volatile as the 1-month rate. Moreover, volatilities of the interest rates have a hump-shaped structure as noted in Dai et al. [17]. The standard deviation increases from 2.45% for the 1-month rate to 2.60% for the 6-month rate, and then declines to 2.32% for the 5-year rate. Also note that all interest rates are very persistent with high autocorrelation coefficients. The 6-month and 5-year rate are plotted in Fig. 3.1.

We report in Table 3.2 the results from the standard regressions regarding the expectation hypothesis, which states that a long-term interest rate is just the average of the expected short-term interest rate over the life of the long-term bond. The regression used to test this hypothesis is

$$i(R_{t+j}^i - R_t^{i+j}) = \alpha + \beta_{ij}[j(R_t^{i+j} - R_t^j)] + \varepsilon_{t+j}^i$$

where R_t^k is the k-period interest rate at time t. Under the null hypothesis of the expectation theory, $\beta_{ij} = 1$. However, as it is well known, most regressions produce estimates of β_{ij} that are significantly less than 1, and often are negative. Table 3.2 confirms this stylized fact. The estimates of β_{ij} are either insignificantly different

---

[15] To make our study comparable, we consider the roughly same sample period as that in Bansal and Zhou [5].

Table 3.1 Interest rates summary statistics

	1M	3M	6M	1Y	2Y	3Y	4Y	5Y [a]
Mean	0.0594	0.0638	0.0659	0.0681	0.0702	0.0717	0.0729	0.0735
Std. dev.	0.0245	0.0258	0.0260	0.0252	0.0246	0.0238	0.0235	0.0232
Maximum	0.1614	0.1603	0.1652	0.1581	0.1564	0.1556	0.1582	0.1500
Minimum	0.0164	0.0171	0.0178	0.0192	0.0237	0.0296	0.0336	0.0367
Skewness	1.4278	1.3717	1.3122	1.1737	1.1288	1.1283	1.1003	1.0565
Kurtosis	5.4659	5.1336	4.9150	4.4157	4.1226	4.0313	3.9196	3.7344
Auto corr	0.947	0.971	0.971	0.970	0.976	0.978	0.979	0.981

[a] 1M, 3M, 6M indicate 1-, 3- and 6-month interest rates respectively. 1Y, 2Y, ..., 5Y indicate 1-, 2-, ...and 5-year interest rate respectively

from 0 or significantly negative. It is interesting to note, however, the yield spread between 5- and 4-year rate predicts the future (4 years ahead) 1-year rate with correct sign and the expectation hypothesis can not be rejected. In fact, as maturity increases, the estimate of β_{ij} tends to increase from negative to positive, suggesting that the expectation hypothesis might hold in longer terms.

Table 3.3 contains correlation coefficients between the expected excess bond returns and a business cycle dummy variable, BC. NBER dates of business cycles are used to distinguish between expansions ($BC = 1$) and recessions ($BC = 0$). Excess bond returns are obtained as the differences between holding-period returns (1-month) on long-term bonds (1, 2, ..., 5-year bonds respectively) and the 1-month

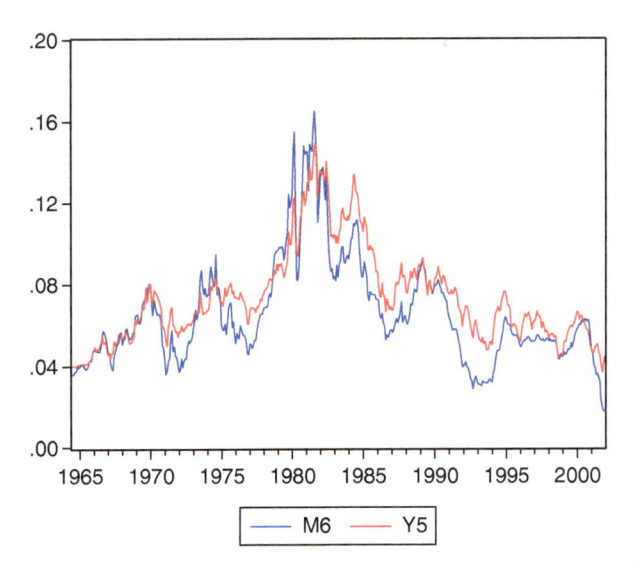

Fig. 3.1 Historical interest rates and the business cycle. The figure plots the 6-month (series M6) and 5-year (series Y5) interest rates during 1964–2001. NBER business cycle recessions are indicated the *shaded area*

Table 3.2 Expectation-hypothesis regression [a]

	i+j=0.5	i+j=1	i+j=2	i+j=3	i+j=4	i+j=5
j=0.25	−0.8656 (0.3766)					
j=0.5		−0.6564 (0.4669)				
j=1			−0.8769 (0.3932)	−1.2513 (0.4697)	−1.6585 (0.5270)	−1.6085 (0.6084)
j=2				−0.3395 (0.5879)	−0.8568 (0.6337)	−1.0011 (0.7539)
j=3					0.0554 (0.4221)	−0.0619 (0.5583)
j=4						0.6853 (0.4681)

[a] This table reports the estimate of β_{ij} in regression $i(R_{t+j}^i - R_t^{i+j}) = \alpha + \beta_{ij}[j(R_t^{i+j} - R_t^j)] + \varepsilon_{t+j}^i$ where R_t^k is the k-year interest rate at time t. Under the null hypothesis of the expectation theory, $\beta_{ij} = 1$. Numbers in parentheses are Newy-White standard errors

interest rate.[16] We can see from Table 3.3 that the correlation coefficients are all negative, which is consistent with the counter-cyclical behavior of risk premiums as documented in Fama and French [27]. Alternatively we can regress the ex-post excess bond returns on the business cycle dummy and the yield spread of the previous period. We include the yield spread in the regression because empirical studies have suggested that yield spreads or forward rates contain information about the state variables that drive the interest rates. Again, the regression coefficients on the business cycle dummy variable are all negative and significant, confirming the counter-cyclical property of bond risk premiums. Interestingly, if we don't include the business cycle dummy variable in the regressions, estimates of the coefficient on the yield spread are all positive and significant (not reported in Table 3.3), indicating that yield spreads do forecast bond returns. Once we include the business cycle dummy in the regressions, however, the dummy variable completely drives out the predicating power of yield spreads for bond returns.

3.3.2 Estimation Procedure

The econometric methodology we adopt to estimate the term structure model in Proposition 2 is Efficient Method of Moments (EMM) proposed in Bansal et al. [6]

---

[16] Continuously compounded bond returns are $H_{t+\Delta t} \equiv -(\tau - \Delta t)R_{t+\Delta t}(\tau - \Delta t) + \tau R_t(\tau)$, where $R_t(\tau)$ is the yield on a τ-year bond at time t. Since we don't have data on $R_{t+\Delta t}(\tau - \Delta t)$, we approximate it by $R_{t+\Delta t}(\tau)$ for $\tau \gg \Delta t$, where $\Delta t = 1$ month. Also note that $\text{Corr}(E_t(H_{t+\Delta t}), BC_t) = \text{Corr}(H_{t+\Delta t}, BC_t)$ under rational expectations.

Table 3.3 Correlation among bond returns and the business cycle [a]

	RETY1	RETY2	RETY3	RETY4	RETY5	BC(-1)
RETY1	1					
RETY2	0.9423	1				
RETY3	0.9057	0.9632	1			
RETY4	0.8515	0.9328	0.9592	1		
RETY5	0.8506	0.9289	0.9553	0.9655	1	
BC(−1)	−0.1986	−0.1703	−0.1518	−0.1277	−0.1218	1
$\hat{\alpha}_i$	0.8966	1.4411	0.0419	0.3453	0.3283	
	(0.4320)	(1.2241)	(1.5286)	(1.8086)	(1.8807)	
$\hat{\beta}_i$	−0.0335**	−0.0596**	−0.1153**	−0.1202**	−0.1270**	
	(0.0096)	(0.0213)	(0.0298)	(0.0383)	(0.0443)	
$\hat{\gamma}_i$	−1.0916	0.1590	2.4134	2.7410	2.8611	
	(0.9559)	(1.3883)	(1.7150)	(2.0863)	(2.3112)	
Adjusted R^2	0.048	0.041	0.042	0.036	0.033	

[a] The first six rows of this table report the sample correlation coefficients among the excess bond returns and the business cycle. RETY1, RETY2, ..., RETY5 are the *ex-post* holding-period (1-month) returns on 1-, 2-, ..., 5-year bonds minus the 1-month interest rate, respectively. BC is the dummy variable for the business cycle with BC=1 indicating an expansion and BC=0 indicating a recession. The last four rows report the OLS regression $RETYi_t = c + \alpha_i SP_{i,t-1} + \beta_i BC_{t-1} + \gamma_i [SP_{i,t-1} \times BC_{t-1}] + \varepsilon_{i,t}$, where RETY$i$ is the holding-period excess return on a i-year bond, SP$_i$ is the yield spread between the i-year bond and the 1-month Bill. Numbers in parentheses are Newy-West standard errors. An ** indicates the estimate is significant at 5 % level

and Gallant and Tauchen [31, 33].[17] We assume that there are two distinct regimes ($N = 2$) for $S(t)$. Therefore (3.17) and (3.18) define a system of three differential equations that must be solved simultaneously. Under regime shifts, the number of parameters of the model increases quickly with each additional factor. In this paper, we estimate a two-factor version of the model, and fit the model to the 6-month and the 5-year interest rates as in Bansal and Zhou [5].

Under EMM procedure, the empirical conditional density of the observed interest rates is first estimated by an auxiliary model that is a close approximation to the true data generating process. Gallant and Tauchen [33] suggests a semi-nonparametric (SNP) series expansion as a convenient general purpose auxiliary model. As pointed out by Bansal and Zhou [5], one advantage of using the semi-nonparametric specification for the auxiliary model is that it can asymptotically converge to any smooth distributions (see also Gallant and Tauchen [32]), including the density of Markov regime-switching models. The dimension of this auxiliary model can be selected by, for example, the Schwarz's Bayesian Information Criterion (BIC).[18] Table 3.4 reports the estimation results of the preferred auxiliary semi-nonparametric model.

---

[17] Bansal and Zhou [5] and Bansal et al. [7] are excellent examples of applying EMM to estimate the term structure model under regime shifts. Dai and Singleton [14] also provides extensive discussions of estimating affine term structure models using EMM procedure.

[18] As for model selection for regime switching models (or general Hidden Markov models), Scott [45] gave an excellent review on limitations of various criteria, including BIC and AIC.

Table 3.4 Parameter estimates of SNP density [a]

Parameter	Estimates	Standard error
a(0,0)	1.00000	0.00000
a(0,1)	0.30987	0.07457
a(1,0)	0.05612	0.04619
a(0,2)	−0.35596	0.06619
a(1,1)	0.10687	0.02843
a(2,0)	−0.04299	0.03182
a(0,3)	−0.02352	0.01820
a(3,0)	−0.01374	0.01023
a(0,4)	0.03257	0.01022
a(4,0)	0.02216	0.00514
$\mu(1,0)$	−0.07309	0.01232
$\mu(2,0)$	−0.05375	0.00800
$\mu(1,1)$	0.81742	0.03382
$\mu(1,2)$	0.15274	0.03288
$\mu(2,1)$	0.00221	0.03763
$\mu(2,2)$	0.95765	0.03690
R(1,0)	0.01843	0.00295
R(2,0)	0.15853	0.01250
R(3,0)	0.18509	0.01616
R(1,1)	0.17234	0.02936
R(2,1)	0.11702	0.05355
R(1,2)	0.12163	0.02951
R(2,2)	0.09551	0.05472
R(1,3)	0.02379	0.03465
R(2,3)	0.10629	0.05286
R(1,4)	0.04649	0.02601
R(2,4)	0.07097	0.04875
R(1,5)	0.05068	0.02554
R(2,5)	0.02903	0.04716

[a] This table reports point estimates as well as their standard errors of the parameters in the preferred SNP model according to BIC (BIC=−1.2488, AIC=−1.3786). $a(i, j)$ are parameters of the Hermit polynomial function. $\mu(i, j)$ are parameters of the VAR conditional mean. $R(i, j)$ are parameters of the ARCH standard deviation of the innovation z. See Gallant and Tauchen [33] or Bansal and Zhou [5] for more detailed interpretations of these parameters

The score function of the auxiliary model are then used to generate moment conditions for computing a chi-square criterion function, which can be evaluated through simulations given the term structure model under consideration. A nonlinear optimizer is used to find the parameter setting that minimizes the criterion function. Gallant and Tauchen [31] shows that such estimation procedure yields fully efficient estimators if the score function of the auxiliary model encompasses the score functions of the model under consideration.

Without further normalization, however, Dai and Singleton [14] shows that affine models as the one in Proposition 2 are under-identified. Therefore we restrict in EMM estimation that $\Theta_1(S)$ and $\lambda_{1,D}$ are lower triangular matrixes. We also restrict

3 An Econometric Model of The Term Structure of Interest Rates

Table 3.5 Parameter estimates of the term structure model [a]

		Regime 1	Regime 2
$\Theta_0(S)$	$\Theta_{0,1}$	0.0050(0.0021)	0.0277(0.0025)
	$\Theta_{0,2}$	0.0002(0.0038)	−0.0003(0.0014)
$\Theta_1(S)$	$\Theta_{1,11}$	−0.1743(0.0464)	−0.2150(0.0049)
	$\Theta_{1,12}$	0	0
	$\Theta_{1,21}$	0.0198(0.0376)	−0.0887(0.0065)
	$\Theta_{1,22}$	−0.8476(0.1480)	−0.6940(0.1946)
$\Sigma_0(S)$	$\sqrt{\sigma_{0,1}}$	0	0
	$\sqrt{\sigma_{0,2}}$	0.0054(0.0006)	0.0091(0.0002)
Σ_1	Σ_1^{11}	0.0018(0.0001)	0.0018
	Σ_1^{12}	0	0
	Σ_1^{21}	0	0
	Σ_1^{22}	0	0
$\Theta_0^Q(S)$	$\Theta_{0,1}^Q$	0.0032(0.0002)	0.0015(0.0009)
	$\Theta_{0,2}^Q$	0.0002	−0.0003
Θ_1^Q	$\Theta_{1,11}^Q$	−0.0184(0.0057)	−0.0184
	$\Theta_{1,12}^Q$	0	0
	$\Theta_{1,21}^Q$	0.5467(0.0084)	0.5467
	$\Theta_{1,22}^Q$	−0.1203(0.0024)	−0.1203
	h_0	−1.6458(0.0318)	−1.2675(0.0320)
	ϕ	−0.5403(5.0976)	−0.3500(2.3850)
$\chi^2 = 23.42$		zvalue = 5.03	d.o.f. = 6

[a] This table reports the EMM estimation result of the term structure model in Sect. 2.5. Numbers in parentheses are standard errors. If an estimate is reported without a standard error, it means that the parameter is not estimated, but fixed at that particular value. Regime-dependent parameters are identified by their dependence on S. Parameter definitions can be found in Sect. 2.5. In particular, $\Theta_0^Q(S)$ and Θ_1^Q are the risk-neutral counterparts of $\Theta_0(S)$ and $\Theta_1(S)$ respectively. This simply implies that $\lambda_{0,D} = \Theta_0(S) - \Theta_0^Q(S)$ and $\lambda_{1,D} = -\Theta_1^Q$. Note that Θ_1^Q (hence $\lambda_{1,D}$) is *not* regime-dependent

$\psi_1 = (1,0)'$ and fix $\psi_0(S) = 0$ so that the instantaneous short-term interest rate $r_t = x_{1,t}$. In other words we restrict that the first state variable is simply the short-term interest rate. Because of the relation between the regime-switching risk and the transition probabilities as discussed in Sect. 2.6, we assume that the transition probabilities are not time-varying, therefore we don't force the regime-switching risk to be priced. In other words, we fix $h_1(u,S) = 0$. Under these restrictions, the model has 29 parameters. After initial estimation, we further fix at zero those parameters whose estimates are close to 0 and statistically insignificant, and re-estimate the model. The final results are reported in Table 3.5.

3.3.3 Discussions

Many empirical models of interest rates that incorporate regime switching are motivated by the recurrent shifts between different phases of the business cycle experienced by the aggregate economy. It is therefore interesting to see how the latent regimes in our model of the term structure of interest rates correspond to the business cycle. Following the approach in Bansal and Zhou [5], we first compute interest rates of different maturities conditional on each regime, $\hat{R}(t,\tau|S_t)$, using the estimated term structure model reported in Table 3.5. An estimate of S_t is then obtained by choosing the regime that minimizes the differences between the actually observed interest rates $R(t,\tau)$ and $\hat{R}(t,\tau|S_t)$ or the pricing errors, that is, $\hat{S}_t = \arg\min \sum_\tau |R(t,\tau) - \hat{R}(t,\tau|S_t)|$. The estimated regimes are plotted in Fig. 3.2 together with the business cycle expansions and recessions identified by NBER. Consistent with the result in Bansal and Zhou [5], the figure clearly shows that the regimes underlying the dynamics of the term structure of interest rates are intimately related to the fluctuations of the aggregate economy. Our model is able to identify all six recessions in the sample period. The result is also consistent with the findings from some earlier empirical studies, such as Estrella and Mishkin [25] and Chauvet and Potter [12] among others, that the yield curve has a significant predicative power for the turning point of the business cycle.

The point estimates reported in Table 3.5 confirm that regime-switching indeed seems to be an important feature of interest rate dynamics. For example, according to the estimated parameters, the first factor has a lower long-run level (2.8%) and

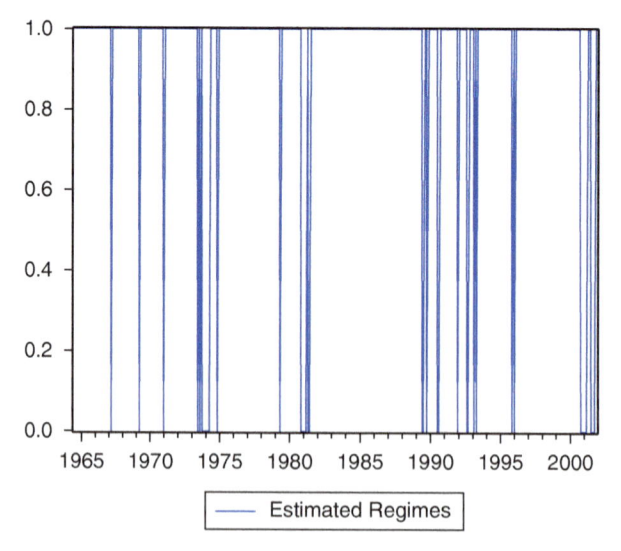

Fig. 3.2 Estimated term structure regimes during 1964–2001. The figure plots the estimated interest rates during 1964–2001. The two regimes are coded as 1 and 0. The *shaded areas* are economic recessions dated by NBER

smaller mean-reverting coefficient (0.1743), hence higher persistence, in Regime 1 than in Regime 2 where the long-run level is 12.89 % with a mean-reverting coefficient 0.2150. Since the volatility of the factor is specified to be proportional to its level, these estimates imply that the factor exhibits higher volatility in Regime 2 than in Regime 1. The estimates of $\sigma_{0,2}$ in Regime 1 and Regime 2 suggest that the second factor also has higher volatility in Regime 2 than in Regime 1. These results are consistent with early findings (Ang and Bekaert [3], for example) with regard to the persistence and volatility of interest rates across different regimes. We plot in Figs. 3.3 and 3.4 the estimated mean yield curve together with the observed average yield curve in both regimes. We also plot in Fig. 3.5 the standard deviations of the estimated interest rates together with the corresponding sample standard deviations in both regimes. We can see that our model fits the yield curve data reasonably well. The yield curve is upward sloping on average in Regime 1 and is flat or slightly downward sloping in Regime 2. Moreover, interest rates are less volatile in Regime 1 than in Regime 2. One stylized fact of the yield curve is that interest rate volatility doesn't seem to attenuate as maturity increases. Figure 3.5 shows that this non-attenuating volatility is mainly a Regime 1 phenomenon where the volatility is relatively low. On the other hand, in the high-volatility regime (Regime 2), interest rate volatility does decline significantly as maturity increases.

In Table 3.5, $\Theta_0^Q(S)$ and Θ_1^Q are the risk-neutral counterparts of $\Theta_0(S)$ and $\Theta_1(S)$. They imply that the coefficients of the market price of diffusion risk are $\lambda_{0,D}(S) = \Theta_0(S) - \Theta_0^Q(S)$ and $\lambda_{1,D} = -\Theta_1^Q$. Note that in order to obtain a closed-form solution of the terms structure of interest rates, we have restricted Θ_1^Q, hence

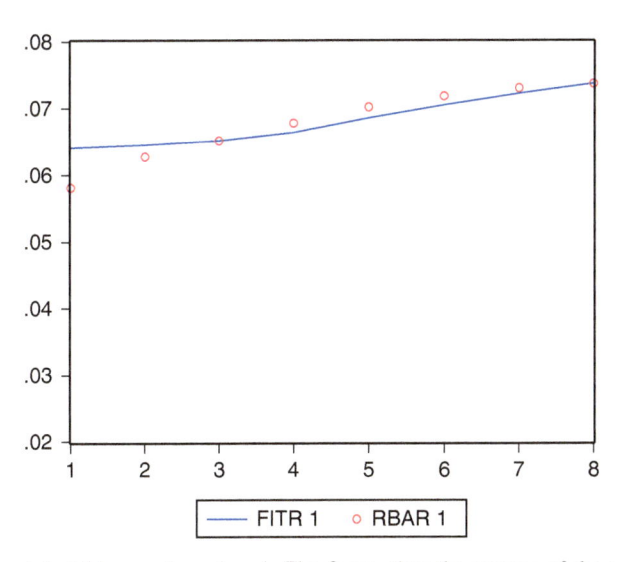

Fig. 3.3 Estimated yield curve in regime 1. The figure plots the average of the estimated yield curve (FITR1) regime 1. RBAR1 are the average interest rates during 1964–2001 in regime 1. Interest rate maturity ranges from 1 month to 5 years

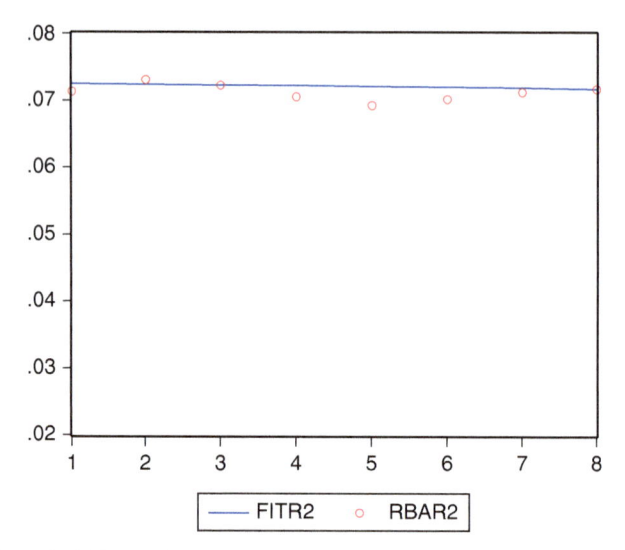

Fig. 3.4 Estimated yield curve in regime 2. The figure plots the average of the estimated yield curve (FITR2) regime 2. RBAR2 are the average interest rates during 1964–2001 in regime 2. Interest rate maturity ranges from 1 month to 5 years

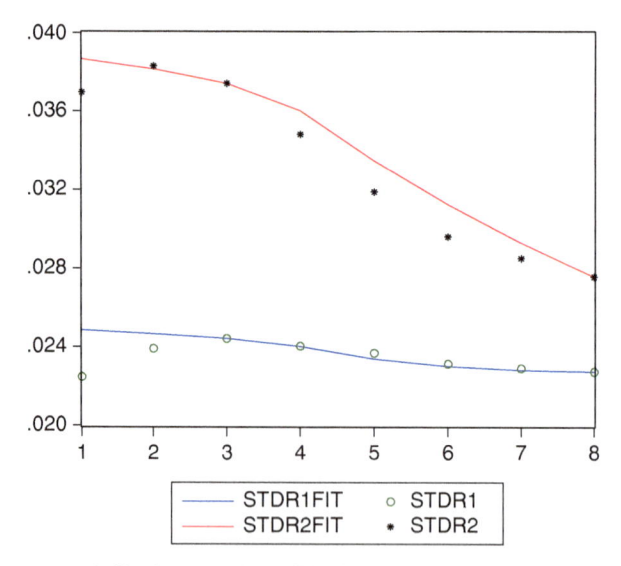

Fig. 3.5 Interest rate volatility in two regimes. The figure plots the standard deviations of the fitted yield curve (STDR1FIT, STDR2FIT) in regime 1 and 2 respectively. STDR1 and STDR2 are the sample standard deviations of the interest rates in the two regimes. Interest rate maturity ranges from 1 month to 5 years

$\lambda_{1,D}$, to be independent of regimes. Nonetheless, our model still allows $\lambda_{0,D}$, hence the market price of risk, to change across regimes, and the estimates of $\Theta_0^Q(S)$ and $\Theta_0(S)$ confirm that the market price of risk in Regime 1 is indeed very different from that in Regime 2. $\lambda_{0,D} = 0.0018$ in Regime 1 and $\lambda_{0,D} = 0.0262$ in Regime 2. Bansal and Zhou [5] also finds evidence that the market price of risk is regime-dependent and shows that it is this feature of the market price of risk that accounts for the improved empirical performance of their model over the existing ones. The difference between the model in Bansal and Zhou [5] and the present one is that Bansal and Zhou [5] restricts $\lambda_{0,D}$ to be zero and allows $\lambda_{1,D}$ to be regime-dependent, and as a result, a closed-form solution of the term structure of interest rate can only be obtained using log-linear approximations.

The negative regression coefficients reported in Table 3.2 strongly reject the expectation hypothesis of the term structure of interest rates. Researchers have often pointed to the presence of time-varying risk premiums as the main cause for this stylized fact about interest rates (see, for example, Dai and Singleton [15]). The model of the term structure of interest rates in the present paper has a very flexible specification of bond risk premiums. As Equation (3.20) shows, the risk premium is time-varying first because it is a function of the underlying state variable $X(t)$ as in the standard affine models. Regime-switching, however, makes the coefficients of the risk premium vary across regimes, therefore adds another source of time-variation. Moreover, our model allows for the regime-switching risk to be priced, therefore introduces a new component to bond risk premiums that is also time-varying. To see how this flexible specification of bond risk premiums helps account for the stylized fact of the term structure of interest rates, we use the estimated model to simulate interest rates of various maturities and run the same expectation-hypothesis regressions as those reported in Table 3.2. The results are included in Table 3.6. We can see that our model is able to replicate the negative regression coefficients typically found in the literature.

Table 3.6 Expectation-hypothesis regression using simulated interest rates [a]

	i+j=0.5	i+j=1	i+j=2	i+j=3	i+j=4	i+j=5
j=0.25	−1.9169 (0.1323)					
j=0.5		−1.7966 (0.1033)				
j=1			−1.6602 (0.0836)	−1.8859 (0.0868)	−2.1152 (0.0909)	−2.3428 (0.0958)
j=2				−1.2804 (0.0680)	−1.4769 (0.0696)	−1.6721 (0.0717)
j=3					−1.0817 (0.0625)	−1.2511 (0.0.0636)
j=4						−0.9612 (0.0600)

[a] This table reports the estimate of β_{ij} in regression $i(R_{t+j}^i - R_t^{i+j}) = \alpha + \beta_{ij}[j(R_t^{i+j} - R_t^j)] + \varepsilon_{t+j}^i$ where R_t^k is the k-year interest rate at time t *using the simulated interest rates* according to the estimated term structure model in Sect. 2.5. Under the null hypothesis of the expectation theory, $\beta_{ij} = 1$. Numbers in parentheses are Newy-White standard errors

Another stylized fact about the yield curve is that bond risk premiums are typically counter-cyclical as reported in Table 3.3. For 1- to 5-year bonds, the risk premiums are all negatively correlated with the business cycle dummy variable. Moreover, simple regressions of the bond risk premiums on the business cycle dummy variable and the yield spread also produce significantly negative coefficients on the business cycle dummy variable in all cases. The dynamic model of term structure of interest rates in the present paper allows us to estimate the instantaneous expected excess return on a long-term bond based on Eq. (3.20). We can then compute the same correlation coefficients as well as the simple regressions as those in Table 3.3 using the estimated bond risk premiums. The results are reported in Table 3.7. We can see that the estimated risk premiums of different bonds are highly and positively correlated as in the data. The correlation coefficients between the estimated risk premium and the business cycle dummy variable, however, are all negative and are similar in magnitude to those found in the data. For example, for the 2-year bond, the correlation coefficient is -0.1703 in the data. Using the estimated risk premiums, the correlation coefficient is -0.1680. We also regress the estimated risk premiums on the business cycle dummy variable and the yield spread. As in the data, the regression coefficients on the business cycle dummy variable are all negative and significant (except for the 1-year bond) in Table 3.7. A difference, though, is that the yield spread seems to retain its predictive power for bond returns even in the presence of the business cycle dummy variable when the estimated bond risk

Table 3.7 The estimated bond risk premiums and the business cycle[a]

	RPY1	RPY2	RPY3	RPY4	RPY5	BC
RPY1	1					
RPY2	0.9776	1				
RPY3	0.9446	0.9925	1			
RPY4	0.9159	0.9799	0.9969	1		
RPY5	0.8947	0.9687	0.9917	0.9987	1	
BC	−0.1290	−0.1680	−0.1872	−0.1980	−0.2043	1
$\hat{\alpha}_i$	0.1663**	0.3978**	0.6884*	1.0270*	1.4032	
	(0.0799)	(0.1999)	(0.3630)	(0.5638)	(0.7937)	
$\hat{\beta}_i$	−0.0028	−0.0100**	−0.0210**	−0.0352**	−0.0517**	
	(0.0017)	(0.0043)	(0.0078)	(0.0120)	(0.0169)	
$\hat{\gamma}_i$	−0.0550	−0.1059	−0.1544	−0.1996	−0.2441	
	(0.0883)	(0.2210)	(0.4313)	(0.6232)	(0.8774)	
Adjusted R^2	0.038	0.051	0.057	0.060	0.062	

[a] The first six rows of this table report the correlation coefficients among the estimated bond risk premiums and the business cycle. RPY1, RPY2, ..., RPY5 are the instantaneous expected excess return on 1-, 2-, ..., 5-year bonds, respectively, given in (3.20) in Sect. 2.6. BC is the dummy variable for the business cycle with BC=1 indicating an expansion and BC=0 indicating a recession. The last four rows report the OLS regression $RPYi_t = c + \alpha_i SP_t + \beta_i BC_t + \gamma_i [SP_t \times BC_t] + \varepsilon_{i,t}$, where RPY$i$ is the risk premium on the i-year bond, SP is the yield spread between the 5-year bond and the 1-month Bill. Numbers in parentheses are Newy-West standard errors. An * indicates the estimate is significant at 10 % level. An ** indicates the estimate is significant at 5 % level.

3 An Econometric Model of The Term Structure of Interest Rates

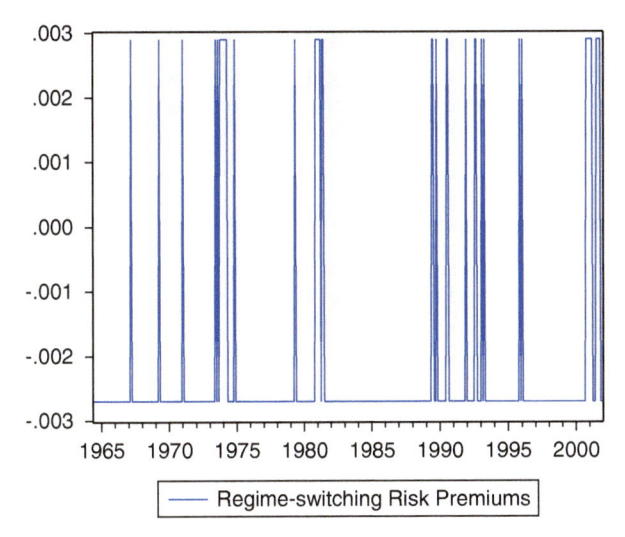

Fig. 3.6 Regime-switching risk premiums. The figure plots the estimated regime-switching risk premiums on the 5-year bond during 1964–2001. The *shaded areas* are economic recessions dated by NBER

premiums are used in the regression (Table 3.7), whereas the business cycle dummy variable completely drives out the predicating power of yield spreads in the data (Table 3.3)

One interesting question about the term structure models under regime shifts is that whether or not the regime-switching risk is priced. In our model, the market price of regime-switching risk, $\lambda_S(u,S,X)$, is given by $\lambda_S(u,S,X) = 1 - e^{\phi(u,S)}/e^{h_0(u,S)}$. From the estimates in Table 3.5, $\lambda_S = 1 - e^{-0.5403}/e^{-1.6458} = -2.0207$ in Regime 1 and $\lambda_S = 1 - e^{-0.3500}/e^{-1.2675} = -1.5030$ in Regime 2. With the estimate of the market price of regime-switching risk, the risk premium associated with regime-switching shocks can be obtained by the second term in (3.20). Figure 3.6 plots the estimated regime-switching risk premiums during the sample period. These values seem economically important and suggest that the regime-switching risk is indeed priced by bond investors. The standard errors of the estimate of ϕ in both regimes, however, are very big (5.0976 and 2.3850 respectively). In fact, among all the estimated parameters, ϕ is the least accurately estimated one. Notice that $\lambda_S = 0$ when $\phi(u,S) = h_0(u,S)$. With the larger standard errors, we are in fact not able to reject that $\phi(u,S) = h_0(u,S)$, or $\lambda_S = 0$ in both regimes. The uncertainty regarding the regime-switching risk premiums may have reflected the caveat of affine models under regime shifts as we discussed above. In order to obtain a closed-form solution of the term structure of interest rates in an affine model, we have to restrict the risk-neutral regime-switching probabilities to be constant. This implies that the model would force the regime-switching risk to be priced if the regime-switching probabilities are allowed to be time-varying under the physical probability measure. In this paper we choose not to impose a non-zero market price of regime-switching risk a prior by using a less general specification of the regime-switching probabilities. Our empirical results suggest that more studies are needed in order to get a better assessment of the regime-switching risk premiums.

3.4 Conclusion

Using an analytically simple representation of Markov regime shifts, this paper develops and estimates a continuous-time affine model of the term structure of interest rates under the risk of regime-switching. The model elucidates the dynamic effects of regime shifts on the yield curve and bond risk premiums. The empirical results show that the model is able to account for many salient features of the term structure of interest rates and confirm that regime-switching indeed seems to be an important property of interest rate movements. There are still some uncertainties regarding the magnitude of the regime-switching risk premiums that warrant the development of more general dynamic models of the term structure of interest rates under regime shifts.

In the current model, regimes, though a latent variable to econometricians, are assumed to be observable to bond investors. A natural extension is to assume that the regimes are not observable to bond investors either, and that the investors must learn the regimes through other observable state variables. One example is that the regimes may represent different stances of the monetary policy and bond investors must infer from different signals about the true intentions of the central bank.

It should be noted that the model developed in the current paper is an empirical one. The regimes identified by the model lack clear structural interpretations. Another extension of the present paper is to incorporate the model of the term structure of interest rates into a well specified macroeconomic model with regime shifts. With such a structural model, we will be able to identify and interpret different regimes in terms of macroeconomic fundamentals. These extensions are left for future studies.

Acknowledgements We would like to thank seminar participants at Federal Reserve Bank of Kansas City, North American Summer Meetings of Econometric Society and Missouri Valley Economics Association Annual Meetings for helpful comments. A part of the research is done while the first author is visiting Federal Reserve Bank of Kansas City. He is grateful for their hospitalities. The second author gratefully acknowledges financial support of National Science Foundation under DMS-1228244 and University Missouri Research Board.

References

1. Aït-Sahalia, Y., Kimmel, R.: Estimating affine multifactor term structure models using closed-form likelihood expansions. J. Financ. Econ. **98**, 113–144 (2010)

2. Ang, A., Bekaert, G.: Regime switches in interest rates. J. Bus. Econ. Stat. **20**(2), 163–182 (2002)
3. Ang, A., Bekaert, G.: Short rate nonlinearities and regime switches. J. Econ. Dyn. Control **26**(7–8), 1243–1274 (2002)
4. Ang, A., Bekaert, G., Wei, M.: The term structure of real rates and expected inflation. J. Financ. **63**, 797–849 (2008)
5. Bansal, R., Zhou, H.: Term structure of interest rates with regime shifts. J. Financ. **57**, 1997–2043 (2002)
6. Bansal, R., Gallant, R., Tauchen, G.: Nonparametric estimation of structural models for high-frequency currency market data. J. Econom. **66**, 251–287 (1995)
7. Bansal, R., Tauchen, G., Zhou, H.: Regime-shifts, risk premiums in the term structure, and the business cycle. J. Bus. Econ. Stat. **22**, 396–409 (2004)
8. Bielecki, T., Rutkowski, M.: Multiple ratings model of defaultable term structure. Math. Financ. **10**, 125–139 (2000)
9. Bielecki, T., Rutkowski, M.: Modeling of the defaultable term structure: conditional Markov approach. Working paper, The Northeastern Illinois University (2001)
10. Boudoukh, J., Richardsona, M., Smithb, T., Whitelaw, R.F.: Regime shifts and bond returns. Working paper, Arizona State University (1999)
11. Cecchetti, S., Lam, P., Mark, N.: The equity premium and the risk-free rate: matching the moments. J. Monet. Econ. **31**, 21–45 (1993)
12. Chauvet, M., Potter, S.: Forecasting recessions using the yield curve. J. Forecast. **24**, 77–103 (2005)
13. Cox, J., Ingersoll, J., Ross, S.: A theory of the term structure of interest rates. Econometrica **53**, 385–407 (1985)
14. Dai, Q., Singleton, K.: Specification analysis of affine term structure models. J. Financ. **55**, 1943–1978 (2000)
15. Dai, Q., Singleton, K.: Expectation puzzles, Time-varying risk premia, and affine models of the term structure. J. Financ. Econ. **63**, 415–441 (2002)
16. Dai, Q., Singleton, K.: Term structure dynamics in theory and reality. Rev. Financ. Stud. **16**, 631–678 (2003)
17. Dai, Q., Singleton, K., Yang, W.: Regime shifts in a dynamic term structure model of the U.S. treasury bond yields. Rev. Financ. Stud. **20**, 1669–1706 (2007)
18. Duarte, J.: Evaluating alternative risk preferences in affine term structure models. Rev. Financ. Stud. **17**, 379–404 (2004)
19. Duffee, G.: Term premia and interest rate forecasts in affine models. J. Financ. **57**, 405–443 (2002)
20. Duffie, D., Kan, R.: A yield-factor model of interest rates. Math. Financ. **6**, 379–406 (1996)
21. Duffie, D., Pan, J., Singleton, K.: Transform analysis and asset pricing for affine jump-diffusions. Econometrica **68**, 1343–1376 (2000)
22. Elliott, R.J. et al.: Hidden Markov Models: Estimation and Control. Springer, New York (1995)

23. Elliott, R.J., Mamon, R.S.: A complete yield curve descriptions of a Markov interest rate model. Int. J. Theor. Appl. Financ. **6**, 317–326 (2001)
24. Elliott, R.J., Siu, T.K.: On Markov-modulated exponential-affine bond price formulae. Appl. Math. Financ. **16**(1), 1–15 (2009)
25. Estrella, A., Mishkin, F.S.: Predicting U.S. recessions: financial variables as leading indicators. Rev. Econ. Stat. **80**, 45–61 (1998)
26. Evans, M.: Real risk, inflation risk, and the term structure. Econ. J. **113**, 345–389 (2003)
27. Fama, E.F., French, K.R.: Business conditions and expected returns on stocks and bonds. J. Financ. Econ. **25**, 23–49 (1989)
28. Ferland, R., Gauthier, G., Lalancette, S.: A regime-switching term structure model with observable state variables. Financ. Res. Lett. **7**, 103–109 (2010)
29. Filardo, A.: Business cycle phases and their transition dynamics. J. Bus. Econ. Stat. **12**, 299–308 (1994)
30. Futami, H.: Regime switching term structure model under partial information. Int. J. Theor. Appl. Financ. **14**, 265–294 (2011)
31. Gallant, R., Tauchen, G.: Which moment to match? Econom. Theory **12**, 657–681 (1996)
32. Gallant, R., Tauchen, G.: Reprojecting partially observed system with application to interest rate diffusions. J. Am. Stat. Assoc. **93**, 10–24 (1998)
33. Gallant, R., Tauchen, G.: Efficient method of moments. Working paper, University of North Carolina (2001)
34. Garcia, R., Perron, P.: An analysis of the real interest rate under regime shifts. Rev. Econ. Stat. **78**(1), 111–125 (1996)
35. Gray, S.: Modeling the conditional distribution of interest rates as a regime-switching process. J. Financ. Econ. **42**, 27–62 (1996)
36. Hamilton, J.D.: Rational expectations econometric analysis of changes in regimes: an investigation of the term structure of interest rates. J. Econ. Dyn. Control **12**, 385–423 (1988)
37. Hamilton, J.D.: A new approach to the economic analysis of nonstationary time series and the business cycle. Econometrica **57**, 357–384 (1989)
38. Harrison, M., Kreps, D.: Martingales and arbitrage in multiperiod security markets. J. Econ. Theory **20**, 381–408 (1979)
39. He, W., Wang, J.G., Yan, J.A.: Semimartingale theory and stochastic calculus. Science Press, Beijing/CRC, Boca Raton (1992)
40. Landen, C.: Bond pricing in a hidden Markov model of the short rate. Financ. Stoch. **4**, 371–389 (2000)
41. Last, G., Brandt, A.: Marked point processes on the real line. Springer, New York (1995)
42. Le, A., Dai, Q., Singleton, K.: Discrete-time dynamic term structure models with generalized market prices of risk. Rev. Financ. Stud. **23**, 2184–2227 (2010)
43. Lewis, K.K.: Was there a 'Peso problem' in the US term structure of interest rates: 1979–1982? Int. Econ. Rev. **32**, 159–173 (1991)
44. Protter, P.: Stochastic intergration and differential equations, 2nd edn. Springer, Berlin (2003)

45. Scott, S.: Bayesian methods for hidden Markov models: recursive computing in the 21st Century. J. Am. Stat. Assoc. **97**, 337–351 (2002)
46. Siu, T.K.: Bond pricing under a Markovian regime-switching jump-augmented Vasicek model via stochastic flows. Appl. Math. Comput. **216**(11), 3184–3190 (2010)
47. Siu, T.K.: Regime switching risk: to price or not to price? Int. J. Stoch. Anal. **2011**, 1–14 (2011)
48. Sola, M., Driffill, J.: Testing the term structure of interest rates using a vector autoregression with regime switching. J. Econ. Dyn. Control **18**, 601–628 (1994)
49. Timmermann, A.: Moments of Markov switching models. J. Econom. **96**, 75–111 (2000)
50. Wu, S., Zeng, Y.: A general equilibrium model of the term structure of interest rates under regime-siwthcing risk. Int. J. Theor. Appl. Financ. **8**, 839–869 (2005)
51. Xiang, J., Zhu, X.: A regime-switching NelsonCSiegel term structure model and interest rate forecasts. J. Financ. Econom. **11**, 522–555 (2013)
52. Yin, G.G., Zhang, Q.: Continuous-time Markov chains and applications: a singular perturbation approach. Springer, Berlin (1998)

Chapter 4
The LIBOR Market Model: A Markov-Switching Jump Diffusion Extension

Lea Steinrücke, Rudi Zagst, and Anatoliy Swishchuk

Abstract This paper demonstrates how the LIBOR Market Model of Brace et al. (Math Financ 7(2):127–147, 1997) and Miltersen et al. (J Financ 52(1):409–430, 1997) may be extended in a way that not only takes into account sudden market shocks without long-term effects, but also allows for structural breaks and changes in the overall economic climate. This is achieved by substituting the simple diffusion process of the original LIBOR Market model by a Markov-switching jump diffusion. Since interest rates of different maturities are modeled under different (forward) measures, we investigate the effects of changes between measures on all relevant quantities. Using the Fourier pricing technique, we derive pricing formula for the most important interest rate derivatives, caps/caplets, and calibrate the model to real data.

4.1 Introduction

Of all derivatives traded on over-the-counter (OTC) markets, contracts on interest rates clearly take the most prominent role, as in terms of notional value, they account to around 78 % of all contracts traded (BIS Quarterly Review, June 2012). When

---

L. Steinrücke (✉)
Berlin Doctoral Program in Economics and Management Science, Humboldt Universität,
Spandauer Strasse 1, D-10099 Berlin, Germany
e-mail: lea.steinruecke@hu-berlin.de

R. Zagst
Chair of Mathematical Finance, Technische Universität München, Parkring 11,
D-85748 Garching, Germany
e-mail: zagst@tum.de

A. Swishchuk
Department of Mathematics and Statistics, University of Calgary, 612 Campus Place NW,
2500 University Drive NW, Calgary, AB T2N 1N4, Canada
e-mail: aswish@ucalgary.ca

compared to other financial markets, the interest-rate segment takes a somewhat special role, as researchers have to face the special challenge of working with a continuum of bond maturities and hence an infinite number of underlyings. As a consequence, certain questions arise that do not require consideration in commodity or stock markets, in particular, when it comes to the identification of no-arbitrage conditions. For a long time, starting in the 1970s, short rate models were used as the main tool for the pricing and hedging of interest-rate products. Even though they still remain important in certain applications, their role in the derivatives market has been gradually replaced – first by the instantaneous forward rates approach of Heath et al. [24], and, more recently in 1997, by the so-called LIBOR market model (LMM). It was in particular the latter, introduced by Miltersen et al. [30] and Brace et al. [10], that revolutionized the pricing of interest-rate derivatives. By turning towards the consideration of simple rather than instantaneous forward rates, it was finally possible to analytically price some of the most widely traded interest-rate products, caps and floors, with a Black (1976)-type formula. The new approach to modeling was especially based on exploring the relation between the simple rates and the bond prices. Shortly after, Jamshidian [26] adapted the approach to the swap market and the pricing of swaptions.

Since its introduction in 1997, the LMM has experienced an unprecedented raise in popularity and has become the most popular pricing approach among practitioners. The model has, however, been criticized for not being suited to adequately reproduce the market-observed prices of interest-rate derivatives. In particular, the presumption that the LIBOR dynamics can be modeled as diffusion processes with deterministic coefficients has been challenged. Rebonato [33], for instance, has pointed out that log-normal dynamics are incapable of reproducing heavy tails, jumps or non-flat volatility surfaces. In response to these shortcomings, a large amount of extensions has been introduced over the course of years. Inter alia, it was proposed to replace the simple log-normal processes of the original model by displaced diffusions [28], Lévy processes [15], generalized jump diffusions [22, 5], Markov-switching geometric Brownian motions [18], processes with stochastic volatility [2, 4] and general semimartingales [27]. Even extensions accounting for default risk were brought forth [14, 13].

In the approach presented here, two of the most promising concepts are merged: generalized jump diffusions and Markov-switching processes. By doing so, we create an extension to the original LMM that is suitable to incorporate sudden market shocks into the interest-rate dynamics and, at the same time, accounts for changes in the overall economic climate. Jump diffusions, on the one hand, are immediate generalizations to ordinary diffusion processes which are expanded by a component accounting for sudden up- or downward movements in the market. As demonstrated by Belomestny and Schoenmakers [5], modeling the simple interest rate through jump diffusions is not only suited to reflect sudden jumps observed in the market dynamics, but also allows to successfully capture the non-flat implied volatility surfaces typically observed in the interest-rate derivatives markets. As these jumps should be thought of as random shocks, they typically do not have an impact on the overall economic environment. Nonetheless, Rebonato and Joshi [35] and

Rebonato [34] present considerable evidence indicating that there are in fact different economic phases. This observation is incorporated into the model by the introduction of a Markov-switching feature: All jump diffusion parameters are assumed to be dependent on an underlying finite-state space Markov chain moving according to the overall economic development.

In the following chapter, we have chosen a structure that should enable even those readers that are not too familiar with market models and/or Markov-switching jump diffusions to follow the main ideas and intuitions behind the model. Keeping this in mind, the proofs presented are rather intuitive sketches intended to convey the underlying ideas rather than mathematically rigorous arguments. All considerations are nonetheless based on mathematically impeccable concepts. Those that are interested in the exact details are referred to Steinrücke et al. [40], where all underlying proofs are demonstrated. There, the concept is furthermore extended to the swap market model and the pricing of swaptions.

The chapter is divided into six sections. Section 4.2 is pointed at getting the reader acquainted with the fundamental tools needed when working with the LIBOR market model – Girsanov's Theorem, the Doléans-Dade exponential and the Change-of-Numéraire Technique. Next, Sect. 4.3 gives a quick introduction to the log-normal LIBOR market model of Brace et al. [10] and Miltersen et al. [30], followed by Sect. 4.4, where the framework for the Markov-switching jump diffusion extension to the original model is introduced. It is explained how the dynamics of different LIBOR rates can be interrelated and the special role of the Markov chain under measure changes is investigated. Then, in Sect. 4.5, it is demonstrated how caps/caplets, one of the most important families of interest-rate derivatives, can be priced within a Markov-switching jump diffusion framework. Last, but not least, Sect. 4.6 gives an idea of how the proposed extension can be successfully calibrated to market data. Section 4.7 finally wraps up the main results and gives an outlook to possible future research.

4.2 Mathematical Preliminaries

Given the finite time horizon $T^* > 0$, let the interest-rate market be modeled on the complete stochastic basis $(\Omega, \mathscr{F}, \mathbb{F}, \mathbb{P})$, i.e., $(\Omega, \mathscr{F})$ is a measurable space, $\mathbb{F}$ is a filtration on $\mathscr{F}$ satisfying the usual conditions of right-continuity and completeness, and $\mathbb{P}$ denotes the physical measure of the market. The convenience of considering a market with an only finite time horizon is that any local martingale can be treated as a martingale [7]. It is furthermore assumed that the market is frictionless with bank account $(B_t)_{t \in [0,T^*]}$, $B_0 = 1$, and zero-coupon bonds $(B(t,T))_{t \in [0,T]}$ trading for every maturity $0 \leq T \leq T^*$. The bonds of different maturities are the so-called primary traded instruments of the interest-rate market. Also, the ad-hoc assumption is made that all processes involved are specified in such a way that all operations to be performed (differentiation in the T-variable, differentiation in the t-variable under the integral sign and interchange of order of integration) are well-defined.

While the pricing in stock markets is largely dependent on changing the $\mathbb{P}$-dynamics of a given stock to the dynamics under some risk-neutral measure (assuming it exists), pricing in the interest-rate market involves changes to measures associated with different numéraires than the simple bank account B. For this, the *Change-of-Numéraire Technique* is needed (see, e.g., [11] or [42]):

Theorem 4.1 (Change-of-Numéraire Technique). *Let $(Y(t))_{t\in[0,T^*]}$ be a primary traded asset of the market and $\mathcal{Q}$ an equivalent martingale measure (EMM) under which $(B_t^{-1}Y(t))_{t\in[0,T^*]}$ follows a martingale. Let $A = (A(t))_{t\in[0,T^*]}$ and $E = (E(t))_{t\in[0,T^*]}$ be two arbitrary numéraires, satisfying that the discounted processes $(B_t^{-1}A(t))_{t\in[0,T^*]}$ and $(B_t^{-1}E(t))_{t\in[0,T^*]}$ are both $\mathcal{Q}$-martingales. Then, the following holds:*

- *There exists an equivalent probability measure $\mathcal{Q}^A$,*

$$\left.\frac{d\mathcal{Q}^A}{d\mathcal{Q}}\right|_{\mathcal{F}_t} := \frac{A(t)}{A(0)B_t}, \quad t \in [0,T^*],$$

such that $(A(t)^{-1}Y(t))_{t\in[0,T^*]}$ is a $\mathcal{Q}^A$-martingale.

- *The Radon-Nikodým derivative of $\mathcal{Q}^E$ with respect to $\mathcal{Q}^A$ is given as*

$$\left.\frac{d\mathcal{Q}^E}{d\mathcal{Q}^A}\right|_{\mathcal{F}_t} := \frac{E(t)}{A(t)} \cdot \frac{A(0)}{E(0)}, \quad \forall t \in [0,T^*].$$

- *For any contingent claim $D = D(T)$ with underlying Y, the time-t-price is given as*

$$B_t \mathbb{E}_{\mathcal{Q}}\left[\frac{D}{B_T}\bigg|\mathcal{F}_t\right] = A(t)\mathbb{E}_{\mathcal{Q}^A}\left[\frac{D}{A(T)}\bigg|\mathcal{F}_t\right] = E(t)\mathbb{E}_{\mathcal{Q}^E}\left[\frac{D}{E(T)}\bigg|\mathcal{F}_t\right].$$

The "numéraires of choice" for our purposes are the bonds $(B(t,T))_{t\in[0,T]}$ with maturities $0 < T \leq T^*$, for which it is natural to assume that $B(t,T) > 0$. The measure $\mathcal{Q}^T$ associated with the numéraire $(B(t,T))_{t\in[0,T^*]}$ is named *T-forward measure*.

By nature of the upcoming extension, the main role in the following considerations will be played by so-called *jump diffusion processes*. Let $\mathcal{H}$ be an arbitrary probability measure and $(E,\mathcal{E})$ a measurable space which satisfies certain technical properties. A jump diffusion $Z = (Z(t))_{t\in[0,T^*]}$ is the solution to the stochastic differential equation

$$dZ(t) = Z(t-)dY(t)$$
$$= Z(t-)\left(\alpha(t)dt + \delta(t)dW^{\mathcal{H}}(t) + \int_E \gamma(t,z)\left(\mu - \nu^{\mathcal{H}}\right)(dt,dz)\right), \quad (4.1)$$

where $W^{\mathcal{H}}$ is a multi-dimensional standard $\mathcal{H}$-Brownian motion, μ is an integer-valued random measure on the mark space $[0,T^*] \times E$, $\nu^{\mathcal{H}}$ is the $\mathcal{H}$-compensator measure associated with μ,[1] and the coefficient functions α, δ and γ satisfy certain predictability and integrability assumptions (see, e.g., Jacod and Shiryaev [25] or Protter [31]). The solution to (4.1) is given by the so-called *stochastic exponential* or *Doléans-Dade* exponential $Z = \mathscr{E}(Y)$, and satisfies

$$Z(t) = Z(0) \cdot \exp\Big(\int_0^t \big(\alpha(s) - \frac{1}{2}\|\delta(s)\|^2\big)ds + \int_0^t \delta(s)' dW^{\mathscr{H}}(s)$$
$$+ \int_0^t \int_E \ln(1+\gamma(s,z))\big(\mu - \nu^{\mathscr{H}}\big)(ds,dz)$$
$$+ \int_0^t \int_E \big[\ln(1+\gamma(s,z)) - \gamma(s,z)\big]\nu^{\mathscr{H}}(ds,dz) \Big).$$

It can be shown, that any jump diffusion is a special semimartingale, i.e., it allows for the representation

$$Z = Z(0) + M + A, \qquad (4.2)$$

where $Z(0)$ is finite and $\mathscr{F}_0$-measurable, M a local martingale, A a process of finite variation and $M(0) = A(0) = 0$. In extension to ordinary semimartingales, A is in addition predictable and decomposition (4.2) is unique [25]. The fact of Z being a special semimartingale allows for the use of a variety of results related to the theory on special semimartingales and especially the corresponding version of Itō's Lemma. The other important theorem that will be frequently used is the following version of *Girsanov's Theorem*. Note that measure changes do not have an impact on the jump measure, but only on the compensator ν.

Theorem 4.2 (Girsanov's Theorem). *Let $(\Omega, \mathscr{F}, \mathbb{F}, \mathscr{H})$ be a complete stochastic basis with $\mathscr{H}$ an arbitrary probability measure. Furthermore, let $W^{\mathscr{H}}(t)$ be a d-dimensional $\mathscr{H}$-Brownian motion and μ an integer-valued random measure with mark space $([0,T^*] \times E, \mathscr{B}([0,T^*]) \otimes \mathscr{E})$ and $\mathscr{H}$-compensator $\nu^{\mathscr{H}}(dt,dz) = \lambda^{\mathscr{H}}(t) k^{\mathscr{H}}(t,dz) dt$. $\lambda^{\mathscr{H}}$ and $k^{\mathscr{H}}$ denote the predictable jump intensity and the marker distribution, respectively, and it is assumed that $k^{\mathscr{H}}(t,A)$ is predictable $\forall A \in \mathscr{E}$. Also, let θ be a d-dimensional predictable process and $\Phi(t,z)$ a nonnegative predictable function satisfying the usual integrability assumptions. Define the process $Z(t)$ by $Z(0) = 1$ and*

$$\frac{dZ(t)}{Z(t-)} = \theta(t) dW^{\mathscr{H}}(t) + \int_E (\Phi(s,z) - 1)\big(\mu(ds,dz) - \nu^{\mathscr{H}}(ds,dz)\big)$$

---

[1] Intuitively speaking, the $\mathscr{H}$-compensator measure $\nu^{\mathscr{H}}$ contains all the distributional information related to the random measure μ under $\mathscr{H}$. In more precise terms: Assuming that μ satisfies certain regularity conditions, ν is defined as the a.s. unique predictable random measure with the following property: For any predictable stochastic process $f : \Omega \times [0,T^*] \times E \to \mathbb{R}$ with $|f| * \mu$ an increasing, locally integrable process, the process M, $M(\omega,t) := \int_0^t \int_E f(\omega,s,z)\mu(\omega,ds,dz) - \int_0^t \int_E f(\omega,s,z)\nu(\omega,ds,dz)$ is a (local) martingale with respect to $\mathscr{H}$ [38].

It is assumed that $\mathbb{E}^{\mathscr{H}}[Z(t)] = 1$ for all $0 \leq t \leq T$. Define the equivalent measure $\mathscr{R} \sim \mathscr{H}$ by the Radon-Nikodým derivative $d\mathscr{R}/d\mathscr{H}|_{\mathscr{F}_t} = Z(t)$. Then, the following holds:

(i) The process $W^{\mathscr{R}}$ is a $\mathscr{R}$-Brownian motion,

$$dW^{\mathscr{R}}(t) = dW^{\mathscr{H}}(t) - \theta(t)\,dt.$$

(ii) The a.s. unique predictable $\mathscr{R}$-compensator of μ is given as

$$v^{\mathscr{R}}(dt,dz) := \Phi(t,z)\,v^{\mathscr{H}}(dt,dz).$$

The corresponding jump intensity and the marker distribution are

$$\lambda^{\mathscr{R}}(t) = \phi(t)\,\lambda^{\mathscr{H}}(t) \quad \text{and} \quad k^{\mathscr{R}}(t,dz) = Z_E(z)\,k^{\mathscr{H}}(t,dz),$$

respectively, where $\phi(t) := \int_E \Phi(t,z)\,k^{\mathscr{H}}(t,dz)$, and $Z_E(z) := \Phi(t,z)/\phi(t)$ for $\phi(t) > 0$, $Z_E(z) = 1$ otherwise.

4.3 The Log-Normal LIBOR Framework

A frequent approach[2] to modeling interest-rate markets is based on the consideration of *instantaneous forward interest rates*. Most intuitively, these can be understood as those interest rates that can be locked in today to guarantee a certain future spot rate. The instantaneous forward rate $f(t,T)$ at time t for the future time point T is defined by the limit

$$f(t,T) = \lim_{\delta \to 0} \frac{1}{\delta} \ln \frac{B(t,T)}{B(t,T+\delta)}.$$

One can easily construct a portfolio that allows to replicate the desired forward rate via simple means (see [39], p. 423). Zero-bond prices can be easily recovered from instantaneous forward rates by the well-known formula

$$B(t,T) = \exp\left(-\int_t^T f(t,s)\,ds\right). \tag{4.3}$$

By investigating the instantaneous forward rates and their relation to the bond dynamics, Heath et al. [24] were able to derive an arbitrage-free framework for the stochastic evolution of the whole yield curve. In detail, the authors assumed that under a given measure $\mathbb{P}$ and for a fixed maturity $t \leq T \leq T^*$, each instantaneous forward rate $f(.,T)$ evolves as a diffusion process.

---

[2] The introductory Sect. 4.3 is mainly based on Brigo and Mercurio [11], Filipovic [20], Rebonato [33] and Zagst [42].

While models for instantaneous forward rates have since been increasingly extended and amended (see, e.g., the jump diffusion model in a semimartingale setting by Björk et al. [6, 7]), the approach bears the intrinsic problem of working with infinitely many interest rates which are not directly observable in the market. Even more severe, the most basic interest-rate derivatives on the market (caps and swaptions) cannot be evaluated via a closed-form pricing formula. It was mainly these problems that gave rise to a new approach to modeling, where simple instead of instantaneous forward rates are considered: The LIBOR market model.

4.3.1 An Introduction to the LIBOR Market Model: The Log-Normal Dynamics

In order to define forward LIBOR rates, first a *tenor structure* $0 = T_0 < T_1 < \ldots < T_N$ is fixed, with constant *tenor* $\delta \equiv T_{i+1} - T_i$, $i = 1,\ldots,N-1$. δ is a fraction of a year (usually $\delta = 1/4$ or $\delta = 1/2$). The *forward LIBOR* or *simple rate* $L_i(t) := L(t, T_i, T_{i+1})$ at time t with *maturity* T_i and *expiry* T_{i+1} is the simple interest an investor can lock in at t for the future interval $[T_i, T_{i+1}]$ and is given by the relation

$$1 + \delta \cdot L_i(t) = \frac{B(t, T_i)}{B(t, T_{i+1})}. \tag{4.4}$$

This is indeed the simple forward interest rate for the interval $[T_i, T_{i+1}]$, as it can be easily replicated by the following payoff scheme: On the one hand, at time t, the investor goes long both a bond $B(t, T_i)$ with maturity T_i and a forward contract on the simple interest rate L_i on one monetary unit for the interval $[T_i, T_{i+1}]$. The initial value of this portfolio is $B(t, T_i)$. At time T_i, the bond payoff of 1 is invested at the forward-contract secured interest rate L_i and yields payoff $1 + \delta L_i(t)$ at time T_{i+1}. On the other hand, at time t, the investor goes short $1 + \delta L_i(t)$ zero coupon bonds with maturity T_{i+1}, yielding a portfolio with initial value $-B(t, T_{i+1})(1 + \delta L_i(t))$. The payoff of the portfolio at time T_{i+1} is $-(1 + \delta L_i(t))$. By no-arbitrage, it follows that the initial portfolio values have to coincide, $B(t, T_i) = (1 + \delta L_i(t))B(t, T_{i+1})$, which, in turn, immediately implies (4.4). Taking the expiry of one LIBOR rate as the maturity of the next then leaves us with an array of $N-1$ LIBOR rates.[3]

The natural question that arises is how the dynamics of each forward LIBOR rate in (4.4) can be modeled. In the case considered by Brace et al. [10], the only source of randomness in the market is a d-dimensional standard $\mathcal{Q}$-Brownian motion $W^{\mathcal{Q}}$, and the market filtration $\mathbb{F}$ is the augmented and completed version of the filtration $\mathbb{F}^{W^{\mathcal{Q}}}$. The authors showed that the LIBOR rate setting can be directly embedded into the framework of instantaneous forward rates: By (4.3),

---

[3] Most authors substitute the accurate term "forward LIBOR rate" for $L_i(t) = L(t, T_i, T_{i+1})$ by the more convenient shortened expression "LIBOR rate". Strictly speaking, this is only appropriate when $t = T_i$, but since no great confusion should be expected, we will also follow this convention.

$$L_i(t) = \frac{1}{\delta}\left[\exp\left(\int_{T_i}^{T_{i+1}} f(t,s)\,ds\right) - 1\right], \qquad (4.5)$$

and the LIBOR rate dynamics may be straight-forwardly derived from the underlying instantaneous forward rate model. Having mentioned this, one might be tempted to think that the two approaches of instantaneous and simple forward rates can be used interchangeably, but this is a fallacious conclusion. In the words of Glasserman and Kou [22], we are dealing with more than a question about the choice of variables. It could be noted, e.g., that (4.5) involves an integral and hence prevents forthright inversion of the expression.

Nonetheless, the embedding is very useful to show that the model is arbitrage-free and that there exists a spot martingale measure $\mathcal{Q}$, under which all bonds discounted with the money market account B are martingales. Furthermore, one can show that each LIBOR rate follows a martingale under an appropriately chosen forward measure. To verify the latter claim, observe that the right-hand side $F_i(t) := B(t,T_i)/B(t,T_{i+1})$ in (4.4) can in fact be interpreted as the price of a forward contract on a bond with maturity T_i, with the expiry of the contract being T_{i+1}. The bonds involved, $B(t,T_i)$ (serving as underlying) and $B(t,T_{i+1})$ (serving as numéraire), are primary traded assets of the market, and hence, $B(t,T_i)/B_t$ and $B(t,T_{i+1})/B_t$ are $\mathcal{Q}$-martingales. Let $\mathcal{Q}^{i+1}$ denote the T_{i+1}-forward measure associated with numéraire $B(t,T_{i+1})$. Then, by the Change-of-Numéraire Theorem 4.1, $F_i(t)$ is a $\mathcal{Q}^{i+1}$-martingale, with the Radon-Nikodým derivative being given as

$$\eta_{i+1}(t) := \left.\frac{d\mathcal{Q}^{i+1}}{d\mathcal{Q}}\right|_{\mathcal{F}_t} = \frac{B(t,T_{i+1})}{B_t} \cdot \frac{1}{B(0,T_{i+1})} \qquad \text{for all } 0 \le t \le T_i.$$

Since each forward LIBOR rate L_i differs from F_i only by the addition and multiplication of constants 1 and δ, it follows immediately that L_i is a $\mathcal{Q}^{i+1}$-martingale as well. As a result, the martingale representation theorem for Brownian markets (see, e.g., Zagst [42], p. 31) implies that each forward LIBOR rate L_i can be modeled as a geometric Brownian motion with drift 0 under the respective forward measure $\mathcal{Q}^{i+1}$,

$$dL_i(t) = L_i(t) \cdot \sigma_i(t)'\,dW^{i+1}(t), \qquad L_i(0) = l_i, \qquad (4.6)$$

where W^{i+1} is assumed to be a d-dimensional $\mathcal{Q}^{i+1}$-Brownian motion, σ_i a predictable d-dimensional vector function satisfying $\int_0^{T_i} \|\sigma_i(s)\|^2\,ds < \infty$ $\mathcal{Q}^{i+1}$-a.s. and l_i is determined according to (4.4) evaluated at 0. The solution to the SDE in (4.6) is given by the stochastic exponential

$$L_i(t) = L_i(0) \cdot \exp\left(\int_0^t \sigma_i(s)'\,dW^{i+1}(s) - \frac{1}{2}\int_0^t \|\sigma_i(s)\|^2\,ds\right). \qquad (4.7)$$

Given the exponential form, the i-th LIBOR rate is non-negative, whenever $L_i(0) \ge 0$, which can be ensured by an initial term structure of the zero-coupon bonds $B(0,T_i)$, $i = 1,\ldots,N$ which is positive and non-increasing in the maturity, $0 < B(0,T_N) \le \ldots \le B(0,T_1)$.

4.3.2 Pricing of Caps and Floors in the Log-Normal LMM

The main reason for the raving success of the LMM almost immediately following its publication can be found in the implications which it had on the evaluation of some of the most important interest-rate instruments, caps and floors (being composed of caplets and floorlets, respectively), among others. The model finally allowed for a closed-form evaluation of these instruments in terms of Black's formula [8].

An *interest-rate caplet (floorlet)* is an instrument that protects its owner against too high (low) interest rates. At time T_i, the interest rate L_i is fixed for the period $[T_i, T_{i+1}]$. The owner of the caplet (floorlet) receives a payment at time T_{i+1} that is equal to the amount in which the interest $\delta \cdot L_i(T_i)$ exceeds (falls below) the pre-specified strike δK,

$$\delta \cdot (L_i(T_i) - K)^+ \qquad \left[\delta \cdot (K - L_i(T_i))^+ \right]. \tag{4.8}$$

A caplet (floorlet) can thus be seen as an interest-rate market equivalent to what a European call (put) option is on the stock market. An *interest-rate cap (floor)* is then a strip of caplets over a collection of time periods $[T_i, T_{i+1}]$, $T_0 < T_1 \ldots < T_N$, where at the end of each period the buyer of the contract receives a payment, if the interest rate fixed at the beginning of the period exceeds (falls below) the fixed, pre-specified strike price K. The time-t payoff of the cap equals the time-t payoff of all remaining caplets,

$$\sum_{i=1,\ldots,n-1; T_i \geq t} \delta \frac{B_t}{B_{T_{i+1}}} (L_i(T_i) - K)^+ .$$

Note that the strike K is the same for all caplets of which the cap is composed. Since a change in sign is the only difference between the payoff of cap/caplets and floor/floorlets, it suffices to henceforth concentrate on the former.

By the usual principle of risk-neutral valuation and the Change-of-Numéraire Theorem 4.1, the price at time t of entering into a caplet contract on the i-th LIBOR rate L_i equals

$$\begin{aligned}\text{Caplet}_i(t) &= \mathbb{E}_{\mathcal{Q}} \left[\frac{B_t}{B_{T_{i+1}}} \cdot \delta \cdot (L_i(T_i) - K)^+ \Big| \mathscr{F}_t \right] \\ &= \delta \cdot B(t, T_{i+1}) \cdot \mathbb{E}_{\mathcal{Q}^{i+1}} \left[(L_i(T_i) - K)^+ \big| \mathscr{F}_t \right]. \end{aligned} \tag{4.9}$$

Under dynamics (4.7), the pricing relation (4.9) can be evaluated using a Black-Scholes type argument (see [37], p. 18):

Theorem 4.3 (Black's Formula for Caplets and Floorlets). *The time-t price of a caplet on the i-th LIBOR rate L_i with strike K is given by*

$$\text{Caplet}_i(t) = \delta \cdot B(t, T_{i+1}) \cdot \left[L_i(t) \cdot \mathcal{N}(\tilde{d}_1) - K \cdot \mathcal{N}(\tilde{d}_2) \right], \tag{4.10}$$

where $\mathscr{N}$ denotes the cumulative distribution function of the standard normal distribution and

$$\tilde{d}_{1,2} = \frac{\ln\left(\frac{L_i(t)}{K}\right) \pm \frac{1}{2}\int_t^{T_i} \|\sigma_i(s)\|^2 \, ds}{\sqrt{\int_t^{T_i} \|\sigma_i(s)\|^2 \, ds}}. \tag{4.11}$$

The time-t price of a floorlet on the i-th LIBOR rate L_i with strike K is given by

$$Floorlet_i(t) = \delta \cdot B(t, T_{i+1}) \cdot \left[K \cdot \mathscr{N}(-\tilde{d}_2) - L_i(t) \cdot \mathscr{N}(-\tilde{d}_1)\right].$$

As the cap over the time period $[0, T_n]$, $0 < n \leq N-1$, is nothing more than a series of caplets with the same underlying strike K, its price is given as

$$\text{Cap}(t) = \sum_{i=1,\dots,n-1; T_i \geq t} \text{Caplet}_i(t).$$

4.4 The Markov-Switching Jump Diffusion (MSJD) Extension of the LMM

Figure 4.1 depicts the movements of the implied volatilities of seven USD ATM caps with different maturities between 1 and 10 years in the time frame 2003/01/01–2012/06/22. There are two major observations that can be made: On the one hand, there seem to be market phases, where prices tend to be more volatile than during other times. On the other hand, one can observe sudden jumps occurring at certain time points, which cannot be properly explained in a diffusion model of the type (4.6), as large displacement are very unlikely to occur for normally distributed

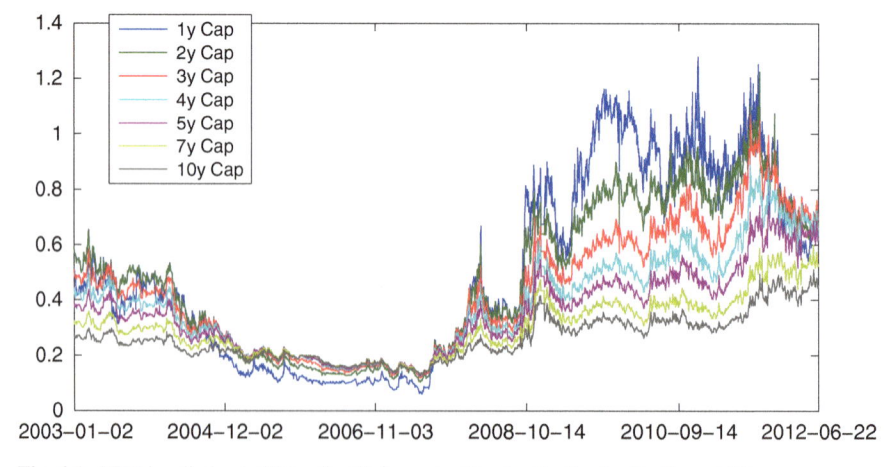

Fig. 4.1 ATM implied volatilities (in %) for maturities 1y, 2y, 3y, 4y, 5y, 7y and 10y

increments. These observations motivate the following proposed extension to the log-normal LMM, where the diffusion process in (4.6) is substituted by a process that is suited to overcome the observed shortcomings. In detail, the ordinary diffusion is replaced by a so-called Markov-switching jump diffusion (MSJD) process. As the name foretells, this process is a jump diffusion whose coefficient functions are deterministic conditional on the state of a given underlying finite-state Markov chain representing the overall market movement. It is worthwhile noting that in the special case, where the jump part equals 0 and the Markov chain can only take on one state, the MSJD extension coincides with the original log-normal LMM.

4.4.1 Presenting the Extended Framework

In extension to the log-normal LMM, let the bond structure again be positive and non-increasing in the maturity. Furthermore, assume, as before, the existence of a bank account $(B_t)_{t \in [0,T^*]}$ and a risk-neutral measure $\mathcal{Q}$ with respect to which all discounted bonds $(B(t,T)/B_t)_{t \in [0,T]}$ follow martingales, $0 < T \leq T^*$. In addition to before, let X be a continuous, time-homogeneous, finite Markov chain, taking values in the standard basis $E = \{e_1, \ldots, e_M\}$ of $\mathbb{R}^M$. The infinitesimal generator of X with respect to the terminal measure $\mathcal{Q}^N$ is given as $\mathcal{A}$. The filtration generated by X is denoted by $\mathbb{F}^X$. It can be shown [16] that X has a semimartingale representation,

$$X_t = X_0 + \int_0^t \mathcal{A}' X_s ds + M_t \qquad (4.12)$$

where $M = (M_t)_{t \geq 0}$ is a right-continuous, square-integrable $\mathbb{R}^M$-valued martingale with respect to $(\mathbb{P}, \mathbb{F}^X)$. Also, let μ be an integer-valued random jump measure defined on the mark space $[0, T^*] \times \mathbb{R}^k$, which is taken to be of finite activity i.e. $\mu\left([0,t] \times \mathbb{R}^k\right) < \infty$ for all $t \in [0, T^*]$.

In extension to (4.6), we propose to model each LIBOR rate L_i, $i = 1, \ldots, N-1$, as a Markov-switching jump diffusion, where the diffusion dynamics of (4.6) are not only augmented by a jump part, but additionally all parameter functions are dependent on the underlying Markov chain X. In detail, we assume that every L_i is governed by the SDE

$$\frac{dL_i(t)}{L_i(t-)} = \sigma_i(t, X_{t-})' dW^{i+1}(t) + \int_{\mathbb{R}^k} \psi_i(t, X_{t-}, z) \left(\mu - v_{X_{t-}}^{i+1}\right) (dt, dz), \qquad (4.13)$$

with W^{i+1} a d-dimensional Brownian motion and $v_{X_{t-}}^{i+1}$ the predictable $\mathcal{Q}^{i+1}$-compensator of μ. σ_i denotes the regime-dependent volatility and ψ_i the regime-dependent jump function associated with the jump term. The objects involved (i.e., processes, measures and compensators) are to satisfy the subsequent assumptions:

(I) X is the only source of randomness for the volatilities and jump functions. For all $i \in \{1, \ldots, N-1\}$, these are defined as

$$\sigma_i(t) = \sigma_i(t, X_{t-}) = \sum_{j=1}^{M} \langle X_{t-}, e_j \rangle \sigma_i(t, e_j), \qquad t \in [0, T_i],$$

$$\psi_i(t, z) = \psi_i(t, X_{t-}, z) = \sum_{j=1}^{M} \langle X_{t-}, e_j \rangle \psi_i(t, e_j, z), \qquad t \in [0, T_i]$$

with $\langle \cdot, \cdot \rangle$ denoting the usual scalar product, and volatilities and jump functions satisfying $\sum_{j=1}^{M} \int_0^{T_i} \|\sigma_i(s, e_j)\|^2 ds < \infty$, for $i = 1, \ldots, N-1$ and $\sum_{j=1}^{M} \int_0^{T_i} \int_{\mathbb{R}^k} |\psi_i(s, j, z)| v_j^i(ds, dz) < \infty$, for $i = 1, \ldots, N-1$. For $t > T_i$, we set $\sigma_i(t, X_{t-}) \equiv 0$ and $\psi_i(t, X_{t-}, z) \equiv 0$, for all $i = 1, \ldots, N-1$. The introduction of such an underlying driving Markov chain is a convenient and commonly used tool to incorporate endogenous structural breaks into the interest rate dynamics (see, e.g., [23], [12], or, more recently, [17]).

(II) Conditional on the Markov chain X, the $\mathcal{Q}^N$-Wiener process W^N and the $\mathcal{Q}^N$-compensated jump measure $\mu - v_{X_{t-}}^N$ are independent. Similar to (I), this is a somewhat standard assumption and has, e.g., been employed by Belomestny and Schoenmakers [5].

(III) The $\mathcal{Q}^N$-compensator $v^N(dt, dz)$ of μ is the predictable compensator associated with a homogeneous Markov-switching marked Poisson process,

$$v_{X_{t-}}^N(dt, dz) = k^N(X_{t-}, dz) \lambda^N(X_{t-}) dt = \sum_{j=1}^{M} \langle X_{t-}, e_j \rangle k_j^N(dz) \lambda^N(e_j) dt$$

with $\lambda^N(e_j)$ being the jump intensity and $k_j^N(dz)$ the conditional distribution of the markers in state e_j (see, e.g., Björk et al. [7], Eberlein et al. [14], or Belomestny and Schoenmakers [5]).

While the intuition behind requirements (II) and (III) will be elaborated further in Sect. 4.4.2, there is already a remark in place both with respect to requirement (I) and the question of no-arbitrage:

Remark 4.1 (Model Specification/No-Arbitrage Conditions).

- Observe that volatilities, jump functions and compensators are dependent on X_{t-} rather than on X_t in order to ensure predictability. In combination with the requirement that μ is integer-valued, this is a necessary assumption to define a special semimartingale of the type (4.1), as introduced in Sect. 4.2. The solution to the stochastic differential equation in (4.13) is accordingly given by the Doléans-Dade exponential

$$L_i(t) = L_i(0) \cdot \exp\Big(-\frac{1}{2}\int_0^t \|\sigma_i(s, X_{s-})\|^2 ds + \int_0^t \sigma_i(s, X_{s-})' dW^{i+1}(s)$$
$$+ \int_0^t \int_{\mathbb{R}^k} \ln(1 + \psi_i(s, X_{s-}, z)) \left(\mu - v_{X_{s-}}^{i+1}\right)(ds, dz)$$
$$+ \int_0^t \int_{\mathbb{R}^k} [\ln(1 + \psi_i(s, X_{s-}, z)) - \psi_i(s, X_{s-}, z)] v_{X_{s-}}^{i+1}(ds, dz)\Big).$$

4 The LIBOR Market Model: A Markov-Switching Jump Diffusion Extension

In combination with the non-decreasing, positive structure of bond prices, this immediately yields the non-negativity of all LIBOR rates, $L_i \geq 0$, $1 \leq i \leq N-1$.

- The other important issue that requires some consideration is the question if the model introduced above is free of arbitrage. Fortunately, one may show that no-arbitrage does in fact hold, for instance by embedding of the MSJD-extension into the generalized HJM framework of Björk et al. (1997). Absence of arbitrage then follows from the no-arbitrage condition in the instantaneous forward rate model, for which no-arbitrage conditions are already known, and the existence of a spot martingale measure $\mathscr{Q}$ can be ensured. The detailed proof on the embedding can be found in Steinrücke et al. [40].

4.4.2 The Measure Changes and Its Consequences

Oftentimes, interest-rate products are not dependent on only one LIBOR rate, but rather multiple ones of different maturities. In order to be able to evaluate these instruments, it is necessary to see how the dynamics of LIBOR rates with different maturities can be interrelated. We investigate the consequences of measure changes for the Wiener processes and the compensators first, before turning towards examining the Markov chain and its generator under different measures.

4.4.2.1 The Measure Changes, the Wiener Process and the Compensator

In order to see how Wiener processes and compensator measures under different forward measures are related to each other, we employ the Change-of-Numéraire Theorem 4.1 in combination with Girsanov's Theorem 4.2. Observe first that the Radon-Nikodým derivative $\eta_{i+2,i+1}(t)$ associated with a measure change from $\mathscr{Q}^{i+2}$ to $\mathscr{Q}^{i+1}$ is given by

$$\eta_{i+2,i+1}(t) := \left. \frac{d\mathscr{Q}^{i+1}}{d\mathscr{Q}^{i+2}} \right|_{\mathscr{F}_t} = \frac{B(0,T_{i+2})}{B(0,T_{i+1})} \cdot \frac{B(t,T_{i+1})}{B(t,T_{i+2})} = \frac{B(0,T_{i+2})}{B(0,T_{i+1})} \cdot [1+\delta L_{i+1}(t)]$$

and, following from (4.13), can be easily shown to have dynamics

$$\frac{d\eta_{i+2,i+1}(t)}{\eta_{i+2,i+1}(t-)} = \frac{\delta L_{i+1}(t-)}{1+\delta L_{i+1}(t-)} \\ \times \left[\sigma_{i+1}(t,X_{t-})' dW^{i+2}(t) + \int_{\mathbb{R}^k} \psi_{i+1}(t,X_{t-},z)\left(\mu - v_{X_{t-}}^{i+2}\right)(dt,dz) \right] \quad (4.14)$$

for $i = N-2, \ldots, 1$. Iteratively, using Girsanov's Theorem, it follows that the $\mathscr{Q}^{i+1}$-Wiener process and the $\mathscr{Q}^{i+1}$-compensator $v_{X_{t-}}^{i+1}$ may be expressed in terms of their $\mathscr{Q}^N$-counterparts as follows,

$$W^{i+1}(t) = -\int_0^t \sum_{j=i+1}^{N-1} \frac{\delta L_j(s-)}{1+\delta L_j(s-)} \sigma_j(s, X_{s-}) ds + W^N(t), \tag{4.15}$$

$$v_{X_{t-}}^{i+1}(dt, dz) = \prod_{j=i+1}^{N-1} \left(1 + \frac{\delta L_j(t-) \psi_j(t, X_{t-}, z)}{1+\delta L_j(t-)}\right) v_{X_{t-}}^N(dt, dz). \tag{4.16}$$

Girsanov's Theorem 4.2 allows us even to make a statement about how jump intensity and the distribution of the markers change. When changing from $\mathscr{Q}^N$ to $\mathscr{Q}^{N-1}$, e.g., the relation between intensities and marker distributions is given as follows,

$$\lambda^{N-1}(t, X_{t-}) = \lambda^N(X_{t-}) \int_{\mathbb{R}^k} [1 + \gamma_{N-1}(t, X_{t-}) \psi_{N-1}(t, X_{t-}, z)] k^N(X_{t-}, dz)$$

$$k^{N-1}(X_{t-}, dz) = \frac{1 + \gamma_{N-1}(t, X_{t-}) \psi_{N-1}(t, X_{t-}, z)}{\int_{\mathbb{R}^k} [1 + \gamma_{N-1}(t, X_{t-}) \psi_{N-1}(t, X_{t-}, z)] k^N(X_{t-}, dz)} k^N(X_{t-}, dz)$$

where $\gamma_{N-1}(t, X_{t-}) = \delta \frac{L_{N-1}(t-)}{1+L_{N-1}(t-)}$. By inserting (4.15) and (4.16) into (4.13), the dynamics of each LIBOR rate L_i, $i = 1, \ldots, M$ can be expressed in terms of $\mathscr{Q}^N$,

$$\frac{dL_i(t)}{L_i(t-)} = -\sum_{j=i+1}^{N-1} \frac{\delta L_j(t-)}{1+\delta L_j(t-)} \sigma_i(t, X_{t-})' \sigma_j(t, X_{t-}) dt + \sigma_i(t, X_{t-})' dW^N(t)$$

$$- \int_{\mathbb{R}^k} \psi_i(t, X_{t-}, z) \left(\prod_{j=i+1}^{N-1} \left(1 + \frac{\delta L_j(t-)}{1+L_j(t-)} \cdot \psi_i(t, X_{t-}, z)\right) - 1\right) v_{X_{t-}}^N(dt, dz)$$

$$+ \int_{\mathbb{R}^k} \psi_i(t, X_{t-}, z) (\mu - v_{X_{t-}}^N)(dt, dz),$$

with W^N being a $\mathscr{Q}^N$ Brownian motion and $v_{X_{t-}}^N$ the $\mathscr{Q}^N$-compensator of μ. As complicated as this expression might look, the representation is necessary in order to allow for the pricing of more evolved financial products than caplets.

Examining (4.15) and (4.16), we can deepen our understanding of requirements (II) and (III):

Remark 4.2 (Measure Changes and Model Assumptions (II) and (III)).
At first sight, it was not immediately obvious, why assumption (II) was only specified for the terminal measure $\mathscr{Q}^N$. Assuming a similar independence structure under all other measures $\mathscr{Q}^2, \ldots, \mathscr{Q}^{N-1}$ would in fact be quite convenient, but unfortunately, this feature cannot be transferred from the terminal measure to any other forward measure. In fact, since in representations (4.15) and (4.16), both $W^{i+1}(t)$ and $v_{X_{t-}}^{i+1}$ contain terms $L_{i+1}, \ldots, L_{N-1}$, which in turn contain integral terms involv-

4 The LIBOR Market Model: A Markov-Switching Jump Diffusion Extension   99

ing W^N and $v^N_{X_{t-}}$, no such independence may be assumed for Wiener process and the compensated measure under any of the measures $\mathscr{Q}^{i+1}$ apart from $\mathscr{Q}^N$.

There is a problem of similar nature arising for requirement (III). We specified the terminal compensator $v^N_{X_{t-}}$ to be the compensator of a time-homogeneous Markov switching measure associated with a marked Poisson process, with Markov-switching jump intensity $\lambda^N(X_{t-})$ and marker distribution $k^N_{X_{t-}}(dz)$. This means, that conditional on the state of the Markov chain, the compensator is deterministic. From (4.16), it is immediately obvious, why we could only make this specification for the terminal measure, as this property is not preserved under the measure change, when switching to any other measure $\mathscr{Q}^{i+1}$, $i = 1,\ldots,N-2$, due to the factors $\delta L_i/(1+\delta L_i)$, $j = i+1,\ldots,N$.

Both points mentioned turn out to be particularly inconvenient when it comes to pricing in Sect. 4.5. Not all is lost, however, since one may follow a proposal of Belomestny and Schoenmakers [5] and approximate (4.15) and (4.16) by freezing the L_j's at 0,

$$dW^{i+1}(t) \approx d\tilde{W}^{i+1}(t) = \sum_{j=i+1}^{N-1} \frac{\delta L_j(0)}{1+\delta L_j(0)} \sigma_j(t,X_{t-})\,dt + dW^N(t), \quad (4.17)$$

$$v^{i+1}_{X_{t-}}(dt,dz) \approx \tilde{v}^{i+1}_{X_{t-}}(dt,dz) := \prod_{j=i+1}^{N-1}\left(1 + \frac{\delta L_j(0)\,\psi_j(t,X_{t-},z)}{1+\delta L_j(0)}\right) v^N_{X_{t-}}(dt,dz). \quad (4.18)$$

This immediately yields as a consequence that $\tilde{v}^{i+1}_{X_{t-}}$ is state-dependent deterministic and the independence between compensator and Wiener process may be preserved.

4.4.2.2 The Measure Changes and the Markov Chain

The other question that needs to be considered is in which way the measure change influences the Markov chain. Since the dynamics of the i-th LIBOR rate are given with respect to the new measure $\mathscr{Q}^{i+1}$, we are interested in also expressing the Markov chain with respect to this measure rather than staying with $\mathscr{Q}^N$. So far, we have used the simple, no-index notation $\mathscr{A}$ for the infinitesimal generator of X. This is a reasonable choice, since we can show that the infinitesimal generator, if first given with respect to the terminal measure $\mathscr{Q}^N$, is not affected by measure changes towards any other forward measure $\mathscr{Q}^{i+1}$. This fact is captured in the following proposition for measure changes from $\mathscr{Q}^{i+1}$ to $\mathscr{Q}^i$:

Proposition 4.1 (Markov Chain under the Measure Change). *The measure change from $\mathscr{Q}^{i+1}$ to $\mathscr{Q}^i$ has no influence on the infinitesimal generator of the Markov chain X.*

Proof. For details, see Steinrücke et al. [40]. The idea is the following: By definition, the infinitesimal generator under the T_i-forward measure $\mathscr{Q}^i$ is given as

$$\mathscr{A}^i f(x) = \lim_{t \downarrow 0} \frac{\mathbb{E}^{\mathscr{Q}^i}[f(X_t)|X_0=x] - f(x)}{t} =: \lim_{t \downarrow 0} \frac{\mathbb{E}^{\mathscr{Q}^i}_x[f(X_t)] - f(x)}{t} \quad (4.19)$$

for any bounded, Borel-measurable function f, for which this limit exists. By application of the Change-of-Numéraire Theorem 4.1, (4.19) may be rewritten as

$$\mathscr{A}^i f(x) = \lim_{t \downarrow 0} \underbrace{\left[\frac{\mathbb{E}^{\mathscr{Q}^{i+1}}_x\left[\frac{d\mathscr{Q}^i}{d\mathscr{Q}^{i+1}}\big|_{\mathscr{F}_t} \cdot f(X_t) \right] - \mathbb{E}^{\mathscr{Q}^{i+1}}_x[f(X_t)]}{t} \right]}_{R_i(X_t)} + \mathscr{A}^{i+1} f(x). \quad (4.20)$$

Using the law of iterated expectation and the martingale property of the Radon-Nikodým derivative with initial value 1, one may show that $R_i(X_t) = 0$. Hence, $\mathscr{A}^i = \mathscr{A}^{i+1}$, and the proposition holds. □

Remark 4.3 (The Infinitesimal Generator and the Change from $\mathscr{Q}$ to $\mathscr{Q}^{i+1}$).
The same statement about the invariability of the generator of the Markov chain is true when changing from the spot martingale measure $\mathscr{Q}$ to $\mathscr{Q}^{i+1}$. The claim can be proved in a similar fashion as in Proposition 4.1.

4.5 Pricing in the MSJD Framework

Having specified and elaborated the dynamics of the MSJD extension to the LMM, the next step is to turn to the pricing of caplets/caps. For convenience, recall that each LIBOR rate is following dynamics

$$\frac{dL_i(t)}{L_i(t-)} = \sigma_i(t, X_{t-})' dW^{i+1}(t) + \int_{\mathbb{R}^k} \psi_i(t, X_{t-}, z)\left(\mu - v^{i+1}_{X_{t-}}\right)(dt, dz), \quad (4.21)$$

where σ_i and ψ_i are state-dependent predictable functions, W^{i+1} is a d-dimensional $\mathscr{Q}^{i+1}$-Brownian motion and $v^{i+1}_{X_{t-}}$ is the $\mathscr{Q}^{i+1}$-compensator measure of the integer-valued random measure μ. There are different possibilities of how volatilities, jump functions and compensators in the MSJD framework can be specified. Since the calibration of the model involves the fitting of parameters related to a wide range of LIBOR rates with different maturities, a first step is to limit the dimension to $d=1$ and $k=1$.[4] In the particular case considered here, it is furthermore assumed that both volatilities and jump functions are regime-dependent constant, i.e.,

---

[4] At least in the case, where no jumps are considered, this is a justifiable assumption, when it comes to caplet pricing. Similar to pricing formulas (4.10) and (4.11) developed in the log-normal LMM, our considerations will show that prices depend only on the norm of σ_i, that is $\|\sigma_i(t, X_{t-})\|$, and not on $\sigma_i(t, X_{t-})$ itself. As underlined, e.g., by Filipovic [20], p. 213, there is thus no gain in flexibility for caplet pricing by introducing additional dimension into the model. Note, nonetheless, that this is no longer true for swaption pricing.

$$\sigma_i(t, X_{t-}) = \sigma_i(X_{t-}), \psi_i(t, X_{t-}, z) = \psi_i(X_{t-}, z) \ \forall \ i = 1, \ldots, N-1. \tag{4.22}$$

This specification appears a bit crude, but has the advantage that it allows for explicit pricing. In any case, the assumption should be seen in the light of changes in X being a direct indicator for overall changing market conditions and can be interpreted in accordance with assumption (I).

Let $\mathscr{Q}$ be an appropriately chosen spot martingale measure, under which all discounted bonds $B(t, T_1)/B_t, \ldots, B(t, T_N)/B_t$ follow martingales. Like in the introductory section on the log-normal LMM, the price of a caplet can be determined as

$$\text{Caplet}_i(t) = \delta \cdot B(t, T_{i+1}) \cdot \mathbb{E}_{\mathscr{Q}^{i+1}}\left[(L_i(T_i) - K)^+ \big| \mathscr{F}_t\right]. \tag{4.23}$$

Already when not taking into consideration a Markov-switching feature, it is usually not possible to analytically price caplets/caps in a jump-diffusion framework. In this case, many authors employ the techniques of either Laplace- or Fourier-transforms. It turns out, that the Laplace transform is indeed one possible path to follow, when working with the yet more complicated MSJD models. To this end, let $t = 0$ for simplicity. Caplet price (4.23) may also be expressed as

$$\text{Caplet}_i(0) = \delta \cdot B(0, T_{i+1}) \cdot K \cdot \mathbb{E}_{\mathscr{Q}^{i+1}}\left[\left(e^{Y_i(T_i) - k_i} - 1\right)^+\right],$$

with $Y_i(t) := \ln(L_i(t)/L_i(0)) = \ln(L_i(t)) - \ln(L_i(0))$ and $k_i = \ln(K/L_i(0))$. Following the derivation in Raible [32], the caplet's price can be determined via the Laplace-transform

$$\text{Caplet}_i(0) = Ke^{xk_i} \delta B(0, T_{i+1}) \frac{1}{\pi} \int_0^\infty \text{Re}\left(e^{iuk_i} \frac{\phi_i(ix - u, T_i)}{x^2 + x - u^2 + iu(2x + 1)}\right) du, \tag{4.24}$$

for some $x < -1$, where $\phi_i(u, t) = \mathbb{E}_{\mathscr{Q}^{i+1}}[\exp(iuY_i(t))]$ denotes the characteristic function of Y_i under the corresponding T_{i+1} forward measure. In order to be able to evaluate (4.24), it is necessary to find a closed-form expression for $\phi_i(ix - u, T_i)$ for all $1 \leq i \leq N - 1$.

4.5.1 Determining the Characteristic Function of Y_{N-1}

The crucial point in the pricing of and calibration to caplets is to start with the LIBOR rate $L_{N-1}(t)$ of longest maturity T_{N-1} whose dynamics are given as

$$\frac{dL_{N-1}(t)}{L_{N-1}(t-)} = \sigma_{N-1}(t, X_{t-}) dW^N(t) + \int_{\mathbb{R}} \psi_{N-1}(t, X_{t-}, z)\left(\mu - v_{X_{t-}}^N\right)(dt, dz). \tag{4.25}$$

As observed before, this LIBOR rate takes a special role among the other rates, because it is the only rate, where the jump distribution is known and the intensity is constant when conditioned on the state of the underlying Markov chain,

$$v^N_{X_{t-}}(dt,dz) = \lambda^N(X_{t-})k^N(X_{t-},dz)\,dt.$$

In order to derive the characteristic function of $Y_{N-1}(t) = \ln(L_{N-1}(t)/L_{N-1}(0))$, one observes that the dynamics of Y_{N-1} are given by the Doléans-Dade exponential

$$L_{N-1}(t) = L_{N-1}(0)\cdot\exp\Big(-\frac{1}{2}\int_0^t \sigma^2_{N-1}(s,X_{s-})ds + \int_0^t \sigma_{N-1}(s,X_{t-})dW^N(s)$$
$$-\int_0^t\int_{\mathbb{R}} \psi_{N-1}(s,X_{s-},z)v^N_{X_{s-}}(ds,dz)$$
$$+\int_0^t\int_{\mathbb{R}} \ln(1+\psi_{N-1}(s,X_{s-},z))\mu(ds,dz)\Big).$$

It then immediately follows that the dynamics of $Y_{N-1}(t) = \ln(L_{N-1}(t)/L_{N-1}(0))$ are given as

$$dY_{N-1}(t) = -\frac{1}{2}\sigma^2_{N-1}(t,X_{t-})dt + \sigma_{N-1}(t,X_{t-})dW^N(t)$$
$$-\int_{\mathbb{R}} \psi_{N-1}(t,X_{t-},z)v^N_{X_{t-}}(dt,dz) + \int_{\mathbb{R}} \ln(1+\psi_{N-1}(t,X_{t-},z))\mu(dt,dz).$$
(4.26)

For X being in a fixed state, say $X_{t-} \equiv e_j$, we write $Y_{N-1}(t,j)$ and (4.26) reads

$$dY_{N-1}(t,j) = -\frac{1}{2}\sigma^2_{N-1}(t,e_j)dt + \sigma_{N-1}(t,e_j)dW^N(t)$$
$$-\int_{\mathbb{R}} \psi_{N-1}(t,e_j,z)v^N_j(dt,dz) + \int_{\mathbb{R}} \ln(1+\psi_{N-1}(t,e_j,z))\mu(dt,dz).$$
(4.27)

The characteristic function of $Y_{N-1}(t,j)$ can be easily determined:

Proposition 4.2 (Characteristic Function of $Y_{N-1}(.,j)$). *The characteristic function $\phi_{N-1}(u,t,j) = \mathbb{E}_{\mathcal{Q}^N}[\exp(iuY_{N-1}(t,j))]$ of $Y_{N-1}(.,j)$ is given as*

$$\phi_{N-1}(u,t,j) = \exp\Big(\int_0^t \zeta_{N-1}(s,e_j,u)\,ds\Big),$$

with

$$\zeta_{N-1}(s,e_j,u) := -\frac{u^2}{2}\sigma^2_{N-1}(s,e_j) - \frac{1}{2}iu\sigma^2_{N-1}(s,e_j) - iu\lambda^N(e_j)\int_{\mathbb{R}} \psi_{N-1}(s,e_j,z)k^N(e_j,dz)$$
$$+\lambda^N(e_j)\int_{\mathbb{R}} [\exp(iu\ln(\psi_{N-1}(s,e_j,z)+1)) - 1]k^N(e_j,dz).$$

Proof. Observing that the first and third term in (4.27) are deterministic, it suffices to analyze the second and fourth term, which are, by definition, independent. The claim then follows from the general form of characteristic functions for Brownian motions and marked Poisson processes. □

Clearly, it also follows then that the characteristic function $\phi_{N-1}(u,t)$ of Y_{N-1} is given as

$$\phi_{N-1}(u,t) = \mathbb{E}_{\mathcal{Q}^N}\left[\exp\left(iuY_{N-1}(t)\right)\right] = \mathbb{E}_{\mathcal{Q}^N}\left[\mathbb{E}_{\mathcal{Q}^N}\left[\exp\left(iuY_{N-1}(t)\right)|\mathcal{F}_t^X\right]\right]$$

$$= \mathbb{E}_{\mathcal{Q}^N}\left[\exp\left(\int_0^t \left[-\frac{u^2}{2}\sigma_{N-1}^2(s,X_{s-}) - \frac{1}{2}iu\sigma_{N-1}^2(s,X_{s-})\right.\right.\right.$$

$$\left.- iu\lambda^N(X_{s-})\int_{\mathbb{R}} \psi_{N-1}(s,X_{s-},z)k^N(X_{s-},dz)\right.$$

$$\left.\left.\left.+ \lambda^N(X_{s-})\int_{\mathbb{R}}\left[\exp\left(iu\ln\left(\psi_{N-1}(s,X_{s-},z)+1\right)\right)-1\right]k^N(X_{s-},dz)\right]ds\right)\right],$$

by the law of iterated expectation. Set

$$Z_{N-1}(t,u) = \exp\left(\int_0^t \left[-\frac{u^2}{2}\sigma_{N-1}^2(s,X_{s-}) - \frac{1}{2}iu\sigma_{N-1}^2(s,X_{s-})\right.\right.$$

$$\left.- iu\lambda^N(X_{s-})\int_{\mathbb{R}} \psi_{N-1}(s,X_{s-},z)k^N(X_{s-},dz)\right.$$

$$\left.\left.+ \lambda^N(X_{s-})\int_{\mathbb{R}}\left[\exp\left(iu\ln\left(\psi_{N-1}(s,X_{s-},z)+1\right)\right)-1\right]k^N(X_{s-},dz)\right]ds\right).$$

Then, *given that volatility and jump function are regime-dependent constant*, $\phi_{N-1}(u,t) = \mathbb{E}_{\mathcal{Q}^N}[Z_{N-1}(t,u)]$ may be evaluated as follows:

Proposition 4.3 (Characteristic Function of Y_{N-1}). *Let X be a Markov chain with infinitesimal generator $\mathscr{A}$ taking its values in the M-dimensional state space $E = \{e_1,\ldots,e_M\}$. Furthermore, let $\sigma_i(t,X_{t-}) = \sigma_i(X_{t-})$ and $\psi_i(t,X_{t-},u) = \psi_i(X_{t-},u)$ of each state be non-time-dependent. Then, the characteristic function $\phi_{N-1}(u,t) = \mathbb{E}_{\mathcal{Q}^N}\left[\exp\left(iuY_{N-1}(t)\right)\right]$ of Y_{N-1} is given as*

$$\phi_{N-1}(u,t) = \langle 1_M, \exp\left(\mathscr{C}_{N-1}(u)\cdot t\right)X_0\rangle \quad (4.28)$$

where $1_M \in \mathbb{R}^M$ is the vector consisting only of ones, $\langle .,.\rangle$ the Euclidean scalar product in $\mathbb{R}^M$ and $\mathscr{C}_{N-1}(u)$ given as

$$\mathscr{C}_{N-1}(u) = \mathscr{A}' + \mathrm{diag}\left(\zeta_{N-1}(e_1,u),\ldots,\zeta_{N-1}(e_M,u)\right),$$

and $\zeta_{N-1}(e_j,u) \equiv \zeta_{N-1}(t,e_j,u)$, $j = 1,\ldots,M$, as in Proposition 4.2.

Proof. This proof can conceptually be traced back to an idea as initially introduced by Elliott and Valchev [18]. The proof is in parts based on the considerations in Elliott and Valchev [18]. Set $G_t = X_t \cdot Z_{N-1}(t,u)$. Then, using the semimartingale decomposition for Markov chains and integration by parts, one may show that

$$G_t = X_0 + \int_0^t \left[\mathscr{A}' + \text{diag}\left(\zeta_{N-1}(s,e_1,u),\ldots,\zeta_{N-1}(s,e_M,u)\right)\right] G_s ds + \int_0^t (\ldots) dM_s,$$

where M is a martingale with respect to $\mathbb{F}^X$. Taking expectations yields, under application of Fubini's Theorem,

$$\mathbb{E}_{\mathscr{Q}^{i+1}}[G_t] = X_0 + \int_0^t \underbrace{\left[\mathscr{A}' + \text{diag}\left(\zeta_{N-1}(s,e_1,u),\ldots,\zeta_{N-1}(s,e_M,u)\right)\right]}_{=:B_s} \mathbb{E}_{\mathscr{Q}^{i+1}}[G_s] ds + 0. \quad (4.29)$$

For any time-independent matrix, the Lappo-Danilevskiî condition, i.e. $B_s \int_t^s B_v dv = \int_t^s B_v dv B_s$, always holds. Consequently, by Lemma 4.2.1 in Adrianova [1], the solution to (4.29) is then given as

$$\mathbb{E}_{\mathscr{Q}^{i+1}}[G_t] = \exp\left(\int_0^t B_s ds\right) X_0 = \exp\left(t \cdot \left[\mathscr{A}' + \text{diag}\left(\zeta_{N-1}(s,e_1,u),\ldots,\zeta_{N-1}(s,e_M,u)\right)\right]\right) X_0.$$

As $1_M := (1,\ldots,1)' \in \mathbb{R}^M$ and X is taking its values in $E = \{e_1,\ldots,e_M\}$, we have $\langle 1_M, X_{T_i}\rangle = 1$. This yields the claim. □

Remark 4.4 (Limitations and Possible Extensions). Note that a most convenient simplification of $\Phi_{N-1}(u,t)$ in the form

$$\Phi_{N-1}(u,t) = \left\langle 1_M, \exp\left(\mathscr{A}' \cdot t\right) \cdot \exp\left(\text{diag}\left(\zeta_{N-1}(e_1,u),\ldots,\zeta_{N-1}(e_M,u)\right) \cdot t\right) X_t\right\rangle$$
$$= \left\langle 1_M, C \cdot \text{diag}\left(e^{\lambda_1 \cdot t},\ldots,e^{\lambda_M \cdot t}\right) \cdot C^{-1}\right.$$
$$\left. \times \text{diag}\left(e^{\zeta_{N-1}(e_1,u)\cdot t},\ldots,e^{\zeta_{N-1}(e_M,u)\cdot t}\right) X_t\right\rangle, \quad (4.30)$$

as proposed in Elliott and Wilson [19], with $\lambda_1,\ldots,\lambda_M$ the real eigenvalues of $\mathscr{A}$, and C the matrix consisting of the corresponding eigenvectors $\{c_1,\ldots,c_M\}$, is in general not possible. This follows from the fact that for matrix exponentials, $\exp(A+B) = \exp(A)\exp(B)$ usually does not hold and therefore the second equality sign in (4.30) is in general not true. The second bad news is that whenever the coefficient functions are not constant in time, i.e.

$$\sigma_i(t,X_{t-}) \neq \sigma_i(X_{t-}) \quad \text{and} \quad \psi_i(t,X_{t-},z) \neq \psi_i(X_{t-},z),$$

the Lappo-Danilevskiî condition is usually violated and a solution to (4.29) cannot be derived so easily.

There is, fortunately, some good news as well: There are techniques how the linear homogeneous system (4.29) may be approximated numerically, in the case that the coefficient functions are not state-dependent constant. One possible way is the so-called *Magnus expansion* [29] yielding the characteristic function to be given as

$$\phi_{N-1}(u,t) = \left\langle 1_M, \left[\prod_{k=0}^{n_{T_{N-1}}} \exp\left(\Omega\left(t_k, t_{k-1}\right)\right) \right] \cdot X_t \right\rangle,$$

on an approximation grid covering the whole interval $[0, T_{N-1}]$, and all $\Omega(t_k, t_{k-1})$ are linear combinations of integrals and nested commutators involving the matrix B_s on the corresponding interval $(t_{k-1}, t_k]$. Further details on the employment of the Magnus expansion can be found in Blanes et al. [9].

4.5.2 Determining the Characteristic Function of Y_j, $j = 1, \ldots, N-2$

As the derivation of Proposition 4.3 is dependent on the fact that, conditional on X, the compensator is deterministic, the strategy cannot be straight-forwardly extended to the pricing of caplets on the other LIBOR rates. We may, however, use observation (4.18) in Sect. 4.4.2, and approximate the respective compensators $v_{X_{t-}}^{i+1}$ in terms of the terminal compensator $v_{X_{t-}}^N$. Writing $\tilde{L}_i$ for the approximated LIBOR dynamics under (4.18), one observes that in a situation where ψ_i is constant conditional on the state X_{t-}, $\psi_i(t, X_{t-}, z) = \psi_i(X_{t-}, z)$, the compensator is state-dependent deterministic with regime-dependent constant jump intensity. Consequently, we are back in the setting of the previous proposition and the approximative characteristic function of $\tilde{Y}_i = \ln \tilde{L}_i(t) - \ln \tilde{L}_i(0)$, $i = 1, \ldots, N-2$ can be determined accordingly.

4.6 Calibration

Having developed formulae for the pricing of caplets/caps, the ultimate step is to find parameters that can accurately reproduce the observed market prices for these products. With the LIBOR rate dynamics being given as

$$\frac{dL_i(t)}{L_i(t-)} = \sigma_i(t, X_{t-}) dW^{i+1}(t) + \int_{\mathbb{R}} \psi_i(t, X_{t-}, z) \left(\mu - v_{X_{t-}}^{i+1}\right)(dt, dz), \quad (4.31)$$

the fitting procedure entails determining the parameters specifying the rate matrix $\mathscr{A}$, the volatilities $\sigma_i(t, X_{t-})$, the jump functions $\psi_i(t, X_{t-}, z)$ and the compensators $v_{X_{t-}}^{i+1}(dt, dz)$, $i = 1, \ldots, N-1$, for a given observation period of market data. Even for very simple specifications of volatilities, jump functions and compensators, it is easy to see that the calibration procedure entails determining a wide range of parameters. It is hence reasonable to make some further simplifying assumptions to limit the number of parameters to be estimated. Some straight-forward requirements are, for example, the following:

- The Markov chain takes its values in a state space with only two states, $E = \{e_1, e_2\}$.

- Wiener processes and mark space are one-dimensional.[5]
- Volatilities $\sigma_i(t, X_{t-}) \equiv \sigma_i(X_{t-})$ and jump terms $\psi_i(t, X_{t-}, z) = e^z - 1$ are regime-dependent constant.
- The marker distribution of the $\mathcal{Q}^N$-compensator $v^N_{X_{t-}}$ is Markov-switching Gaussian,

$$v^N_{X_{t-}}(dt, dz) = \lambda^N(X_{t-}) \frac{1}{\sqrt{2\pi v_J^2(X_{t-})}} \exp\left(-\frac{(z - m_J(X_{t-}))^2}{2v_J^2(X_{t-})}\right) dz\, dt.$$

- The $\mathcal{Q}^{i+1}$-compensators $v^{i+1}_{X_{t-}}$, $1 \leq i \leq N-2$, are approximated as

$$\tilde{v}^{i+1}_{X_{t-}}(dt, dz) = \prod_{j=i+1}^{N-1} \left(1 + \frac{\delta L_j(0)(e^z - 1)}{1 + \delta L_j(0)}\right) v^N_{X_{t-}}(dt, dz). \tag{4.32}$$

Under this specification, we successfully calibrated the model in a multi-step procedure based on a combination of bootstrapping techniques, Markov-Chain Monte Carlo methods and non-linear optimization algorithms, which were all implemented in MATLAB. As the calibration procedure entails a number of rather extensive steps, only the results shall be presented here. Further details on the methods employed can be found in Steinrücke et al. [40]. In view of the large amount of parameters characterizing the interest rate dynamics, a joint estimation is unfortunately not possible which in turn lead us to the employment of the hereafter elaborated iterative procedure. The method employed proved to be robust and lead to reasonable results.

4.6.1 The Data

For the calibration of the MSJD extension of the LMM, we employed data over a time period of more than 9 years, encompassing all trading days between 2003/01/01 and 2012/06/22 (2,464 days). The data source was Thomson Reuters. In detail, we considered USD ATM caps with maturities 1y, 2y, 3y, 4y, 5y, 7y and 10y years, quoted in volatilities v^1, v^2, v^3, v^4, v^5, v^7 and v^{10}. As caplet prices are usually not directly quoted on the market, they were derived from the available cap information using a bootstrapping technique based on the Black (1976)-formula for caplet prices. For the caps with given maturities, this yielded prices for 39 caplets with maturities $(0.25, \ldots, 9.75)$ and expiries $(0.5, \ldots, 10)$. Figure 4.2 depicts the (forward) volatility of the caplets with maturities 1y, 1.25y, 1.5y and 1.75y. [6] Figure 4.3 depicts the

---

[5] Given these assumptions, the specified model can be seen as a simple, straight-forward generalization to the one-dimensional LIBOR model of Brace et al. [10], where not only the volatility, but also parameters characterizing jumps and structural breaks in the market can be derived in the calibration procedure. The reduction to one-dimensional spaces allows to significantly reduce the amount of parameters to be fitted. All challenges met and a more detailed explanation on the necessity of employing a multi-step iterative procedure may be found in Steinrücke et al. [40].

[6] Observe that the bootstrapping procedure assumes constant volatilities for caplets in between caps of succeeding maturity. For example, for $t = 0$, the (identical) volatilities σ_{1y} for caplets with maturities 1y, 1.25y, 1.5y and 1.75y are derived based on the formula

corresponding caplet prices. Even from pure eyesight, it is apparent, that there are times of low (in particular in the time frame between ∼2004/12/02 and 2007/08/21) and high fluctuation (in particular in the time frame between ∼2009/12/08 and 2011/06/21) in the market.

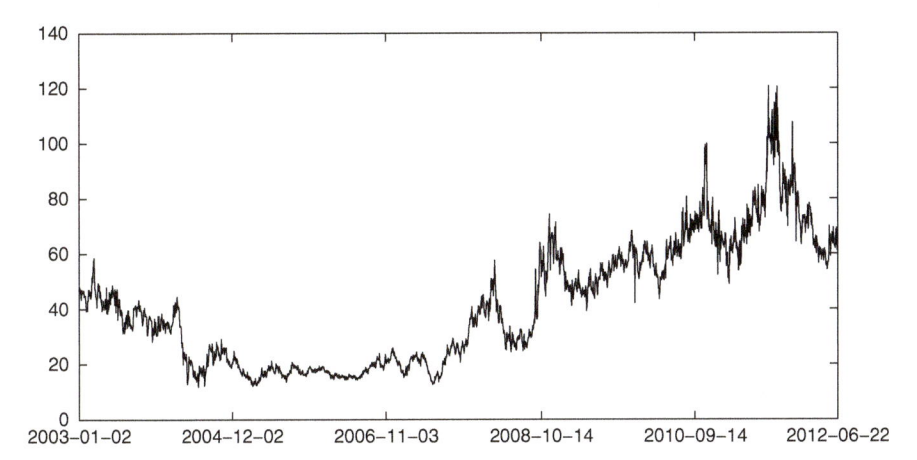

Fig. 4.2 Fitted (forward) caplet volatility (in %) for maturities 1y, 1.25y, 1.5y and 1.75y

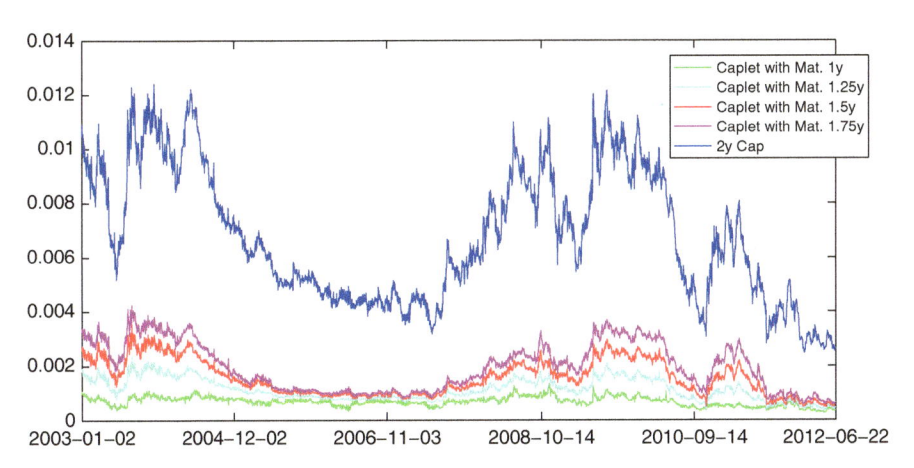

Fig. 4.3 Fitted caplet prices for maturities 1y, 1.25y, 1.5y and 1.75y

$$\text{Cap}(0,2y,K_{2y}) - \text{Cap}(0,1y,K_{1y}) = \text{Caplet}(0,1y,\sigma_{1y},K_{2y}) + \text{Caplet}(0,1.25y,\sigma_{1y},K_{2y})$$
$$+ \text{Caplet}(0,1.5y,\sigma_{1y},K_{2y}) + \text{Caplet}(0,1.75y,\sigma_{1y},K_{2y}),$$

where $\text{Caplet}(0,T_j,\sigma_{1y},K_{2y})$ denotes the price of a caplet at $t=0$ with maturity T_j, expiry T_{j+1} and ATM cap strike K_{2y} for maturity 2y, evaluated by the Black [8]-formula for constant volatility σ_{1y}. For details, see, e.g., Filipovic [20], p.215 seqq.

4.6.2 Discussion of the Results of the Calibration

For the employed calibration routine, all market-observed caplet prices could be perfectly reproduced within the MSJD extension. It may hence be concluded that this extension is a suitable generalization to existing LIBOR market models and can help to account for jumps and overall market trends. The detailed results for the different steps of the calibration are as follows:

4.6.2.1 Most Likely Path and Infinitesimal Generator of the Markov Chain

The most likely state of the Markov chain at each time point was derived based on the market observed cap volatilities, using an MCMC routine for Markov-switching Vasicek processes (Gibbs-sampler with 100,000 draws, burn-in of 20,000). The day-by-day estimates for each volatility are depicted in Fig. 4.4. As the path estimates for different cap volatilities turn out to be very similar, it seemed reasonable to take the average over all estimates and round to the closest integer (1 or 2) in order to receive an overall estimate for the most likely path of the Markov chain. Based on this average path, the rate matrix $\mathscr{A}$ and the corresponding stationary distribution π were estimated to be given as

$$\mathscr{A} = \begin{pmatrix} -10.7910 & 10.7910 \\ 17.9111 & -17.9111 \end{pmatrix} \text{ and } \pi = \begin{pmatrix} 0.6239 \\ 0.3761 \end{pmatrix}.$$

The findings coincide with the intuition that the market is more likely to reside in a 'normal' state than in the excited state as $\pi_1 \gg \pi_2$ (see also Svoboda [41]).

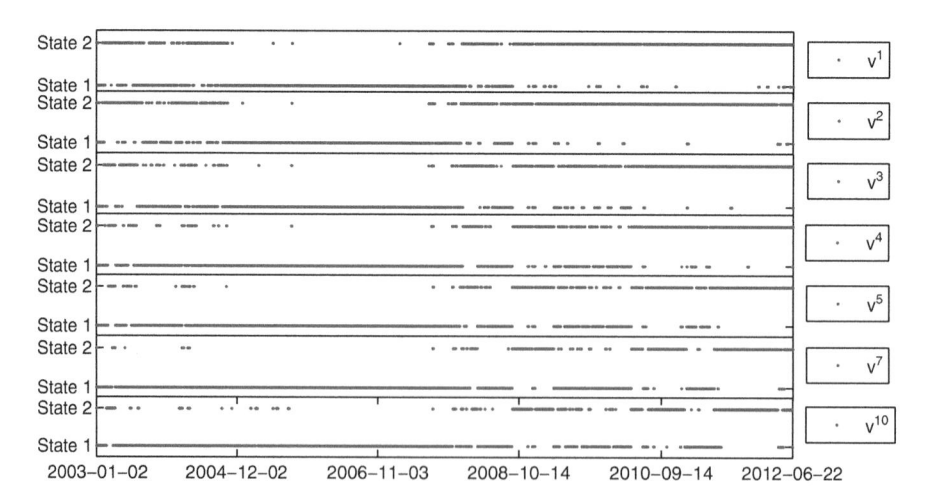

Fig. 4.4 Estimated states of the economy at all days of the time series, for cap volatilities $v^1, \ldots, v^{10}$

4.6.2.2 Parameters Specifying the Compensator Measure with Respect to $\mathbb{P}$

The next step entailed the estimation of the parameters specifying the compensator with respect to the *physical measure* $\mathbb{P}$. These were derived based on the terminal LIBOR rate L_{N-1}, by splitting the time series according to the different states derived in the previous step and using an MCMC algorithm for jump diffusions on each of these time series (again a Gibbs-sampler with 100,000 draws, burn-in of 20,000). Figure 4.5 depicts the logarithmized time series as well as jump times and sizes for both time series displaying the success of the algorithm to identify occasions jump times (counted as incidences over time) and sizes. As the sampled parameters in Table 4.1 display, there is a tendency in state e_1 towards jumps into a positive direction, while jumps in state e_2 have a slightly downwards tendency. This coincides with the intuition that in times of crises and market unrest, a downwards tendency should also manifest itself in the jump sizes. Also, it seems that in times of crises, jumps happen more often, as $\lambda^{\mathbb{P}}(e_2) > \lambda^{\mathbb{P}}(e_1)$. Table 4.1 also displays the standard errors for the sampled parameters.[7]

It should be mentioned that both Gibbs samplers, for the determination of the parameters describing the Markov chain and the jump parameters, display very good sampling behavior. The Markov chain generated by each sampler mixes well, as the parameter space is exploited nicely, and the time to convergence is quick (there is hardly any burn-in time until the samples are drawn from a non-changing range of values). Exemplarily, this can be seen in the trace plot for $\left(\lambda^{\mathbb{P}}(e_1), \lambda^{\mathbb{P}}(e_2)\right)$ in Fig. 4.6. Trace plots for all other parameters exhibit a similar convergece behavior.

Table 4.1 Parameter estimates for the compensator $v^{\mathbb{P}}$

Parameters	Sampled values	Standard error
$(\lambda^{\mathbb{P}}(e_1), \lambda^{\mathbb{P}}(e_2))$	$(0.1823, 0.2263)$	$(8.6105 \cdot 10^{-05}, 3.1956 \cdot 10^{-04})$
$(m_J(e_1), m_J(e_2))$	$(0.0014, -0.0053)$	$(1.2693 \cdot 10^{-05}, 1.8959 \cdot 10^{-05})$
$(v_J^2(e_1), v_J^2(e_2))$	$(0.0026, 0.0026)$	$(1.2317 \cdot 10^{-06}, 2.8182 \cdot 10^{-06})$

4.6.2.3 Volatility Parameters in the Model Specification Without Jumps

In a next step, we derived first volatility estimates for the respective states, by ignoring the jump part in (4.31) and simply considering a Markov-switching log-normal model. Based on market-observed caplet prices of every day and the pricing formula (4.24), we ran a non-linear optimization routine to derive a first estimate of the volatilities $\tilde{\sigma}_i(e_j)$, $i = 1, \ldots, N_1$, $i = 1, 2$. In order to guarantee for a unique solution, we let

---

[7] The variance of the ergodic distribution π of the generated Markov chain X is given as $\Sigma^2 = \text{Var}_\pi(X_1) + 2\sum_{t=2}^{\infty} \text{Cov}_\pi(X_1, X_t)$. For details on the estimation of the standard error, $\hat{\Sigma}/(N - \text{burn-in})$ in the MCMC context, see, e.g., Flegal [21], p. 59 seqq.

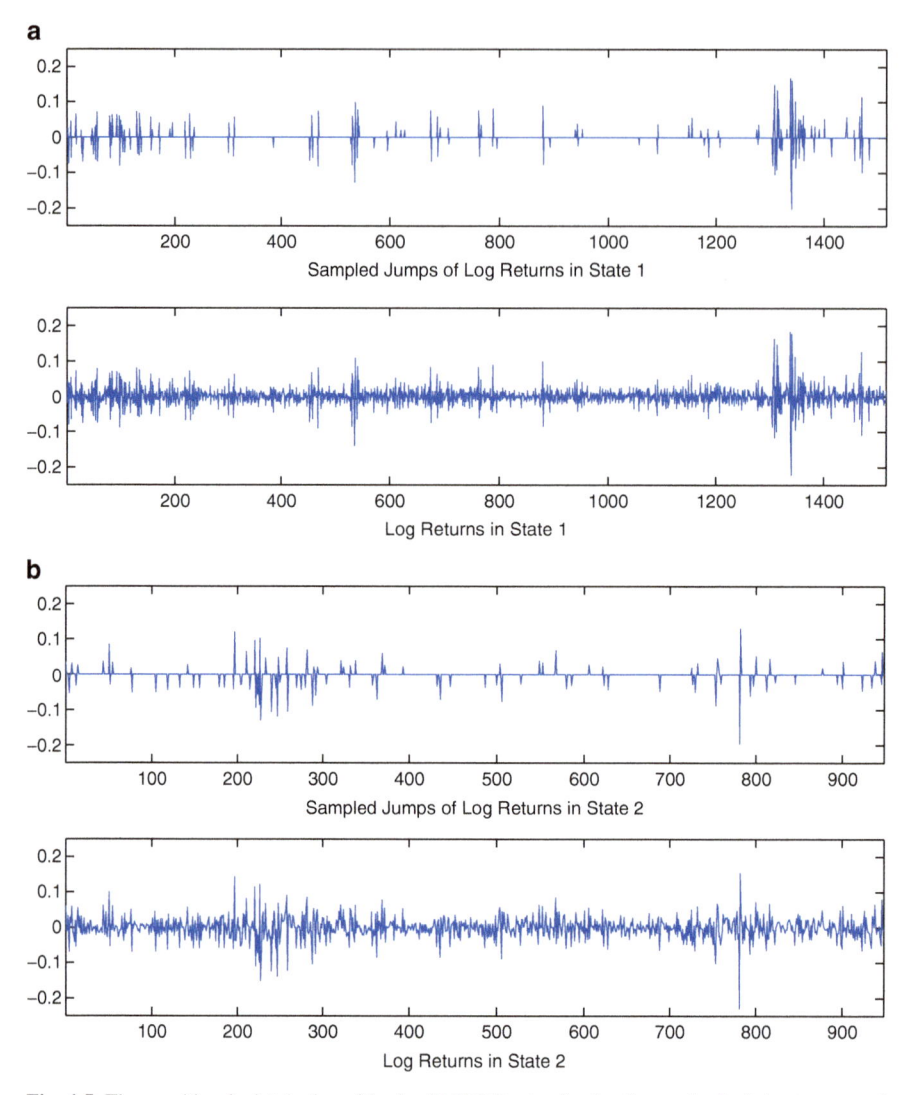

Fig. 4.5 The graphics depict the logarithmized LIBOR rates for the time series both in state e_1 and e_2 (*bottom picture* for (**a**) and (**b**)). The *top picture* in both cases depicts the sampled jump times (counted as incidences over time) and sizes. (**a**) Jump times and sizes for state e_1. (**b**) Jump times and sizes for state e_2

$$\mathcal{Q}^{i+1}(X_t = e_1) \cdot \tilde{\sigma}_i(e_1) + \mathcal{Q}^{i+1}(X_t = e_2) \cdot \tilde{\sigma}_i(e_2) = \sigma_i^{\text{MKT}}, \quad (4.33)$$

which reflects the reasonable assumption that the market-observed volatility is the investors' expected volatility, based on two possible states for each day with corresponding volatilities $\tilde{\sigma}_i(e_1)$ and $\tilde{\sigma}_i(e_2)$. Caplet prices could be perfectly recovered at all times.

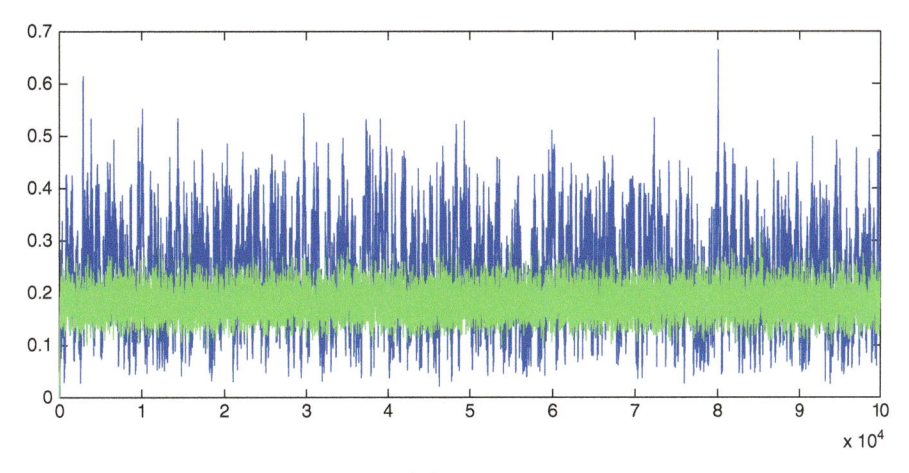

Fig. 4.6 Trace plot of $\lambda(e_1)$ (*green*) and $\lambda(e_2)$ (*blue*)

4.6.2.4 Final Estimates for Volatility and Jump Parameters

In a last step, we derived the final estimates for volatility and jump parameters in a model *with jumps*, by combining the estimates from the previous steps.[8] For simplicity, it was assumed that the jump *size* distribution is unaffected by a measure change from $\mathbb{P}$ to $\mathcal{Q}^N$, $k^N_{X_{t-}} \equiv k^{\mathbb{P}}_{X_{t-}}$. In contrast, the *intensity* was allowed to differ between $\mathbb{P}$ and $\mathcal{Q}^N$. Assuming that the intensities do not change over time, by Girsanov's Theorem 4.2 there exists a $\phi(X_{t-}) \equiv c$ such that $\left(\lambda^N(e_1), \lambda^N(e_2)\right) = c \cdot \left(\lambda^{\mathbb{P}}(e_1), \lambda^{\mathbb{P}}(e_2)\right)$. We found that an estimate of $c = 0.6$ yields very good results for the algorithm minimizing the distance between model and market prices. Setting up some further appropriate constraints, another non-linear optimization routine based on the time series of caplet prices Caplet_{N-1} and $c = 0.6$ was run to determine the volatility σ_{N-1} for each time point of the time series. Finally, the compensators for all other LIBOR rates were approximated through relation (4.32) and σ_i, $i = 1, \ldots, N-2$, were estimated by non-linear optimization based on the respective caplet prices Caplet_i.

All market prices in the investigated observation period were perfectly reproduced. The estimates for the jump parameters can be found in Table 4.2. Observe the similarity of the result to the calibration for the (non Markov-switching) jump diffusion model of Belomestny and Schoenmakers [5], where the jump intensity is estimated as 0.1.

---

[8] In favor of computation time we only considered every tenth day in the available time series for the calibration of the remaining parameters.

Table 4.2 Parameter estimates for the terminal compensator v^N

Parameters	Estimates
$(\lambda^N(e_1), \lambda^N(e_2))$	$(0.1094, 0.1358)$
$(m_J(e_1), m_J(e_2))$	$(0.0014, -0.0053)$
$(v_J^2(e_1), v_J^2(e_2))$	$(0.0026, 0.0026)$

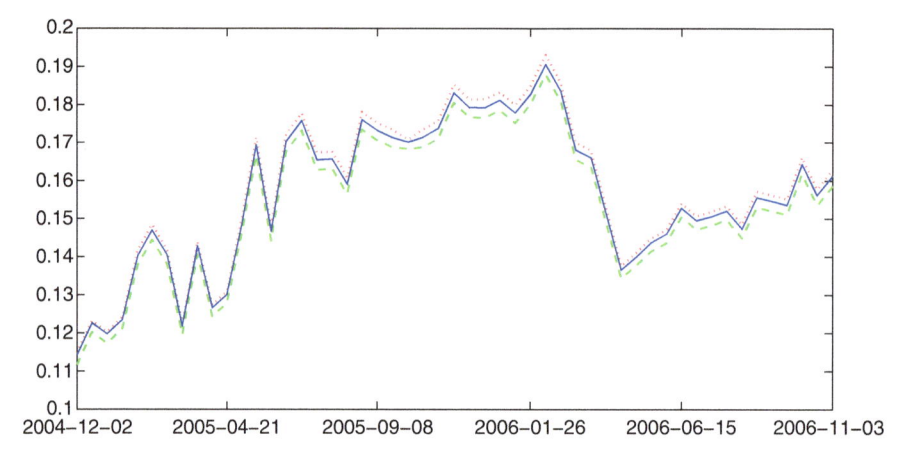

Fig. 4.7 Fitting of $\sigma_{9.75}(e_1)$ (*dotted*) and $\sigma_{9.75}(e_2)$ (*dashed*) and the observed market volatility (*solid*) for the caplet with longest maturity 9.75 (in %). The depicted time frame is 2004/12/02 until 2006/11/03

Final volatility estimates $(\sigma_{9.75}(e_1), \sigma_{9.75}(e_2))$ are depicted in Fig. 4.7. As the volatility estimated are rather close together, only days between 2004/12/02 and 2006/11/03 are shown. Figure 4.8 shows the fitting of the volatilities for the LIBOR rate with maturity 3.75 for the same time frame. Also here, caplet prices Caplet$_{3.75}$ were perfectly reproduced for all time points.

4.7 Conclusion

This chapter has set out to encounter the shortcomings of the log-normal LIBOR market model by replacing the ordinary diffusion of the original model by a Markov-switching jump diffusion process. With LIBOR rates being martingales under their corresponding forward measures, we proposed to model each of the rates as a Markov-switching jump diffusion without drift under the corresponding measure. With measure changes playing the central role within the model, we gave the intuition, why a measure change between forward measures and risk-neutral measure has no effect on the infinitesimal generator of the underlying Markov chain. Wiener processes and Markov-switching compensator, however, *do* follow different dynam-

4 The LIBOR Market Model: A Markov-Switching Jump Diffusion Extension 113

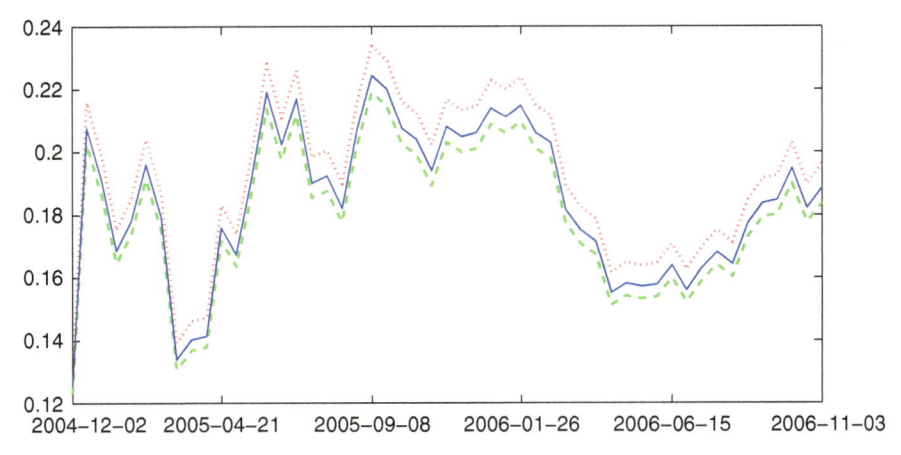

Fig. 4.8 Fitting of $\sigma_{3.75}(e_1)$ (*dashed*) and $\sigma_{3.75}(e_2)$ (*dotted*) and the observed market volatility (*solid*) for the caplet with maturity 3.75 (in %). The depicted time frame is 2004/12/02 until 2006/11/03

ics under different measures and we derived expressions for their interrelations. We observed the particularity that, while assuming that under the terminal measure, Wiener and jump part are independent, this independence is lost when changing to any other measure. Similarly, time-independence of the jump intensity and jump distribution in the Markov-switching compensator cannot be carried over from the terminal measure to any other measure.

While the introduced interest-rate dynamics are rich enough to incorporate both sudden shocks and overall market movement into the model, we showed that they are still simple enough to allow for the pricing of caps/floors. Under the assumption of volatilities being modeled as Markov-switching constants, we showed that the characteristic functions of the logarithmized LIBOR rates needed for pricing with Laplace transforms can be derived in analytical form. Eventually, we demonstrated in the last section that the model can be successfully calibrated to market-observed cap/caplet prices, even though this procedure is related to a considerable amount of effort. In particular, it is necessary to make additional reasonable assumptions to further specify the parameters to be estimated. Ex post, all assumptions made turned out to be reasonable, as market prices could be perfectly recovered in all cases considered.

At the same time, Sect. 4.6 also gave rise to some interesting questions and aspects for further investigation. It appears that the assumption of modeling volatilities as regime-dependent constants is too restrictive, as for the fitted observation times, volatility parameters turned out to be far from being constant when considered over the entire time period. Future research could examine if other volatility specifications can yield more satisfactory results. In our calibration routine, we moreover only considered a Markov chain with two states, helping us to set assumptions for the calibration of the model. However, it is reasonable to assume that a Markov chain with three or four states might be suited even better to represent global economic trends. Incorporating a bigger number of states into the model would require developing new approaches to calibration and could be the focus of further analysis. Also, the choice for modeling of jumps was rather simple. This raises the interesting as well as challenging question of how more complicated jump terms could be included into the approach. Another issue that would require further investigation is if the presented approach is in fact capable of reproducing market-observed volatility skews/smiles, by calibrating the model not only to caplets corresponding to ATM caps, but also to caps of different strikes. We also did not look into hedging issues, as these would have been beyond the scope of this paper, and are therefore left to future research (cf., e.g., Baaquie et al. [3] and Rebonato et al. [36]).

In summary, the MSJD extension to the LIBOR Market Model has been shown to not only considerably enrich the possible dynamics under which interest rates can be modeled, but also provided an appropriate basis for further research and discussion. A possible future focus lies in the calibration of these models.

References

1. Adrianova, L.: Introduction to Linear Systems of Differential Equations. Translations of Mathematical Monographs, vol. 146. American Mathematical Society, Providence (1995)
2. Andersen, L., Brotherton-Ratcliffe, R.: Extended Libor market models with stochastic volatility. J. Comput. Financ. **9**, 1–40 (2005)
3. Baaquie, B., Liang, C., Warachka, M.: Hedging LIBOR derivatives in a field theory model of interest rates. Phys. A Stat. Mech. Appl. **374**, 730–748 (2007)
4. Belomestny, D., Matthew, S., Schoenmakers, J.: Multiple stochastic volatility extension of the libor market model and its implementation, Working paper (2010)
5. Belomestny, D., Schoenmakers, J.: A jump-diffusion Libor model and its robust calibration. Quant. Financ. **11**(4), 529–546 (2011)
6. Björk, T., Di Masi, G., Kabanov, Y., Runggaldier, W.: Towards a general theory of bond markets. Financ. Stoch. **1**, 141–174 (1997)
7. Björk, T., Kabanov, Y., Runggaldier, W.: Bond market structure in the presence of marked point processes. Math. Financ. **7**(2), 211–223 (1997)
8. Black, F.: The pricing of commodity contracts. J. Financ. Econ. **3**, 167–179 (1976)
9. Blanes, S., Cases, F., Oteo, J., Ros, J.: The magnus expansion and xome of its applications. Phys. Rep. **470**, 151–238 (2009)
10. Brace, A., Gatarek, D., Musiela, M.: The market model of interest rate dynamics. Math. Financ. **7**(2), 127–147 (1997)
11. Brigo, D., Mercurio, F.: Interest Rate Models – Theory and Practice. Springer, Berlin/Heidelberg/New York (2006)
12. di Masi, G., Kabanov, Y., Runggaldier, W.: Mean-variance hedging of options on stocks with Markov volatilites. Theory Probab. Appl. **39**(1), 172–182 (1994)

13. Eberlein, E., Grbac, Z.: Rating based Lévy LIBOR model. Mathematical Finance **23**(4), 591–626 (2013)
14. Eberlein, E., Kluge, W., Schönbucher, P.: The Lévy LIBOR model with default risk. J. Credit Risk **2**(2), 3–42 (2006)
15. Eberlein, E., Özkan, F.: The Lévy LIBOR model. Financ. Stoch. **9**(3), 327–348 (2005)
16. Elliott, R.J., Aggoun, L., Moore, J.: Hidden Markov Models: Estimation and Control. Applications of Mathematics, vol. 29. Springer, New York (1994)
17. Elliott, R., Siu, T.: On Markov-modulated exponential-affine bond price formulae. Appl. Math. Financ. **16**(1), 1–15 (2009)
18. Elliott, R.J., Valchev, S.: Libor market model with regime-switching volatility, Working paper No. 228 (2004)
19. Elliott, R.J., Wilson, C.A.: The term structure of interest rates in a Markov setting (2003). Available at: www.fsa.ulaval.ca/nfa2003/papiers/Robert%20J%20Elliott.pdf
20. Filipovic, D.: Term-Structure Models: A Graduate Course. Springer Finance. Springer, Heidelberg/London/New York (2009)
21. Flegal, J.: Monte Carlo standard errors for Markov chain Monte Carlo. Ph.D. thesis, University of Minnesota (2008)
22. Glasserman, P., Kou, S.: The term structure of simple forward rates with jump risk. Math. Financ. **13**(3), 383–410 (2003)
23. Hamilton, J.: Analysis of time series subject to changes in regime. J. Econ. **45**, 39–70 (1990)
24. Heath, D., Jarrow, R., Morton, A.: Bond pricing and the term structure of interest rates: a new methodology for contingent claim valuation. Econometrica **60**, 77–105 (1992)
25. Jacod, N., Shiryaev, A.: Limit Theorems for Stochastic Processes. A Series of Comprehensive Studies in Mathematics, vol. 288, 2nd edn. Springer, Berlin/Heidelberg/New York (2002)
26. Jamshidian, F.: Libor and swap market models and measures. Financ. Stoch. **1**(4), 293–330 (1997)
27. Jamshidian, F.: Libor market model with semimartingales, Working paper, NetAnalytic Ltd. (1999)
28. Joshi, M., Rebonato, R.: A displaced diffusion stochastic volatility libor market model: motivation, definition and implementation. Quant. Financ. **3**(6), 458–469 (2003)
29. Magnus, W.: On the exponential solution of differential equations for a linear operator. Commun. Pure Appl. Math. **7**, 649–673 (1954)
30. Miltersen, K., Sandmann, K., Sondermann, D.: Closed form solutions for term structure derivatives with log-normal interest rates. J. Financ. **52**(1), 409–430 (1997)
31. Protter, P.: Stochastic Integration and Differential Equations. Stochastic Modeling and Applied Probability, vol. 21, 2nd edn. Springer, Berlin/Heidelberg/New York (2005)

32. Raible, S.: Lévy processes in finance: theory, numerics, and empirical facts. Ph.D. thesis, University Freiburg (2000)
33. Rebonato, R.: Modern Pricing of Interest-Rate Derivatives: The LIBOR Market Model and Beyond. Princeton University Press, Princeton (2002)
34. Rebonato, R.: Which process gives rise to the observed dependence of swaption implied volatility of the underlying. Int. J. Theor. Appl. Financ. **6**(4), 419–442 (2003)
35. Rebonato, R., Joshi, M.: A joint empirical/theoretical investigation of the modes of deformation of swaption matrices: implications for the stochastic-volatility libor market model. Int. J. Theor. Appl. Financ. **5**(7), 667–694 (2002)
36. Rebonato, R., McKay, K., White, R.: The SABR/LIBOR Market Model: Pricing, Calibration and Hedging for Complex Interest-Rate Derivatives. Wiley, Chichester (2009)
37. Schoenmakers, J.: Robust Libor Modelling and Pricing of Derivative Products. Chapman & Hall/CRC, BocaRaton/London/New York/Singapore (2005)
38. Schönbucher, P.: Credit Derivatives Pricing Models. Models, Pricing and Implementation. Wiley Finance Series. Wiley, Chichester (2003)
39. Shreve, S.: Stochastic Calculus for Finance II: Continuous Models. Springer Finance, 8th edn. Springer, New York (2008)
40. Steinrücke, L., Zagst, R., Swishchuk, A.: The libor market model: a Markov-switching jump diffusion extension, Working paper (2013)
41. Svoboda, S.: Libor market model with stochastic volatility, First year transfer report, University of Oxford (2005)
42. Zagst, R.: Interest Rate Management. Springer, Berlin (2001)

Chapter 5
Exchange Rates and Net Portfolio Flows: A Markov-Switching Approach

Faek Menla Ali, Fabio Spagnolo, and Nicola Spagnolo

Abstract In this paper we investigate the impact of net bond and equity portfolio flows on exchange rate changes. Two-state Markov-switching models are estimated for Canada, the euro area, Japan and the UK exchange rates *vis-à-vis* the US dollar. Our results suggest that the relationship between net portfolio flows and exchange rate changes is nonlinear for all currencies considered but Canada.

5.1 Introduction

Financial markets deregulation has led to a dramatic increase in international capital mobility across most developed economies. To give an example of the magnitude of such a shift, gross cross-border portfolio investments in equities and bonds for the US were accounting for only 4 % of GDP in 1975, this proportion surged to 100 % in the early of 1990s and has reached 245 % by 2000 [20]. Not surprisingly the recent literature has focused on the impact (causal effect) that the increased capital mobility across-borders has had on exchange rate dynamics especially in the light of the poor performance of macroeconomic models to explain such dynamics (see [24]).

Recent works have shown that the microstructure dynamics of exchange rates, currency order flows, explains a significant proportion of exchange rate movements (e.g., Evans and Lyons [12], Payne [25], Rime et al. [27], and Chinn and Moore [8];

F. Menla Ali • F. Spagnolo (✉)
Department of Economics and Finance, Brunel University London, Uxbridge,
Middlesex UB8 3PH, UK
e-mail: fabio.spagnolo@brunel.ac.uk

N. Spagnolo
Department of Economics and Finance, Brunel University London, Uxbridge,
Middlesex UB8 3PH, UK

Centre for Applied Macroeconomic Analysis (CAMA), Canberra, Australia

among others). Furthermore, Hau and Rey [20], in an influential study, argued that portfolio flows and order flows are closely aligned since both flows are driven by investors' behavior. To this end, the ongoing empirical literature has produced convincing evidence that portfolio investment flows do affect the dynamics of exchange rates to a large extent, using different data sets.

Froot et al. [15], using daily data on cross-border flows for 44 countries, found a positive contemporaneous correlation between net portfolio investment inflows and both equities expressed in US dollar and currency returns, as well as strong positive correlation between net portfolio inflows and lagged equity and currency returns. Furthermore, Brooks et al. [3], using quarterly data from 1988:01 to 2000:03, examined the impact of portfolio and foreign direct investment flows on the yen-dollar and the euro-dollar exchange rates. They found that, while the yen-dollar exchange rate was tracked by mainstream macroeconomic variables such as interest rate differential and current account, the euro-dollar exchange rate was driven primarily by bilateral net portfolio investment flows. More specifically, equity inflows from the euro area towards the US implied a depreciation of the euro against the US dollar.

Siourounis [28], using monthly data for the US exchange rate *vis-à-vis* the UK, Japan, Germany, and Switzerland over the period 1988–2000, provided evidence that equity flows rather than bond flows are tracing the evolution of exchange rates. The study was conducted using an unrestricted vector autoregressive model in which net cross-border capital flows, equity return differentials, exchange rate changes, and interest rate differentials were set endogenously. Furthermore, by employing an equilibrium framework in which exchange rate changes, stock returns, and capital flows are jointly set under incomplete foreign exchange risk trading feature, Hau and Rey [20] showed that the correlation between equity flows and exchange rate changes is significant in 6 out of 17 OECD countries considered. When pooling the data across the countries, the correlation became highly significant. Chaban [6] argued that Hau and Rey's results were not specifically supported in commodity-exporting countries, namely Canada, Australia, and New Zealand. While Chaban argued that commodity prices play a significant role in the transmission of shocks in these countries, Ferreira Filipe [13] showed that differences in country-specific shocks volatility also play a role in these countries and should therefore be accounted for.

Recently Kodongo and Ojah [22], using monthly data over the period 1997:1– 2009:12 for four African countries, namely Egypt, Morocco, Nigeria, and South Africa, showed that the dynamic relationship between real exchange rate changes and international portfolio flows is country-specific and time-dependent. Furthermore, Combes et al. [9], employing a pooled mean group estimator for a sample of 42 emerging and developing economies over the period 1980–2006, showed that public and private inflows, primarily portfolio investment inflows, result in an appreciation of the real effective exchange rate.

The empirical evidence on the effects of portfolio inflows on exchange rate dynamics has been investigated assuming a linear dependence. We argue that this approach is quite limited and do not allow to capture the dynamics observed in the exchange rates over the last few decades. This paper contributes to the existing literature by using a Markov-switching framework which allows us to propose an alternative way of measuring the relationship between net equity and bond portfolio investment flows and exchange rate changes. As shown by Engle and Hamilton [11], Bekaert and Hodrick [1], Engle [10], Caporale and Spagnolo [5], Frömmel et al. [14], Chen [7], Brunetti et al. [4] among others, the Markov regime-switching model is particularly appropriate to model exchange rate dynamics. We use bilateral quarterly data from the US vis-à-vis Canada, the euro area, Japan, and the UK over the period 1990:01–2011:04. To the best of our knowledge, the nonlinear dependence between both variables has not been explored in the literature yet. The nonlinear model employed in this paper separates periods of high and low states of the world for the endogenous variable (exchange rate changes), and therefore allows us to separate the causal effects for periods of exchange rate appreciation and depreciation as well as high volatility and low volatility.

The paper is organized as follows. Section 5.2 describes the econometric model. Section 5.3 provides details on the data set. Section 5.4 discusses the results; and Sect. 5.5 offers some concluding remarks.

5.2 The Model

We propose an alternative way of detecting the causal dynamics between net portfolio flows and exchange rate changes for the US vis-à-vis Canada, the euro area, Japan, and the UK. The regime-switching model considered in this paper[1] allows for shifts in the mean (periods of currency appreciation and depreciation) and in variance (periods of high volatility and low volatility) and is given by:

$$r_t = \mu(s_t) + \sum_{i=1}^{4} \phi_i r_{t-i} + \alpha(s_t) nbf_{t-1} + \beta(s_t) nef_{t-1} + \sigma(s_t)\varepsilon_t, \qquad \varepsilon_t \sim N(0,1) \tag{5.1}$$

$$\mu(s_t) = \sum_{i=1}^{2} \mu^{(i)} \mathbf{1}\{s_t = i\}, \ \sigma(s_t) = \sum_{i=1}^{2} \sigma^{(i)} \mathbf{1}\{s_t = i\}, \ (t \in \mathbb{T})$$

where r_t = changes of exchange rates, nbf_{t-1} = net bond flows, nef_{t-1} = net equity flows. Given that s_t is unobserved, estimation of (5.1) requires restrictions on the probability process governing s_t; it is assumed that s_t follows a first-order, homogeneous, two-state Markov chain. This means that any persistence in the state is completely summarized by the value of the state in the previous period. Therefore, the regime indicators $\{s_t\}$ are assumed to form a Markov chain on $\mathbb{S}$ with transition probability matrix $\mathbf{P}' = [p_{ij}]_{2 \times 2}$, with:

$$p_{ij} = \Pr(s_t = j | s_{t-1} = i), \qquad i,j \in \mathbb{S}, \tag{5.2}$$

---

[1] The model is based on the Markov switching representation proposed by Hamilton [16, 17].

and $p_{i1} = 1 - p_{i2}$ ($i \in \mathbb{S}$), where each column sums to unity and all elements are non-negative. The probability law that governs these regime changes is flexible enough to allow for a wide variety of different shifts, depending on the values of the transition probabilities. For example, values of p_{ii} ($i \in \mathbb{S}$) that are not very close to unity imply that structural parameters are subject to frequent changes, whereas values near unity suggest that only a few regime transitions are likely to occur in a relatively short realization of the process. $\{\varepsilon_t\}$ are i.i.d. errors with $\mathsf{E}(\varepsilon_t) = 0$ and $\mathsf{E}(\varepsilon_t^2) = 1$. $\{s_t\}$ are random variables in $\mathbb{S} = \{1,2\}$ that indicate the unobserved state[2] of the system at time t. It is assumed that $\{\varepsilon_t\}$ and $\{s_t\}$ are independent. Also, note that the independence between the sequences $\{\varepsilon_t\}$ and $\{s_t\}$ implies that regime changes take place independently of the past history of $\{r_t\}$.

We are interested in documenting estimates of the low-high phase exchange rate changes, μ^l and μ^h, and most importantly in investigating the extent to which net bond flows and net equity flows are associated with the low-high phase exchange rate changes. The autoregressive lag dimension is selected according to the Schwarz Bayesian Criterion (SBC), allowing up to four lags. Therefore, the parameters vector of the mean equation (5.1) is defined by $\mu^{(i)}$ ($i = low, high$) and $\sigma^{(i)}$ ($i = low, high$), which are real constants, and the autoregressive terms $\sum_{i=1}^{4} \phi_i$, up to four lags. $\alpha = (\alpha^l, \alpha^h)$ and $\beta = (\beta^l, \beta^h)$ measure the impact of net bond flows and net equity flows respectively on exchange rate changes. The parameter vector is estimated by maximum likelihood. The density of the data has two components, one for each regime, and the log-likelihood function is constructed as a probability weighted sum of these two components. The maximum likelihood estimation is performed using the EM algorithm described by Hamilton [16, 17].

Furthermore, we estimate the linear model commonly used in the literature and consider it as a benchmark. This is given by:

$$r_t = \mu + \sum_{i=1}^{4} \phi_i r_{t-i} + \alpha nb f_{t-1} + \beta ne f_{t-1} + \sigma \varepsilon_t, \qquad \varepsilon_t \sim N(0,1) \qquad (5.3)$$

where the parameters vector of the mean equation (5.2) is defined by the constant parameters $(\mu, \phi_i, \alpha, \beta, \sigma)$.

5.3 Data

The variables employed in this paper are net bond flows (nbf_t), net equity flows (nef_t) and exchange rates (E_t) for the US vis-à-vis Canada, the euro area, Japan, and the UK. Throughout, the US is considered to be the domestic economy. We use quarterly data from 1990:01 to 2011:04. Data on exchange rates are from the IMF's *International Financial Statistics (IFS)*, whereas portfolio investment flows

---

[2] Regime 1 is labelled as the low regime, whereas regime 2 as the high regime.

are sourced from the USA Treasury International Capital (TIC) System.[3] Exchange rate changes are calculated as $r_t = 100 \times (E_t/E_{t-1})$. For the euro area exchange rate, the ECU is considered prior to the introduction of the euro in 1999.

Table 5.1 Descriptive statistics

Variables	Mean	St. dev	Skewness	Ex. Kurtosis	JB[a]
			Canada		
r	0.102	2.766	−0.736	8.1592	104.34[0.000]
nbf	0.355	1.897	3.358	23.535	1692.2[0.000]
nef	0.192	1.653	0.762	6.2477	46.660[0.000]
			Euro area		
r	−0.035	4.773	−0.509	2.9354	3.7741[0.151]
nbf	0.303	1.523	−0.072	4.1495	4.8655[0.087]
nef	0.032	1.547	0.128	3.4093	0.8451[0.655]
			Japan		
r	0.902	4.953	0.501	3.613	5.015[0.081]
nbf	0.972	1.307	−0.185	3.330	0.894[0.639]
nef	−0.403	1.959	−1.186	9.737	184.9[0.000]
			UK		
r[a]	−0.282	4.812	−1.625	8.7961	160.0[0.000]
nbf[a]	0.873	1.450	−2.685	17.829	901.6[0.000]
nef[a]	0.0001	1.781	−0.141	5.8607	29.95[0.000]

[a] r, nbf, and nef indicate exchange rates changes, net bond flows and net equity flows, respectively; JB is the Jarque-Bera test for normality

Net portfolio flows, on the other hand, are constructed as the difference between portfolio inflows and outflows. Inflows and outflows are measured as net purchases and sales of domestic investors (equities and bonds) by foreign residents, and net purchases and sales of foreign assets (equities and bonds) by domestic investors, respectively. Following Heimonen [21], the euro area portfolio flows were calculated aggregating the data for the individual EMU countries (Austria, Belgium-Luxemburg, Finland, France, Germany, Ireland, Italy, the Netherlands, Portugal, and Spain). Furthermore, following Brennan and Cao [2], Hau and Rey [20], and Chaban [6], flows were normalized by the average of their absolute values over the previous four quarters.

Summary statistics for the variables considered are displayed in Table 5.1. The mean quarterly change for exchange rates is positive (US dollar depreciation) for Japan and Canada, while it is negative (US dollar appreciation) for the UK and the euro area. The mean quarterly net bond flows and net equity flows, on the other hand, is positive in all cases but for equity flows in Japan, thereby indicating the existence of net bond and net equity inflows towards the US (net bond and net equity outflows from the counterpart countries). Exchange rate changes, as expected, exhibit higher volatility than net portfolio (bond and equity) flows. Furthermore, skewness and

---

[3] The data are obtained from the US Treasury Department website: http://www.treasury.gov/resource-center/data-chart-center/tic/Pages/country-longterm.aspxhe

excess kurtosis characterize all the series with the exceptions of net bond flows in Japan and all variables of the euro area.[4] Finally, the Jarque-Bera (JB) test statistics shows evidence of no normality in all series but net bond flows in Japan and net equity flows and exchange rate changes in the euro area.

5.4 Empirical Results

The first step of our analysis is to estimate the benchmark model Eq. (5.3) by standard OLS. Results, reported in Table 5.2, indicate that, for all countries, neither equity flows, nor bond flows have statistically significant effects on exchange rate changes. These findings are consistent with those of Hau and Rey [20] and Chaban [6] but contradicts Brooks et al. [3] who found a statistical significant linkage in the euro area.

The null hypothesis of linearity against the alternative of a Markov regime switching cannot be tested directly using a standard likelihood ratio (LR) test. We properly test for multiple equilibria (more than one regime) against linearity using the Hansen's [18, 19] standardized likelihood ratio test. The value of the

Table 5.2 Parameter estimates for linear models[a]

	Japan	Canada	UK	Euro area
μ_1	0.523 (0.635)	−0.014 (0.343)	−0.209 (0.583)	−0.0723 (0.484)
α_1	0.003 (0.437)	0.232 (0.177)	0.023 (0.426)	0.126 (0.323)
β_1	0.119 (0.235)	0.033 (0.210)	−0.042 (0.242)	0.329 (0.313)
ϕ_1	0.238** (0.108)	0.384* (0.110)	0.343* (0.105)	0.246** (0.109)
ϕ_2	−0.309* (0.101)	−0.235** (0.111)	−0.310* (0.106)	−0.206** (0.111)
ϕ_3	0.332* (0.112)	–	–	–
σ_1	4.265	3.024	4.194	4.369
Log Like	−237.933	−212.093	−242.764	−240.485
$LB_{(8)}$	0.685 [0.702]	5.622 [0.466]	6.222 [0.398]	8.143 [0.227]
$LB^2_{(8)}$	0.466 [0.875]	0.479 [0.866]	0.157 [0.995]	0.338 [0.947]
JB	19.68 [0.000]	31.34 [0.000]	33.47 [0.000]	3.470 [0.176]

[a] Standard errors and P-values are reported in (.) and [.], respectively. ***, **, * indicate significance levels at the 1, 5, and 10%, respectively. $LB_{(8)}$ and $LB^2_{(8)}$ are respectively the Ljung-Box test [23] of significance of autocorrelations of eight lags in the standardized and standardized squared residuals. JB is the Jarque-Bera test for normality

---

[4] In the euro area, net bond flows exhibit only excess kurtosis, whereas exchange rate changes exhibit only skewness.

standardized likelihood ratio statistics and related *P*-values (Table 5.3) under the null hypothesis[5] Hansen [18, 19] provide strong evidence in favor of a two-state Markov switching specification. This procedure requires the evaluation of the likelihood function across a grid of different values for the transition probabilities and for each state-dependent parameter. We also test for the presence of a third state (Table 5.3). The results provide strong evidence in favor of a two-state regime-switching specification.

Table 5.3 Markov switching state dimension: Hansen test[a]

Standardized LR test	Linearity vs two-states	Two states vs three-states
	Canada	
LR	4.123	0.2731
$M=0$	(0.001)	(0.643)
$M=1$	(0.002)	(0.677)
$M=2$	(0.003)	(0.690)
$M=3$	(0.009)	(0.715)
$M=4$	(0.011)	(0.722)
	Euro area	
LR	4.021	0.2567
$M=0$	(0.001)	(0.612)
$M=1$	(0.002)	(0.654)
$M=2$	(0.004)	(0.686)
$M=3$	(0.008)	(0.701)
$M=4$	(0.010)	(0.715)
	Japan	
LR	3.985	0.2451
$M=0$	(0.001)	(0.599)
$M=1$	(0.003)	(0.611)
$M=2$	(0.004)	(0.671)
$M=3$	(0.007)	(0.689)
$M=4$	(0.012)	(0.699)
	UK	
LR	3.769	0.2113
$M=0$	(0.001)	(0.587)
$M=1$	(0.002)	(0.599)
$M=2$	(0.003)	(0.645)
$M=3$	(0.006)	(0.661)
$M=4$	(0.013)	(0.688)

[a] The Hansen's standardized Likelihood Ratio test (LR) p-values, in round brackets, are calculated according to the method described in Hansen [18], using 1,000 random draws from the relevant limiting Gaussian processes and bandwidth parameter $M = 0, 1, \ldots, 4$. Test results for the presence of a third state are also reported

---

[5] The p-value is calculated according to the method described in Hansen [19], using 1,000 random draws from the relevant limiting Gaussian processes and bandwidth parameter $M = 0, 1, \ldots, 4$. See Hansen [18] for details.

Table 5.4 Parameter estimates for regime-switching models: Canada[a]

	Model I	Model II	Model III	Model IV
μ_1	−0.024 (0.288)	−1.326*** (0.456)	−0.242 (0.246)	2.202 (1.411)
μ_2	–	1.981** (0.855)	–	−0.341 (0.240)
α_1	0.101 (0.127)	0.214 (0.171)	0.908 (0.778)	−0.254 (0.905)
α_2	5.394*** (1.832)	0.455 (0.559)	0.091 (0.113)	0.118 (0.113)
β_1	0.313* (0.166)	−0.147 (0.276)	−0.678 (0.911)	−1.771* (0.944)
β_2	−2.108** (1.019)	−0.045 (0.312)	0.267* (0.143)	0.278** (0.138)
ϕ_1	–	–	0.235** (0.112)	0.216** (0.105)
ϕ_2	–	−0.267*** (0.111)	–	–
σ_1	2.328*** (0.231)	2.705*** (0.222)	4.667*** (0.786)	4.420*** (0.714)
σ_2	–	–	1.778*** (0.175)	1.781*** (0.165)
P	0.942*** (0.038)	0.946*** (0.040)	0.887*** (0.101)	0.874*** (0.088)
$1-Q$	0.584** (0.235)	0.077 (0.065)	0.043 (0.033)	0.041 (0.029)
Log like	−211.096	−214.209	−203.051	−201.650
AIC	438.192	448.418	426.102	425.3
$LB_{(8)}$	9.061 [0.337]	22.811 [0.088]	11.364 [0.181]	8.542 [0.382]
$LB^2_{(8)}$	0.897 [0.524]	0.498 [0.852]	1.4344 [0.201]	1.073 [0.394]
JB	3.379 [0.184]	31.595 [0.000]	2.9871 [0.224]	3.411 [0.181]

[a] See notes of Table 5.2

Maximum likelihood (ML) estimates are reported in Tables 5.4–5.7. We estimate four nested Markov switching models. Model IV is the general model and allows for shifts in mean, $\mu(s_t)$, variance, $\sigma(s_t)$, bond, $\alpha(s_t)$, and equity, $\beta(s_t)$ parameters; Model III constraints the mean to be not regime dependent, whereas Model II constraints the variance to be constant. Finally Model 1 allows for switches in the bond and equity flows parameters only. Following Psaradakis and Spagnolo [26], we use the Akaike Information Criteria (AIC) to identify the best among the four estimated models. The models selected are Model II (shift in mean) for Canada and the UK, and Model III (shift in variance) for the euro area and Japan.[6] The model selected appears to be well identified: parameters are significant and the standardized resid-

---

[6] We use the AIC to choose the best fitted model among the candidate models considered. For example, in the cases of US/Canada and US/UK, Model II is favored according to the AIC. This implies that in both cases the switching in exchange rate changes is driven primarily by the mean, but not the variance.

5 Exchange Rates and Net Portfolio Flows: A Markov-Switching Approach

Table 5.5 Parameter estimates for regime-switching models: Euro Area[a]

	Model I	Model II	Model III	Model IV
μ_1	−0.173 (0.508)	−4.417*** (0.453)	−0.069 (0.464)	−3.346*** (0.944)
μ_2	–	3.393*** (0.474)	–	3.296*** (0.673)
α_1	−1.126* (0.584)	0.165 (0.349)	−1.662*** (0.427)	0.405 (0.540)
α_2	1.106** (0.469)	0.173 (0.215)	1.070** (0.478)	0.250 (0.244)
β_1	0.147 (0.437)	−0.335 (0.276)	0.095 (0.451)	−0.003 (0.398)
β_2	0.380 (0.502)	0.145 (0.292)	0.359 (0.479)	0.158 (0.373)
ϕ_1	0.195* (0.105)	–	–	–
ϕ_2	−0.321*** (0.110)	−0.431*** (0.086)	−0.225*** (0.109)	−0.260** (0.102)
ϕ_3	–	0.253*** (0.077)	–	–
ϕ_4	−0.150* 0.081	−0.185** (0.072)	–	–
σ_1	3.893*** (0.323)	2.355*** (0.218)	2.891*** (0.462)	3.345*** (0.597)
σ_2	–	–	4.418*** (0.475)	2.468*** (0.404)
P	0.943** (0.068)	0.699*** (0.085)	0.882*** (0.097)	0.757*** (0.106)
$1-Q$	0.078 (0.063)	0.236*** (0.067)	0.081 (0.062)	0.249*** (0.089)
Log like	−235.708	−229.687	−242.444	−239.090
AIC	493.416	483.374	504.888	500.18
$LB_{(8)}$	5.282 [0.382]	6.892 [0.228]	5.465 [[0.706]	3.246 (0.918)
$LB^2_{(8)}$	0.862 [0.553]	0.758 0.640]	0.408 [0.911]	0.775 [0.626]
JB	6.827 [0.032]	8.251 (0.016)	5.321 [0.069]	4.548 [0.102]

[a] See notes of Table 5.2

uals exhibit no signs of linear or non-linear dependence.[7] The periods of high and low exchange rate changes (Canada and UK) and of high and low exchange rate volatility (euro area and Japan) seem to be accurately identified by the filter probabilities. Figures 5.1–5.4 show the plots for exchange rate changes, r_t, along with the corresponding estimated filter probabilities.

More specifically, for Canada the mean value of −1.3 (1.9) in regime low (high) indicates a regime characterized by US dollar appreciation (depreciation) against the Canadian dollar. The probability of staying in regime low (high) is 0.94 (0.92). The filter probabilities (see Fig. 5.1) show a relatively low number of switches consistently with the high regime persistence. There are 52 quarters (61.18 % of the whole sample) where the process is in the low regime and 33 quarters (38.82 %) where

---

[7] The exchange rate lag dimension is chosen according to the SBC (allowing up to four lags), Lags found insignificant are excluded.

Table 5.6 Parameter estimates for regime-switching models: Japan[a]

	Model I	Model II	Model III	Model IV
μ_1	0.902 (0.621)	−3.663*** (0.891)	0.819 (0.646)	−3.688*** (0.828)
μ_2	–	3.679*** (0.621)	–	3.627*** (0.644)
α_1	−1.378 (1.013)	−0.001 (0.595)	−2.158*** (0.652)	−0.002 (0.553)
α_2	0.898* (0.546)	−0.140 (0.382)	0.736 (0.537)	−0.136 (0.395)
β_1	0.894** (0.439)	0.137 (0.371)	0.544 (0.436)	0.133 (0.341)
β_2	−0.220 (0.341)	0.005 (0.199)	0.028 (0.280)	0.007 (0.206)
ϕ_2	−0.280*** (0.104)	−0.519*** (0.080)	−0.296*** (0.114)	−0.503*** (0.083)
σ_1	4.047*** (0.363)	3.125*** (0.255)	2.669*** (0.721)	2.880*** (0.422)
σ_2	–	–	4.496*** (0.447)	3.237*** (0.322)
P	0.828*** (0.128)	0.885*** (0.062)	0.815*** (0.117)	0.883*** (0.063)
$1-Q$	0.106 (0.069)	0.056* (0.031)	0.057 (0.042)	0.056* (0.031)
Log Like	−246.704	−234.806	−246.208	−234.600
AIC	511.408	489.612	512.416	491.2
$Q(8)$	3.425 [0.843]	10.555 [0.159]	4.105 [0.847]	5.156 [0.740]
$ARCH(8)$	0.588 [0.783]	0.527 [0.831]	0.411 [0.909]	0.490 [0.858]
JB	8.989 [0.011]	10.080 [0.006]	6.335 [0.042]	7.706 [0.021]

[a] See notes of Table 5.2

the process is in the high regime. However, equity flows and bond flows are found to be insignificant in either regime. These findings are in line with previous studies, see Chaban [6], and Ferreira Filipe [13]. In the case of the UK, while the mean value associated with the low regime is insignificant, the significant mean, 0.84, of the high regime indicates a US dollar depreciation against the British pound. The probability of staying in a low (high) regime is 0.28 (0.96), with the number of regime switches, being quite frequent when the process is in the low state. There are 4 quarters (4.60 % of the total sample) where the process is in the low regime and 83 quarters (95.40 %) where the process is in the high regime (see Fig. 5.2). Equity flows and bond flows are found to have a significant impact on exchange rate changes in the low regime. These results suggest that both equity and bond inflows towards the US or equity and bond outflows from the UK result in an appreciation of the US dollar against the British pound.

The shift in variance, in the case of the euro area, separates periods of low volatility ($\sigma^l = 2.9$) from high volatility ($\sigma^h = 4.4$). The probability of staying in the low (high) regime is 0.88 (0.92). Both regimes are quite persistent, well capturing the cluster effect. There are 36 quarters (42.35 % of the total sample) where the process is in the low regime and 49 quarters (57.65 %) where the process is in the high

5 Exchange Rates and Net Portfolio Flows: A Markov-Switching Approach

Table 5.7 Parameter estimates for regime-switching models: UK[a]

	Model I	Model II	Model III	Model IV
μ_1	0.127 (0.387)	−1.932 (4.355)	0.026 (0.417)	10.317** (4.944)
μ_2	−	0.847* (0.454)	−	0.303 (0.363)
α_1	−15.231*** (1.430)	−14.173*** (4.976)	−2.179 (1.646)	−25.548** (5.634)
α_2	0.282 (0.291)	−0.219 (0.324)	0.266 (0.313)	0.076 (0.258)
β_1	−1.565* (0.790)	−2.119* (1.201)	−0.310 (1.143)	−3.783** (1.312)
β_2	−0.117 (0.150)	−0.160 (0.188)	−0.079 (0.153)	−0.143 (0.150)
ϕ_1	0.326*** (0.078)	−	0.417*** (0.098)	0.326*** (0.073)
ϕ_2	−0.369*** (0.074)	−	−	−0.256*** (0.091)
ϕ_3	0.175** (0.075)	−	−	−
ϕ_4	−0.146* (0.076)	−	−	−
ϕ_5	0.145** (0.068)	−	−	−
σ_1	2.399*** (0.190)	3.051*** (0.239)	7.244*** (1.250)	4.590*** (1.624)
σ_2	−	−	2.289*** (0.253)	2.410*** (0.214)
P	0.480*** (0.219)	0.280*** (0.245)	0.881*** (0.097)	0.330*** (0.284)
$1-Q$	0.026 (0.018)	0.037* (0.021)	0.025 (0.026)	0.042* (0.024)
Log like	−203.040	−237.161	−227.047	−216.853
AIC	432.08	492.322	474.094	457.706
$LB_{(8)}$	24.28 [0.060]	5.349 [0.719]	10.63 [0.155]	4.931 [0.552]
$LB^2_{(8)}$	0.900 [0.523]	0.717 [0.675]	0.652 [0.730]	0.256 [0.977]
JB	0.402 [0.817]	3.848 [0.146]	0.418 [0.811]	0.187 [0.910]

[a] See notes of Table 5.2

regime. While net equity flows appear to be insignificant, net bond flows are significant in the low regime, suggesting that net bond inflows towards the US (or net bond outflows from the euro area) result in a US dollar appreciation against the euro.

Finally, in the case of Japan, volatility in the high regime ($\sigma^h = 4.5$) is 73% higher than in the low regime ($\sigma^l = 2.6$), with the associated transition probabilities being equal to 0.81 and 0.94 for the low and high regimes, respectively. There are 20 quarters (23.53% of the total observations) where the process is in the low variance regime and 65 quarters (76.47%) where the process is in the high variance regime. Net bond flows are found to have a significant impact on exchange rate changes only in periods of low volatility, in other words inflows towards the US or outflows from Japan result in a US dollar appreciation against the Japanese yen. Our findings contradict what was found by Siourounis [28], where the impact of net

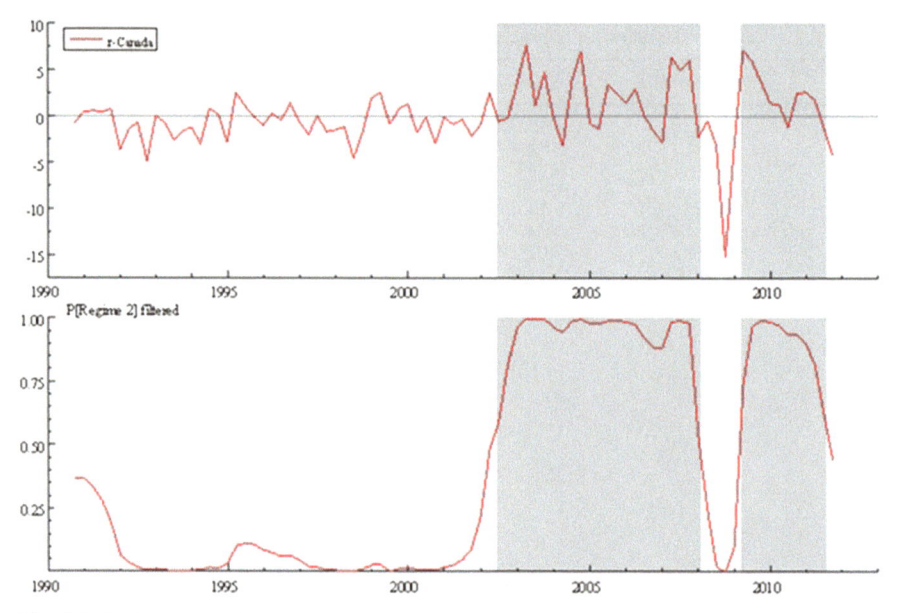

Fig. 5.1 Exchange rate changes, r_t, and the transition probabilities for Canada. Regime 1 denotes the probability of staying in the low (appreciation) regime, while regime 2 indicates the probability of staying in the high (depreciation) regime

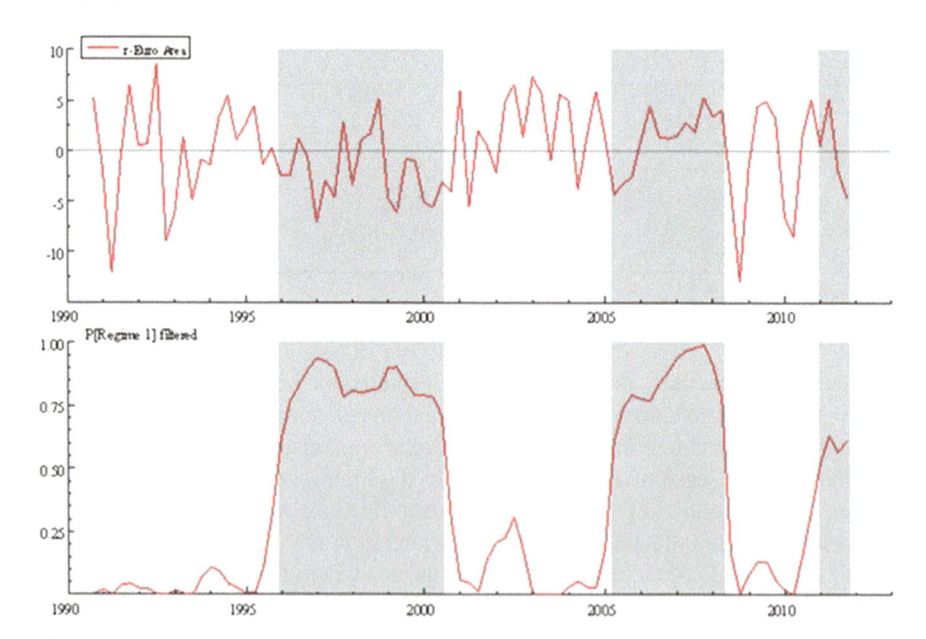

Fig. 5.2 Exchange rate changes, r_t, and the transition probabilities for the Euro area. Regime 1 denotes the probability of staying in the less volatile regime, while regime 2 indicates the probability of staying in the higher volatile regime

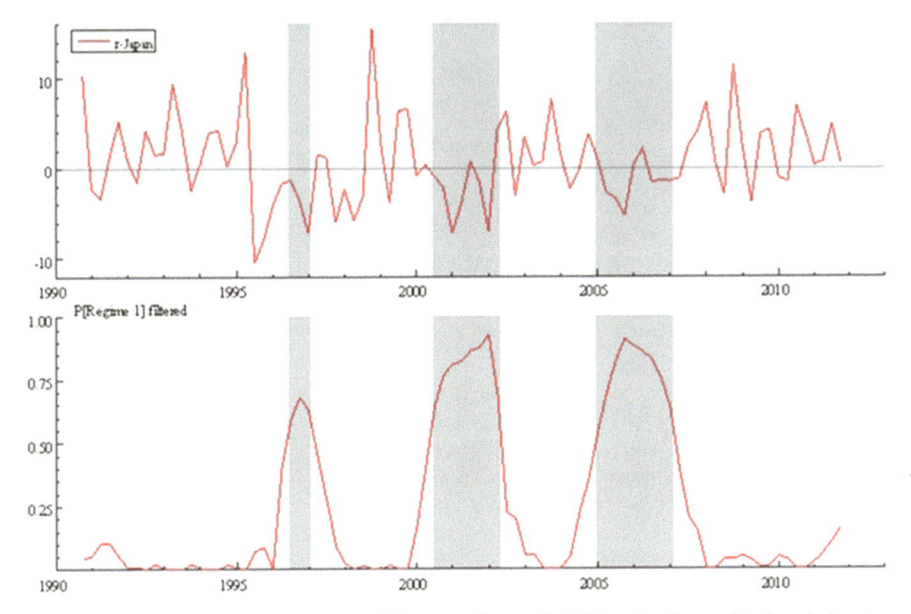

Fig. 5.3 Exchange rate changes, r_t, and the transition probabilities for Japan. Regime 1 denotes the probability of staying in the less volatile regime, while regime 2 indicates the probability of staying in the higher volatile regime

equity flows compared to net bond flows is shown to be rather limited. This applies to both the euro area and Japan. Furthermore, Brooks et al. [3] reported that the yen is likely to be driven by macroeconomic variables rather than portfolio flows. However, using Markov switching specifications, we found that the impact of portfolio flows, primarily bond flows, appears to impact on the US dollar-Japanese yen exchange rate.

5.5 Conclusions

In this paper we have provided some empirical evidence on the causal relationship between net equity and bond portfolio flows and exchange rate changes for the last two decades. The focus of this paper is on the nonlinear casual dynamics and the methodology adopted differentiates this study from most other contributions to the literature. Our argument is that investors behave differently in a bullish market compared to a bearish one. Therefore linear models proposed in the literature are not rich enough to accommodate those different behaviors. Our empirical results show that there is evidence of a nonlinear relationship in three (euro area, Japan and UK) of the four countries under examination. Canada was the only case where net portfolio inflows were found not to impact on exchange rate dynamics. This result is in line with previous studies on commodity exporting countries ([6, 13]). The debate on

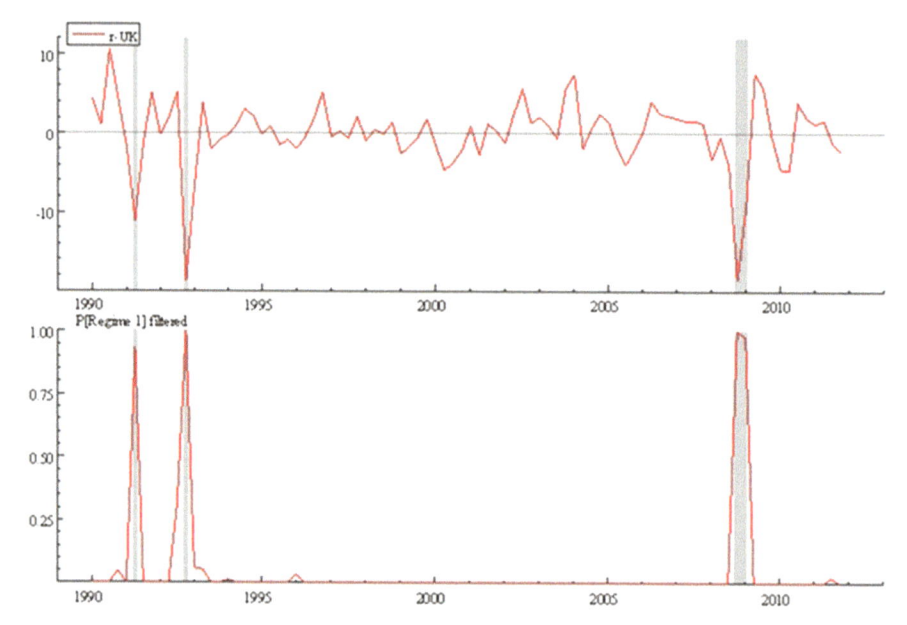

Fig. 5.4 Exchange rate changes, r_t, and the transition probabilities for UK. Regime 1 denotes the probability of staying in the low (appreciation) regime, while regime 2 indicates the probability of staying in the high (depreciation) regime

the linkages between portfolio inflows and exchange rate appreciation/depreciation is still open, but our findings indicate that careful consideration should be given to the often neglected nonlinearities involved. Interesting areas which might be considered for possible future research include studies of the behavior of developed and emerging countries exchange rates and an analysis of the out of sample forecasting performance of our proposed nonlinear approach.

References

1. Bekaert, G., Hodrick, R.J.: On biases in the measurement of foreign exchange risk premiums. J. Int. Money Financ. **12**, 115–138 (1993)
2. Brennan, M.J., Cao, H.H.: International portfolio investment flow. J. Financ. **52**, 1851–1880 (1997)
3. Brooks, R., Edison, H., Kumar, M.S., Slok, T.: Exchange rates and capital flows. Eur. Financ. Manag. **10**, 511–533 (2004)
4. Brunetti, C., Scotti, C., Mariano, R.S., Tan, A.H.H.: Markov switching GARCH models of currency turmoil in southeast Asia. Emerg. Mark. Rev. **9**, 104–128 (2008)

5. Caporale, G.M., Spagnolo, N.: Modelling east Asian exchange rates: a Markov-switching approach. Appl. Financ. Econ. **14**, 233–242 (2004)
6. Chaban, M.: Commodity currencies and equity flows. J. Intern. Money Financ. **28**, 836–852 (2009)
7. Chen, S.S.: Revisiting the interest rate–exchange rate nexus: a Markov-switching approach. J. Dev. Econ. **79**, 208–224 (2006)
8. Chinn, M.D., Moore, M.: Order flow and the monetary model of exchange rates: evidence from a novel data set. J. Money Credit Bank. **43**, 1599–1624 (2011)
9. Combes, J-L., Kinda, T., Plane, P.: Capital flows, exchange rate flexibility, and the real exchange rate. J. Macroecon. **34**, 1034–1043 (2012)
10. Engle, C.: Can the Markov switching model forecast exchange rates? J. Int. Econ. **36**, 151–165 (1994)
11. Engle, C., Hamilton, J.D.: Long swings in the dollar: are they in the data and do markets know it? Am. Econ. Rev. **80**, 689–713 (1990)
12. Evans, M., Lyons, R.K.: Order flow and exchange rate dynamics. J. Political Econ. **110**, 170–180 (2002)
13. Ferreira F.S.: Equity order flow and exchange rate dynamics. J. Empir. Financ. **19**, 359–381 (2012)
14. Frömmel, M., MacDonald, R., Menkhoff, L.: Markov switching regimes in a monetary exchange rate model. Econ. Model. **22**, 485–502 (2005)
15. Froot, K.A., O'Connell, P.G.O., Seasholes, M.S.: The portfolio flows of international investors. J. Financ. Econ. **59**, 151–193 (2001)
16. Hamilton, J.D.: A new approach to the economic analysis of non-stationary time series and the business cycle. Econometrica. **57**, 357–384 (1989)
17. Hamilton, J.D.: Analysis of time series subject to changes in regime. J. Econom. **45**, 39–70 (1990)
18. Hansen, B.E.: The likelihood ratio test under nonstandard conditions: testing the Markov switching model of GNP. J. Appl. Econom. **7**, 61–82 (1992)
19. Hansen, B.E.: Erratum: the likelihood ratio test under nonstandard conditions: testing the Markov switching model of GNP. J. Appl. Econom. **11**, 195–198 (1996)
20. Hau, H., Rey, H.: Exchange rates, equity prices, and capital flows. Rev. Financ. Stud. **19**, 273–317 (2006)
21. Heimonen, K.: The Euro-dollar exchange rate and equity flows. Rev. Financ. Econ. **18**, 202–209 (2009)
22. Kodongo, O., Ojah, K.: The dynamic relation between foreign exchange rates and international portfolio flows: evidence from Africa's capital markets. Int. Rev. Econ. Financ. **24**, 71–87 (2012)
23. Ljung, G.M., Box, G.E.P.: On a measure of lack of fit in time series models. Biometrika. **65**, 297–303 (1978)
24. Meese, R., Rogoff, K.: Empirical exchange rate of the seventies: do they fit out of sample? J. Int. Econ. **14**, 3–24 (1983)
25. Payne, R.: Informed trade in spot foreign exchange markets: an empirical investigation. J. Int. Econ. **61**, 307–329 (2003)

26. Psaradakis, Z., Spagnolo, N.: On the determination of number of regimes in Markov-switching autoregressive models. J. Time Ser. Anal. **24**, 237–252 (2003)
27. Rime, D., Sarno, L., Sojli, E.: Exchange rate forecasting, order flow and macroeconomic information. J. Int. Econ. **80**, 72–88 (2010)
28. Siourounis, G.: Capital flows and exchange rates: an empirical analysis. Available at: http://papers.ssrn.com/sol3/papers.cfm?abstract_id=572025 (2004)

Chapter 6
Hedging Costs for Variable Annuities Under Regime-Switching

Parsiad Azimzadeh, Peter A. Forsyth, and Kenneth R. Vetzal

Abstract A general methodology is described in which policyholder behaviour is decoupled from the pricing of a variable annuity based on the cost of hedging it, yielding two weakly coupled systems of partial differential equations (PDEs): the pricing and utility systems. The utility system is used to generate policyholder withdrawal behaviour, which is in turn fed into the pricing system as a means to determine the cost of hedging the contract. This approach allows us to incorporate the effects of utility-based pricing and factors such as taxation. As a case study, we consider the Guaranteed Lifelong Withdrawal and Death Benefits (GLWDB) contract. The pricing and utility systems for the GLWDB are derived under the assumption that the underlying asset follows a Markov regime-switching process. An implicit PDE method is used to solve both systems in tandem. We show that for a large class of utility functions, the pricing and utility systems preserve homogeneity, allowing us to decrease the dimensionality of the PDEs and thus to rapidly generate numerical solutions. It is shown that for a typical contract, the fee required to fund the cost of hedging calculated under the assumption that the policyholder withdraws at the contract rate is an appropriate approximation to the fee calculated assuming optimal consumption. The costly nature of the death benefit is documented. Results are presented which demonstrate the sensitivity of the hedging expense to various parameters.

P. Azimzadeh • P.A. Forsyth (✉)
Cheriton School of Computer Science, University of Waterloo, Waterloo, ON N2L 3G1, Canada
e-mail: pazimzad@uwaterloo.ca;paforsyt@uwaterloo.ca

K.R. Vetzal
School of Accounting and Finance, University of Waterloo, Waterloo, ON N2L 3G1, Canada
e-mail: kvetzal@uwaterloo.ca

6.1 Introduction

Variable annuities are unit-linked insurance products. These products are a class of insurance vehicles that provide the buyer with particular guarantees without requiring them to sacrifice full control over the funds invested. These funds are usually invested in a collective investment vehicle such as a mutual fund and the writer's position is secured by the deduction of a proportional fee applied to each investors' account.

We propose a method for pricing such contracts when the value of the underlying investment follows a Markovian regime-switching process. Regime-switching was introduced by Hamilton [10], while its application to long-term guarantees was popularized by Hardy [11], who demonstrated its effectiveness by fitting to the S&P 500 and the Toronto Stock Exchange 300 indices. Regime-switching has thus been suggested as a sensible model for pricing variable annuities [30, 19, 3, 33, 23, 14] due to their long-term nature. An alternative to this model is stochastic volatility [13]. However, it could be argued that due to the long-term nature of these guarantees, it is more useful to choose a model which allows for the incorporation of a long-term economic perspective. A regime-switching process has parameters which are economically meaningful, and it is straightforward to adjust these parameters to incorporate economic views. This is perhaps more difficult for a stochastic volatility model, which is typically calibrated to short term option prices. Furthermore, the adoption of stochastic volatility requires an additional dimension in the corresponding partial differential equation (PDE) while the regime-switching model adds complexity proportional to the number of regimes considered, and as a result is computationally less intensive. Moreover, it is straightforward (in the regime-switching framework) to allow for different levels of the risk-free interest rate across regimes. The alternative of incorporating an additional stochastic interest rate factor would add an extra dimension to the PDE, with the associated costs of complexity.

We demonstrate our methodology by considering a specific variable annuity: the Guaranteed Lifelong Withdrawal and Death Benefits (GLWDB) contract. The GLWDB is a response to a general reduction in the availability of defined benefit pension plans, allowing the buyer to replicate the security of such a plan via a substitute. The GLWDB is bootstrapped via a lump sum payment to an insurer, $S(0)$, which is invested in risky assets. We term this the *investment account*. Associated with the GLWDB contract are the *guaranteed withdrawal benefit account* and the *guaranteed death benefit account*, hereafter referred to as the withdrawal and death benefits for brevity. We also refer to these as the *auxiliary accounts*. Both auxiliary accounts are initially set to $S(0)$. At a finite set of *withdrawal dates*, the policyholder is entitled to withdraw a predetermined fraction of the withdrawal benefit (or any lesser amount), even if the investment account diminishes to zero. This predetermined fraction is referred to as the *contract withdrawal rate*. If the policyholder wishes to withdraw in excess of the contract withdrawal rate, they can do so upon the payment of a penalty. Typical GLWDB contracts include penalty rates that are decreasing functions of time. Upon death, the policyholder's estate receives the maximum of the investment account and death benefit. These contracts are often bundled with *ratchets* (a.k.a. step-ups), a contract feature that periodically increases one or more of the auxiliary accounts to the investment account, provided that the investment account has grown larger than the respective auxiliary account. Moreover, *bonus* (a.k.a. roll-up) provisions are also often present, in which the withdrawal benefit is increased if the policyholder does not withdraw on a given withdrawal date.

This contract can be considered as part of a greater family of insurance vehicles offering guaranteed benefits that have emerged as a result of a recent trend away from defined benefits [4]. Our approach can easily be extended to include features present in an arbitrary member of this family. There exists a maturing body of work on pricing these contracts. Bauer et al. [2] introduce a general framework for pricing various products in this family. Monte Carlo techniques and numerical integration are employed, and loss-maximizing (from the perspective of the insurer) withdrawal strategies are considered. Holz et al. [12] compute the fair fee for Guaranteed Lifelong Withdrawal Benefit (GLWB) contracts via a Monte Carlo method. Milevsky and Salisbury [21] employ a numerical PDE approach to price the Guaranteed Minimum Withdrawal Benefits (GMWB) contract. Shah and Bertsimas [29] introduce a GLWB model with stochastic volatility and consider static strategies. Kling et al. [17] provide an extension of the variable annuity model under stochastic volatility. Piscopo and Haberman [27] consider a model with stochastic mortality risk.

In the general area of financial derivatives, the traditional approach is to assume that the policyholder acts so as to maximize the value of owning the contract. The no-arbitrage price of the contract is then calculated as the cost to the writer of the contract of establishing a self-financing hedging strategy that is guaranteed to produce at least enough cash to pay off any future liabilities resulting from the policyholder's future decisions with respect to the contract (in the context of the assumed pricing model). Since derivative payoffs are a zero sum game, this is equivalent to establishing a price on the basis of assuming a worst case scenario to the contract writer. We will refer to the assumption of such behaviour by policyholders here as *loss-maximizing* strategies, as they represent worst case outcomes for the insurer. Such strategies produce an upper bound on the fair price of the contract, but it is far from clear that policyholders actually behave in this manner. Instead, for any of a number of reasons, a policyholder may deviate from loss-maximizing behaviour.

In order to account for this, we provide a new approach here in which we decouple policyholder withdrawal behaviour from the contract pricing equations, and generate said behaviour by considering a policyholder's utility. This general approach is applicable to any contract involving policyholder behaviour, and results in two weakly coupled systems of PDEs. In the context of GLWDBs, this allows for the easy modeling of complex phenomena such as risk aversion and taxation. Solving the PDEs backwards in time allows us to employ the Bellman principle to ensure that the policyholder is able to maximize his or her utility. Since our approach incorporates this added generality, we will generally avoid the use of the term "no-arbitrage" below, and instead refer to the cost of hedging. Of course, under the specific case of loss-maximizing behaviour by the policyholder, our cost of hedging coincides with the traditional no-arbitrage price.

In Sect. 6.2, we introduce a system of regime-switching PDEs used to determine the hedging costs of the GLWDB contract. In Sect. 6.3, we introduce a system of regime-switching PDEs used to model a policyholder's utility and describe how this system is used alongside the system introduced in Sect. 6.2 to determine the cost of hedging the guarantee assuming optimal consumption. In Sect. 6.4, we discuss our numerical methodology. In Sect. 6.5, we present results under both the assumption that the policyholder behaves so as to maximize the cost of the guarantee (i.e. the loss-maximizing strategy) and the assumption that the policyholder maximizes utility.

Overall, the contributions of this work are:

- We introduce a general methodology that allows for the decoupling of policyholder behaviour from the cost of hedging the contract.
  - This approach yields two weakly coupled systems of PDEs: the pricing and utility systems.
  - This approach abandons the arguably flawed notion of a policyholder acting only so as to maximize the cost of a guarantee.
- We model the long-term behaviour of the underlying stock index (or mutual fund) by a Markovian regime-switching process.
- We present the pricing and utility systems for the GLWDB contract.
- We show sufficient conditions for the homogeneity of the systems. This result is computationally significant, as it is used to reduce the dimensionality of the systems.
- We find that assuming optimal consumption yields a hedging cost fee that is very close to the fee calculated by assuming that the policyholder follows the static strategy of always withdrawing at the contract rate. This is a result of particular practical importance as it suggests that policyholders will generally withdraw at the contract rate. This substantiates pricing contracts under this otherwise seemingly naïve assumption.
- We find that the inclusion of a death benefit is often expensive. This may account for the failure to properly hedge this guarantee and the subsequent withdrawal of contracts including ratcheting death benefits from the Canadian market.
- We demonstrate sensitivity to various parameters and we consider the adoption of exotic fee structures in which the proportional fee applies not just to the investment account but rather to the greater of this account and one or more of the auxiliary accounts.

6.2 Hedging Costs

We begin by considering a basic model for pricing GLWDBs under which policyholder withdrawal behaviour is determined so as to maximize the value of the guarantee (i.e. the loss-maximizing strategy). We extend previous work by Forsyth and

6.2.1 Derivation of the Pricing Equation

Let $\mathscr{M}(t)$ be defined as the instantaneous rate of mortality per unit interval. The fraction of policyholders still alive at time t is

$$\mathscr{R}(t) = 1 - \int_0^t \mathscr{M}(s)\,ds,$$

where $t = 0$ is the time at which the contract is purchased. Let $S(t)$ be the amount in the investment account of any policyholder of the GLWDB contract who is still alive at time t. Let $W(t)$ and $D(t)$ be the withdrawal and death benefits at time t. Assume that the underlying value of the investment account is described by

$$dS = (\mu - \alpha)S\,dt + \sigma S\,dZ$$

where Z is a Wiener process. The constant α represents the total fee structure of the contract. It is comprised of two terms. First, the underlying investment fund has a proportional management fee α_M. Second, the insurer charges for the cost of hedging the contractual features through a proportional fee α_R, which we will refer below to as the hedging cost fee. The total proportional deduction applied to the investor's account is $\alpha = \alpha_M + \alpha_R$. If we suppose that α_M is fixed, the pricing problem becomes one of finding α_R such that the insurer can follow a hedging strategy which (in principle) can eliminate risk. This will be discussed further in Sect. 6.4.3. S tracks the index $\hat{S}$ which follows

$$d\hat{S} = \mu \hat{S}\,dt + \sigma \hat{S}\,dZ.$$

It is assumed that the insurer is unable to short S for fiduciary reasons.

We proceed by a hedging argument ubiquitous in the literature [31, 5, 3]. Let $U(S,W,D,t)$ be the cost of funding the withdrawal and death benefits at time t years after purchase for investment account value S, withdrawal benefit W, and death benefit D. The value of U is adjusted to account for the effects of mortality. We assume that this contract was purchased at time zero by a buyer aged x_0. Let T be the smallest time at which $\mathscr{R}(T) = 0$ (we assume that such a time exists; i.e. no policyholder lives forever). The insurer has no obligations at time T and hence

$$U(S,W,D,T) = 0. \tag{6.1}$$

The writer creates a replicating portfolio Π by shorting one contract and taking a position of x units in the index $\hat{S}$. That is,

$$\Pi(S,W,D,t) = -U(S,W,D,t) + x\hat{S}.$$

The contractually specified times at which withdrawals and ratchets occur are referred to as *event times*, gathered in the set $\mathscr{T} = \{t_1, t_2, \ldots, t_{N-1}\}$ and ordered by

$$0 = t_1 < t_2 < \ldots < t_{N-1} < t_N = T.$$

Note that time zero (but not $t_N = T$) is also referred to as an event time even if no withdrawals or ratchets are prescribed to occur at time zero.

Following standard portfolio dynamics arguments [see, e.g. 8] and noting that between event times, dU is a function solely of S and t, we can use Itô's lemma to yield

$$d\Pi = -\left[\left(\frac{1}{2}\sigma^2 S^2 \frac{\partial^2 U}{\partial S^2} + (\mu - \alpha) S \frac{\partial U}{\partial S} + \frac{\partial U}{\partial t}\right) dt + \sigma S \frac{\partial U}{\partial S} dZ\right]$$
$$+ x\left[\mu \hat{S} dt + \sigma \hat{S} dZ\right] + \mathscr{R}(t) \alpha_R S dt - \mathscr{M}(t)\left[0 \vee (D-S)\right] dt,$$

where $a \vee b = \max(a,b)$. The term $\mathscr{R}(t) \alpha_R S dt$ represents the fees collected by the hedger, while $\mathscr{M}(t)\left[0 \vee (D-S)\right] dt$ represents the surplus generated by the death benefit as paid out to the estates of deceased policyholders. Taking $x = (S/\hat{S}) \frac{\partial U}{\partial S}$ yields

$$d\Pi = \left(-\frac{1}{2}\sigma^2 S^2 \frac{\partial^2 U}{\partial S^2} + \alpha S \frac{\partial U}{\partial S} - \frac{\partial U}{\partial t} + \mathscr{R}(t) \alpha_R S - \mathscr{M}(t)\left[0 \vee (D-S)\right]\right) dt. \quad (6.2)$$

As this increment is deterministic, by the principle of no-arbitrage, the corresponding portfolio process must grow at the risk-free rate. That is,

$$d\Pi = r\Pi dt = r\left(-U + \frac{S}{\hat{S}} \frac{\partial U}{\partial S} \hat{S}\right) dt. \quad (6.3)$$

Substituting (6.3) into (6.2),

$$\frac{1}{2}\sigma^2 S^2 \frac{\partial^2 U}{\partial S^2} + (r - \alpha) S \frac{\partial U}{\partial S} + \frac{\partial U}{\partial t} - rU - \mathscr{R}(t) \alpha_R S + \mathscr{M}(t)\left[0 \vee (D-S)\right] = 0. \quad (6.4)$$

Let

$$V(S,W,D,t) = U(S,W,D,t) + \mathscr{R}(t) S \quad (6.5)$$

be the cost of funding the entire contract at time t. Substituting into (6.4), we arrive at

$$\frac{1}{2}\sigma^2 S^2 \frac{\partial^2 V}{\partial S^2} + (r - \alpha) S \frac{\partial V}{\partial S} + \frac{\partial V}{\partial t} - rV + \mathscr{R}(t) \alpha_M S + \mathscr{M}(t)(S \vee D) = 0. \quad (6.6)$$

We stress that V satisfies the above PDE only *between* a pair of adjacent event times t_n and t_{n+1}. We discuss the behaviour of V *across* event times (e.g. from t_n^- to t_n^+) in Sect. 6.2.2.

6.2.2 Events

Remark 6.1 (Notation). In order to reduce clutter, we will sometimes refer to $V(S,W,D,t)$ as $V(\mathbf{x},t)$, where $\mathbf{x} = (S,W,D)$. We will often use this notation for other functions of (S,W,D) as well. We refer to a point $\mathbf{x}$ as a *state*.

6.2.2.1 Event Times

Across event times, V is not necessarily continuous as a function of t. We restrict V to be a *càglàd* function of t so that for all $\mathbf{x}$, $V(\mathbf{x},t) = \lim_{s \uparrow\uparrow t} V(\mathbf{x},s)$ and $V(\mathbf{x},t^+) = \lim_{s \downarrow\downarrow t} V(\mathbf{x},s)$ exist. Whenever $t \in \mathcal{T}$, $V(\mathbf{x},t)$ and $V(\mathbf{x},t^+)$ can be regarded as the price of the contract "immediately before" and "immediately after" the event time, respectively.

6.2.2.2 Withdrawal Strategy

We isolate the withdrawal strategy by introducing a function $\gamma(\mathbf{x},t)$ describing the policyholder's actions at state $\mathbf{x}$ and $t \in \mathcal{T}$.

- $\gamma(\mathbf{x},t) = 0$ indicates that the policyholder does not withdraw anything.
- $\gamma(\mathbf{x},t) \in (0,1]$ indicates a nonzero withdrawal less than or equal to the *contract withdrawal amount*, the maximum amount one can withdraw without incurring a penalty.
- $\gamma(\mathbf{x},t) \in (1,2]$ indicates withdrawal at more than the contract withdrawal amount.

$\gamma(\mathbf{x},t) = 2$ is referred to as a *full surrender*, as it corresponds to the scenario in which the policyholder withdraws the entirety of their investment account, while $\gamma(\mathbf{x},t) \in (1,2)$ is referred to as a *partial surrender*.

Remark 6.2 (Abstract strategy). We stress that we have not yet made any assumptions about policyholder behaviour. The decoupling of policyholder behaviour from the hedging cost equations is the guiding philosophy of this work, and allows us to model complex phenomena visible to the policyholder, but not necessarily visible to the writer. To be more precise, we assume that the insurer can observe the policyholder's strategy, though not the factors which determine that strategy. The robustness of this approach is made concrete via the model developed in Sect. 6.3, which considers the effects of taxation and nonlinear utility functions on a policyholder's withdrawal strategy.

Denote the cost of funding the contract at state $\mathbf{x}$ and event time $t \in \mathcal{T}$ assuming the policyholder performs action $\lambda \in [0,2]$ by

$$v(\mathbf{x},t,\lambda) = V\left(\mathbf{f}(\mathbf{x},t,\lambda),t^+\right) + \mathcal{R}(t)f(\mathbf{x},t,\lambda) \tag{6.7}$$

where f represents cash flow from the writer to the policyholder and $\mathbf{f}\colon \mathbb{R}^3 \times \mathscr{T} \times [0,2] \to \mathbb{R}^3$ describes the state of the contract after the event. The cash flow is adjusted to account only for the fraction of holders still alive at time t, $\mathscr{R}(t)$. The actual (observed) cost of funding the contract is obtained simply by passing the withdrawal strategy employed by the policyholder γ to v. That is,

$$V(\mathbf{x},t) = v(\mathbf{x},t,\gamma(\mathbf{x},t)). \tag{6.8}$$

We cast a withdrawal event in the form (6.7) by considering the three cases enumerated above (i.e. $\lambda = 0$, $\lambda \in (0,1]$, and $\lambda \in (1,2]$) separately.

In the following, we refer to $\mathscr{T}_{\text{Withdraw}} \subset \mathscr{T}$ as the set of times at which withdrawals are prescribed and $\mathscr{T}_{\text{Ratchet}} \subset \mathscr{T}$ as the set of times at which ratchets are prescribed. We begin by assuming $\mathscr{T}_{\text{Withdraw}} \cap \mathscr{T}_{\text{Ratchet}} = \emptyset$ (i.e. ratchets and withdrawals do not occur simultaneously) and subsequently relax this assumption.

6.2.2.3 Bonus

At a time $t \in \mathscr{T}_{\text{Withdraw}}$, nonwithdrawal is indicated by $\lambda = 0$. If the policyholder chooses not to withdraw, the withdrawal benefit is amplified by $1 + B(t)$, where $B(t)$ is the bonus rate available at t. By the principle of no-arbitrage,

$$v(S,W,D,t,0) = V\left(\underbrace{S,\ W(1+B(t)),\ D,\ t^+}_{\mathbf{f}(\mathbf{x},t,0)}\right).$$

6.2.2.4 Withdrawal Not Exceeding the Contract Rate

At a time $t \in \mathscr{T}_{\text{Withdraw}}$, the contract withdrawal amount for withdrawal benefit W is $G(t)W$, where the *contract withdrawal rate* at time t, $G(t)$, is specified by the contract. The amount withdrawn by the policyholder when $\lambda \in (0,1]$ is $\lambda G(t)W$. We express this type of withdrawal as

$$v(S,W,D,t,\lambda) = V\left(\underbrace{(S-\lambda G(t)W) \vee 0,\ W,\ (D-\lambda G(t)W) \vee 0,\ t^+}_{\mathbf{f}(\mathbf{x},t,\lambda)}\right) + \underbrace{\mathscr{R}(t)\lambda G(t)W}_{f(\mathbf{x},t,\lambda)}.$$

For the particular contract that we are considering, the death benefit is reduced whenever any withdrawals are made.

6.2.2.5 Partial or Full Surrender

At a time $t \in \mathscr{T}_{\text{Withdraw}}$, the amount withdrawn if $\lambda \in (1,2]$ is

$$G(t)W + (\lambda - 1)(1 - \kappa(t))S'$$

where $S' = (S - G(t)W) \vee 0$ is the state of the investment account after a withdrawal at the contract withdrawal amount and $\kappa(t) \in [0,1]$ is the *penalty rate* incurred at t for withdrawing above the contract withdrawal amount. For a typical contract, $\kappa(t)$ is monotonically decreasing in time. We express this type of withdrawal as

$$v(S,W,D,t,\lambda) = V\left(\underbrace{(2-\lambda)S',\ (2-\lambda)W,\ (2-\lambda)D,\ t^+}_{\mathbf{f}(\mathbf{x},t,\lambda)}\right)$$
$$+ \mathscr{R}(t)\underbrace{\left(G(t)W + (\lambda-1)(1-\kappa(t))S'\right)}_{f(\mathbf{x},t,\lambda)}.$$

6.2.2.6 Ratchets

At a time $t \in \mathscr{T}_{\text{Ratchet}}$, the withdrawal benefit is increased to the investment account if the latter has grown larger than the former in value. Note that the value of the withdrawal benefit W can never decrease, unless a penalty has been incurred for withdrawing over the contract withdrawal rate. Although ratchets are not controlled by the policyholder, we can still write a ratchet event in the form (6.7) by

$$v(S,W,D,t,\lambda) = V\left(\underbrace{S,\ S \vee W,\ D,\ t^+}_{\mathbf{f}(\mathbf{x},t,\lambda)}\right)$$

irrespective of the value of λ. We also explore the possibility of a ratcheting death benefit.

6.2.2.7 Simultaneous Events

When multiple events are prescribed to occur at the same time, we simply apply them one after the other. Naturally, without a particular order, the pricing problem is not well-posed: the contract is ambiguous. If a withdrawal and a ratchet are prescribed to occur at the same time, we assume that the withdrawal occurs before the ratchet. As we are solving the PDE backwards in time in order to employ the Bellman principle, these events are applied in reverse order (in backwards time).

6.2.3 Loss-Maximizing Strategies

For all states $\mathbf{x}$ and event times $t \in \mathscr{T}$, assuming v is a continuous function of λ, let

$$\Gamma(\mathbf{x},t) = \arg\max_{\lambda \in [0,2]} [v(\mathbf{x},t,\lambda)] \qquad (6.9)$$

Since we are maximizing (6.7), $\Gamma(\mathbf{x},t)$ is simply the set of all actions that maximize the cost of the contract at $\mathbf{x}$ and t. If the writer is interested in computing the hedging cost for the contract in the worst-case scenario, the withdrawal strategy is assumed to satisfy

$$\gamma(\mathbf{x},t) \in \Gamma(\mathbf{x},t) \qquad (6.10)$$

for all $\mathbf{x}$ and $t \in \mathcal{T}$. Any such strategy is termed a *loss-maximizing withdrawal strategy*.

Remark 6.3 (An unfortunate choice of terms). A loss-maximizing withdrawal strategy is often referred to as an optimal strategy in the literature. The adoption of the term optimal is an arguably unfortunate one, as an optimal strategy is not necessarily "optimal" for the policyholder. We stress that an optimal strategy as typically referred to in the literature is simply one that maximizes losses for the writer, and use instead the term "loss-maximizing" for the remainder of this work in order to avoid confusion.

6.2.4 Regime-Switching

We extend the formulation to include a regime-switching framework in which shifts between states are controlled by a continuous-time Markov chain. Letting $\mathscr{S} = \{1,2,\ldots,M\}$ be the state-space consisting of M regimes, we assume that in regime $i \in \mathscr{S}$, the underlying investment account evolves according to

$$dS = (\mu_i - \alpha)S + \sigma_i S dZ + \sum_{j=1}^{M} S(J_{i \to j} - 1) dX_{i \to j}$$

where

$$dX_{i \to j} = \begin{cases} 1 & \text{with probability } \delta_{i,j} + q_{i \to j} dt \\ 0 & \text{with probability } 1 - (\delta_{i,j} + q_{i \to j} dt) \end{cases}$$

and $\delta_{i,j}$ is the Kronecker delta. Here, $q_{i \to j}$ is the objective ($\mathbb{P}$ measure) rate of transition from regime i to j whenever $i \neq j$ and

$$q_{i \to i} = -\sum_{\substack{j=1 \\ j \neq i}}^{M} q_{i \to j}.$$

$J_{i \to j} \geq 0$ is the relative jump size in S associated with a transition from regime i to j. We take $J_{i \to i} = 1$ for all i so that jumps in the underlying are not experienced unless there is a change in regime. Let $V_i(S,W,D,t)$ be the cost of funding a GLWDB in regime i. Following a combination of the hedging arguments in Sect. 6.2.1 and "the Appendix", we arrive at the system of PDEs

$$\mathscr{L}_i V_i + \sum_{\substack{j=1 \\ j \neq i}}^{M} \left[q_{i \to j}^{\mathbb{Q}} V_j \left(J_{i \to j} S, W, D, t \right) \right] + \frac{\partial V_i}{\partial t} + \mathscr{R}(t) \alpha_M S + \mathscr{M}(t) (S \vee D) = 0 \; \forall i \in \mathscr{S}$$

(6.11)

where

$$\mathscr{L}_i = \frac{1}{2} \sigma_i^2 S^2 \frac{\partial^2}{\partial S^2} + \left(r_i - \alpha - \rho_i^{\mathbb{Q}} \right) S \frac{\partial}{\partial S} - \left(r_i - q_{i \to i}^{\mathbb{Q}} \right).$$

$q_{i \to j}^{\mathbb{Q}}$ is the risk-neutral rate of transition from regime i to j whenever $i \neq j$ and

$$q_{i \to i}^{\mathbb{Q}} = - \sum_{\substack{j=1 \\ j \neq i}}^{M} q_{i \to j}^{\mathbb{Q}}.$$

Furthermore, $\rho_i^{\mathbb{Q}}$ is defined as

$$\rho_i^{\mathbb{Q}} = \sum_{\substack{j=1 \\ j \neq i}}^{M} \left[q_{i \to j}^{\mathbb{Q}} (J_{i \to j} - 1) \right] = \sum_{j=1}^{M} \left[q_{i \to j}^{\mathbb{Q}} J_{i \to j} \right].$$

Equation (6.11) is referred to as the *pricing system*.

The events introduced in the single-regime model are simply applied to each regime separately. That is, the regime-switching analogues of (6.7) and (6.8) are

$$v_i(\mathbf{x}, t, \lambda) = V_i \left(\mathbf{f}(\mathbf{x}, t, \lambda), t^+ \right) + \mathscr{R}(t) f(\mathbf{x}, t, \lambda) \quad (6.12)$$

and

$$V_i(\mathbf{x}, t) = v_i(\mathbf{x}, t, \gamma_i(\mathbf{x}, t)). \quad (6.13)$$

Likewise, the withdrawal strategy becomes regime-dependent. The regime-switching analogue of (6.9) and (6.10) is

$$\gamma_i(\mathbf{x}, t) \in \Gamma_i(\mathbf{x}, t) = \underset{\lambda \in [0,2]}{\arg \max} \left[v_i(\mathbf{x}, t, \lambda) \right]. \quad (6.14)$$

6.3 Optimal Consumption

Using a loss-maximizing strategy yields the largest hedging cost fee. Any other strategy will, by definition, yield a smaller fee. Using the fee generated by a loss-maximizing strategy ensures that the writer can, at least in theory, hedge a short position in the contract with no risk. However, insurers are often interested in using a less conservative method for pricing contracts so as to decrease the hedging cost fee while minimizing their exposure. We now extend the framework introduced in Sect. 6.2 to strategies based on optimal consumption from the perspective of the policyholder. As usual, we first consider the single-regime case and subsequently provide the extension to include regime-switching.

6.3.1 Utility PDE

Let $\overline{V}(S,W,D,t)$ be the mortality-adjusted utility of holding a GLWDB contract at t with investment account value S, withdrawal benefit W and death benefit D. Following standard arguments, we express the evolution of a policyholder's utility by

$$\frac{1}{2}\sigma^2 S^2 \frac{\partial^2 \overline{V}}{\partial S^2} + (\mu - \alpha) S \frac{\partial \overline{V}}{\partial S} + \frac{\partial \overline{V}}{\partial t} - \beta \overline{V} + \mathscr{M}(t) u^B(S \vee D) = 0. \tag{6.15}$$

Here, $u^B(x)$ is the *bequest utility*, the utility received from bequeathing x, and β is the *rate of time preference*. Note that (6.15) depends on the real-world drift μ as opposed to the risk-free rate r. We represent the worthlessness of holding a GLWDB after all death benefits have been paid by

$$\overline{V}(S,W,D,T) = 0. \tag{6.16}$$

The drift-diffusion form (6.15) corresponds to a standard additive utility specification.

6.3.2 Events

As in (6.7) and (6.8), we parameterize an event occurring at $t \in \mathscr{T}$ by writing it in the form

$$\overline{v}(\mathbf{x},t,\lambda) = \overline{V}\left(\mathbf{f}(\mathbf{x},t,\lambda),t^+\right) + \mathscr{R}(t)\overline{f}(\mathbf{x},t,\lambda) \tag{6.17}$$

along with

$$\overline{V}(\mathbf{x},t) = \overline{v}(\mathbf{x},t,\gamma(\mathbf{x},t)). \tag{6.18}$$

$\mathbf{f}$ is defined implicitly for each event type in Sect. 6.2.2. It should be noted that the function $\overline{f}$ does not represent a cash flow, but rather an influx of utility to the holder. That is,

$$\overline{f}(\mathbf{x},t,\lambda) = u^C(f(\mathbf{x},t,\lambda)),$$

where f is defined for each event type in Sect. 6.2.2 and $u^C(y)$ is the *consumption utility*, the utility received from consuming y.

6.3.3 Consumption-Optimal Withdrawal

We refer to a withdrawal strategy that satisfies

$$\gamma(\mathbf{x},t) \in \overline{\Gamma}(\mathbf{x},t) = \arg\max_{\lambda \in [0,2]} [\overline{v}(\mathbf{x},t,\lambda)] \qquad (6.19)$$

(for all states $\mathbf{x}$ and event times $t \in \mathscr{T}$) as a *consumption-optimal withdrawal strategy*. Since we are maximizing (6.17), $\overline{\Gamma}(\mathbf{x},t)$ is simply the set of all actions that maximize the policyholder's utility at $\mathbf{x}$ and t.

It should be noted that we are not interested in the value of the numerical solution to the utility PDE but rather in the withdrawal strategy generated by it. Instead of adopting the optimal withdrawal strategy introduced in Sect. 6.2.3, we feed the withdrawal strategy generated by the policyholder's utility into the pricing problem. Given the Cauchy data $V(\cdot,t_{n+1})$ and $\overline{V}(\cdot,t_{n+1})$:

1. Solve $V(\cdot,t_n^+)$ using (6.6) and Cauchy data $V(\cdot,t_{n+1})$.
2. Solve $\overline{V}(\cdot,t_n^+)$ using (6.15) and Cauchy data $\overline{V}(\cdot,t_{n+1})$.
3. Determine $\gamma(\cdot,t_n)$ s.t. (6.19) is satisfied. In doing so, determine $\overline{V}(\cdot,t_n)$ by (6.17) and (6.18).
4. Use $\gamma(\cdot,t_n)$, (6.7) and (6.8) to determine $V(\cdot,t_n)$.

The propagation of information in this procedure is depicted in Fig. 6.1.

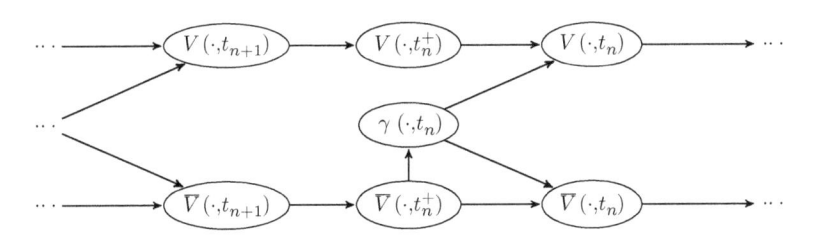

Fig. 6.1 A graph depicting the propagation of information in the pricing procedure

Remark 6.4 (Ensuring uniqueness). Step 3 requires that for each $\mathbf{x}$, we determine $\gamma(\mathbf{x},t_n)$. The expression (6.19) suggests that $\gamma(\mathbf{x},t_n)$ need not be unique. To ensure the uniqueness of V, we need a way to break ties between consumption-optimal strategies. Formally, we substitute condition (6.19) for

$$\gamma(\mathbf{x},t) = c\left(\overline{\Gamma}(\mathbf{x},t)\right)$$

where c is a choice function on the power set of $[0,2]$. For example, a choice function c that selects the smallest element (e.g. $c(\{0,1,2\}) = 0$) corresponds to a policyholder who will always withdraw the least amount possible to break a tie.

6.3.4 Regime-Switching

Assuming the regime-switching model introduced in Sect. 6.2.4, define $\overline{V}_i(S,W,D,t)$ as the mortality-adjusted utility of holding a GLWDB contract at time t years after purchase in regime $i \in \mathscr{S}$. Following standard arguments, we arrive at

$$\frac{\partial \overline{V}_i}{\partial t} + \mathscr{L}_i \overline{V}_i + \sum_{\substack{j=1 \\ j \neq i}}^{M} \left[q_{i \to j} \overline{V}_j (J_{i \to j} S, W, D, t) \right] + \mathscr{M}(t) u_i^B (S \vee D) = 0 \; \forall i \in S \quad (6.20)$$

where

$$\mathscr{L}_i = \frac{1}{2} \sigma_i^2 S^2 \frac{\partial^2}{\partial S^2} + (\mu_i - \alpha) S \frac{\partial}{\partial S} - (\beta_i - q_{i \to i}).$$

Equation (6.20) is referred to as the *utility system*. Note that this system of PDEs does not depend on the risk-neutral rates of transition $q_{i \to j}^{\mathbb{Q}}$ as in Sect. 6.2.4, but instead on the objective ($\mathbb{P}$ measure) rates of transition $q_{i \to j}$. We use the symbols u_i^B and u_i^C to stress that the utility functions can, in general, be regime-dependent.

As in Sect. 6.2.4, events and the corresponding withdrawal strategies become regime-dependent. The regime-switching analogue of (6.17) and (6.18) is

$$\overline{v}_i(\mathbf{x}, t, \lambda) = \overline{V}_i \left(\mathbf{f}(\mathbf{x}, t, \lambda) t^+ \right) + \mathscr{R}(t) u_i^C (f(\mathbf{x}, t, \lambda)) \quad (6.21)$$

and

$$\overline{V}_i(\mathbf{x}, t) = \overline{v}_i(\mathbf{x}, t, \gamma_i(\mathbf{x}, t)). \quad (6.22)$$

Likewise, the regime-switching analogue of (6.19) is

$$\gamma_i(\mathbf{x}, t) \in \overline{\Gamma}_i(\mathbf{x}, t) = \arg\max_{\lambda \in [0, 2]} \left[\overline{v}_i(\mathbf{x}, t, \lambda) \right]. \quad (6.23)$$

In this way, at any event time, the policyholder's utility in regime i (i.e. $\overline{V}_i$) is directly related to the price in regime i (i.e. V_i). In particular, the algorithm presented in Sect. 6.3.3 becomes:

1. Solve $\{V_1(\cdot, t_n^+), \ldots, V_M(\cdot, t_n^+)\}$ using (6.11) and Cauchy data $\{V_1(\cdot, t_{n+1}), \ldots, V_M(\cdot, t_{n+1})\}$.
2. Solve $\{\overline{V}_1(\cdot, t_n^+), \ldots, \overline{V}_M(\cdot, t_n^+)\}$ using (6.20) and Cauchy data $\{\overline{V}_1(\cdot, t_{n+1}), \ldots, \overline{V}_M(\cdot, t_{n+1})\}$.
3. For each regime i,
   a. Determine $\gamma_i(\cdot, t_n)$ such that (6.23) is satisfied. In doing so, determine $\overline{V}_i(\cdot, t_n)$ by (6.21) and (6.22).
   b. Use $\gamma_i(\cdot, t_n)$, (6.12) and (6.13) to determine $V_i(\cdot, t_n)$.

6.3.5 Hyperbolic Absolute Risk-Aversion

We consider policyholder consumption to be governed by hyperbolic absolute risk-aversion (HARA) utility [20]:

$$u_i^C(y; a_i, b_i, p_i) = \lim_{p \to p_i} \frac{1-p}{p} \left(\frac{a_i y}{1-p} + b_i \right)^p. \tag{6.24}$$

We take $u_i^B(y) = h_i u_i^C(y)$, where h_i is termed the *bequest motive*. This is a fairly flexible and general class of utility functions that can be parameterized so that marginal utility is finite at a consumption level of zero. This is potentially of interest in our context since it allows for the possibility that the policyholder will decide to not withdraw any amount at a withdrawal date. Otherwise, with infinite marginal utility at a consumption level of zero, the policyholder will always withdraw some positive amount.

6.4 Numerical Method

6.4.1 Homogeneity

Let $\mathbf{V}$ denote the column vector consisting of $V_1, V_2, \ldots, V_M$. We define $\overline{\mathbf{V}}$ similarly.

Remark 6.5 (Technical assumptions). We assume that all regime-switching jumps are unity (i.e. $J_{i \to j} = 1$ for all i and j), that $\mathbf{V}$ (resp. $\overline{\mathbf{V}}$) is a classical solution (i.e. twice differentiable in the investment account S and once in t on (t_n, t_{n+1}) for all $1 \leq n < N$) satisfying a growth condition to ensure uniqueness (recall that parabolic PDEs do not, in general, admit unique solutions [9]) and that the functions σ_i, r_i, α, $q_{i \to j}^{\mathbb{Q}}$, μ_i, β_i and $q_{i \to j}$ are bounded and continuous. Under these assumptions, it is possible to use the parametrix method [18] to construct a Green's function (denoted by F) representation for $\mathbf{V}$ (resp. $\overline{\mathbf{V}}$) on $t \in (t_n, t_{n+1})$. A more detailed list of these assumptions is provided by Azimzadeh [1]. We further assume that the functions σ_i, r_i, α, and $q_{i \to j}^{\mathbb{Q}}$, μ_i, β_i and $q_{i \to j}$ are independent of S, W and D and exploit this fact in Lemma 6.1.

Definition 6.1 (Homogeneous function). A function $s: X \to Y$ between two cones is said to be homogeneous of order $k \in \mathbb{Z}$ if for all $\eta > 0$ and $\mathbf{x} \in X$, $\eta^k s(\mathbf{x}) = s(\eta \mathbf{x})$. We say $\mathbf{V}$ is homogeneous if for each $i \in \mathscr{S}$, V_i is homogeneous.

Theorem 6.1 (Price homogeneity under loss-maximizing strategy). *Suppose that a loss-maximizing strategy is employed by the policyholder. Then, $\mathbf{V}(\mathbf{x},t)$ is homogeneous of order 1 in $\mathbf{x}$.*

This fact is established via a series of lemmas. Namely, we show that if $\mathbf{V}(\mathbf{x}, t_{n+1})$ is homogeneous in $\mathbf{x}$, so too is $\mathbf{V}(\mathbf{x}, t_n^+)$ (Lemma 6.1). That is, the system (6.11) composed of the operators $\mathscr{L}_1, \mathscr{L}_2, \ldots, \mathscr{L}_M$ preserves homogeneity. Then, we show that if $\mathbf{V}(\mathbf{x}, t_n^+)$ is homogeneous in $\mathbf{x}$, so too is $\mathbf{V}(\mathbf{x}, t_n)$ (Lemma 6.2). That is, homogeneity is preserved across event times under a loss-maximizing strategy. By (6.1) and (6.5), we have $\mathbf{V}(\mathbf{x}, t_N = T) = \mathbf{0}$. Since this is trivially homogeneous, the desired result follows by induction.

Lemma 6.1 (Pricing system homogeneity between event times). *Suppose that for some n with $1 \leq n < N$, $\mathbf{V}(\mathbf{x}, t_{n+1})$ is homogeneous of order 1 in $\mathbf{x}$. Then, for all $t \in (t_n, t_{n+1}]$, $\mathbf{V}(\mathbf{x}, t)$ is homogeneous of order 1 in $\mathbf{x}$.*

Proof. If we let $\tau = t_{n+1} - t$ and

$$g(S, W, D, t) = \mathscr{R}(t) \alpha_M S + \mathscr{M}(t)(S \vee D),$$

we can write (Remark 6.5)

$$\mathbf{V}(S, W, D, t) = \int_0^\infty F\left(\log \frac{S'}{S}, \tau, 0\right) \mathbf{V}(S', W, D, t_{n+1}) \frac{1}{S'} dS'$$
$$+ \int_0^\tau \int_0^\infty F\left(\log \frac{S'}{S}, \tau, \tau'\right) \left(g(S', W, D, t_{n+1} - \tau') \mathbf{1}\right) \frac{1}{S'} dS' d\tau'.$$

where $\mathbf{1}$ is a column vector of ones. The fact that F depends on S' and S only through $\log(S'/S)$ is discussed by Azimzadeh [1] and stems from the assumption that σ_i, r_i, α and $q_{i \to j}^{\mathbb{Q}}$ are independent of S, W and D (Remark 6.5). The substitution $S' = SS''$ yields

$$\mathbf{V}(S, W, D, t) = \int_0^\infty F(\log S'', \tau, 0) \mathbf{V}(SS'', W, D, t_{n+1}) \frac{1}{S''} dS''$$
$$+ \int_0^\tau \int_0^\infty F(\log S'', \tau, \tau') \left(g(SS'', W, D, t_{n+1} - \tau') \mathbf{1}\right) \frac{1}{S''} dS'' d\tau'.$$

Since $\mathbf{V}(\mathbf{x}, t_{n+1})$ and $g(\mathbf{x}, t)$ are both homogeneous in $\mathbf{x}$, it is now straightforward to extend $\mathbf{V}(\mathbf{x}, t)$'s homogeneity to $t \in (t_n, t_{n+1}]$. □

Remark 6.6 (Unit jump size assumption). The assumption that the jump sizes are unity $J_{i \to j} = 1$ is required in order to use the standard Green's function form. However, Lemma 6.1 also holds for the case of non-unit jump sizes, but the proof is somewhat more lengthy.

Lemma 6.2 (Loss-maximizing strategy preserves homogeneity). *Suppose that for some regime $i \in \mathscr{S}$ and for some n with $1 \leq n < N$, $V_i(\mathbf{x}, t_n^+)$ is homogeneous of order 1 in $\mathbf{x}$ and that the policyholder employs a loss-maximizing strategy $\gamma_i(\cdot, t_n)$. Then, $V_i(\mathbf{x}, t_n)$ is homogeneous of order 1 in $\mathbf{x}$.*

Proof. We leave it to the interested reader to show that $\mathbf{f}(\mathbf{x}, t_n, \lambda)$ and $f(\mathbf{x}, t_n, \lambda)$ defined implicitly in Sect. 6.2.2 are homogeneous of order 1 in $\mathbf{x}$. From this and the presumed homogeneity of $V_i(\mathbf{x}, t_n^+)$, it follows that $v_i(\mathbf{x}, t_n, \lambda)$ defined by (6.12) is homogeneous of order 1 in $\mathbf{x}$. Let $\eta > 0$ and $\mathbf{x}$ be arbitrary. By (6.14),

$$\gamma_i(\eta\mathbf{x},t_n) \in \Gamma_i(\eta\mathbf{x},t_n)$$
$$= \underset{\lambda\in[0,2]}{\arg\max}\,[v_i(\eta\mathbf{x},t_n,\lambda)]$$
$$= \underset{\lambda\in[0,2]}{\arg\max}\,\eta\,[v_i(\mathbf{x},t_n,\lambda)]$$
$$= \underset{\lambda\in[0,2]}{\arg\max}\,[v_i(\mathbf{x},t_n,\lambda)]$$
$$= \Gamma_i(\mathbf{x},t_n) \ni \gamma_i(\mathbf{x},t_n).$$

From this, it follows that $v_i(\mathbf{x},t_n,\gamma(\eta\mathbf{x},t_n)) = v_i(\mathbf{x},t_n,\gamma(\mathbf{x},t_n))$. Specifically,

$$V_i(\eta\mathbf{x},t_n) = v_i(\eta\mathbf{x},t_n,\gamma(\eta\mathbf{x},t_n)) = \eta v_i(\mathbf{x},t_n,\gamma(\eta\mathbf{x},t_n)) = \eta v_i(\mathbf{x},t_n,\gamma(\mathbf{x},t_n)) = \eta V_i(\mathbf{x},t_n).$$

◻

The homogeneity of the pricing problem allows us to reduce it from a system of coupled three-dimensional PDEs to a system of coupled two-dimensional PDEs. By Theorem 6.1, for $\eta > 0$,

$$V_i(S,W,D,t) = \frac{1}{\eta}V_i(\eta S,\eta W,\eta D,t).$$

Suppose $W > 0$. Choosing $\eta = W^\star/W$ with $W^\star > 0$ yields

$$V_i(S,W,D,t) = \frac{W}{W^\star}V_i\left(\frac{W^\star}{W}S,W^\star,\frac{W^\star}{W}D,t\right), \tag{6.25}$$

which reveals that we need only solve the problem for two values of the withdrawal benefit: $W^\star$ and zero. We refer to this reduction in dimensionality as a *similarity reduction*.

Theorem 6.2 (Utility homogeneity under consumption-optimal strategy). *Suppose that a consumption-optimal strategy is employed by the policyholder, and that for all regimes $i \in \mathscr{S}$, u_i^B and u_i^C are homogeneous of order p. Then $\mathbf{V}(\mathbf{x},t)$ and $\overline{\mathbf{V}}(\mathbf{x},t)$ are homogeneous of orders 1 and p, respectively, in $\mathbf{x}$.*

The proof of this is almost identical to that of Theorem 6.1, and is hence left to the interested reader. It should be noted that the above assumes that ties in strategies are broken as in Remark 6.4.

Corollary 6.1 (Power law homogeneity). *For all regimes $i \in \mathscr{S}$, take $b_i = 0$ and $p_i = p$ in (6.24) for some constant $p \neq 0$. Suppose that a consumption-optimal strategy is employed by the policyholder. Then, $\mathbf{V}(\mathbf{x},t)$ and $\overline{\mathbf{V}}(\mathbf{x},t)$ are homogeneous of order 1 and p, respectively, in $\mathbf{x}$.*

Proof. This follows directly from Theorem 6.2 and the fact that $u_i^C(x;a,b,p)$ is homogeneous of order p in x and b. ◻

This encompasses a large family of economically relevant functions, namely the power law (a.k.a. isoelastic) utility functions. Under power law utility, we can reduce the system of three-dimensional PDEs to a system of two-dimensional PDEs. As before, we get

$$\overline{V}_i(S,W,D,t) = \left(\frac{W}{W^\star}\right)^p \overline{V}_i\left(\frac{W^\star}{W}S, W^\star, \frac{W^\star}{W}D, t\right),$$

along with (6.25) whenever $W > 0$ and $W^\star > 0$.

6.4.2 Localized Problem and Boundary Conditions

We approximate the original problem, posed on $(S,W,D,t) \in \mathbb{R}^3_{\geq 0} \times [0,T]$, on the truncated domain

$$(S,W,D,t) \in [0, S_{\text{Max}}] \times \mathscr{W} \times [0, D_{\text{Max}}] \times [0,T],$$

where $\mathscr{W} = [0,\infty)$ when a similarity reduction is applied and $\mathscr{W} = [0, W_{\text{Max}}]$ otherwise. We clamp regime-switching jumps that drive the underlying above S_{Max}. That is, we take $\min(J_{i\to j}S, S_{\text{Max}})$ (instead of $J_{i\to j}S$) to be the value of the investment account after a jump from regime i to j. No boundary conditions are needed at $S = 0$, $W = 0$, $D = 0$, $W = W_{\text{Max}}$ and $D = D_{\text{Max}}$. That is, it is sufficient to substitute one of the aforementioned boundary values of S, W or D into (6.11) and (6.20) to retrieve the relevant behaviour. At $S = S_{\text{Max}}$, for each W and D, we impose instead the linearity conditions [32]

$$V_i(S_{\text{Max}}, W, D, t) = C_i(t) S_{\text{Max}} \text{ and } \overline{V}_i(S_{\text{Max}}, W, D, t) = \overline{C}_i(t) S_{\text{Max}} \ \forall i \in \mathscr{S}$$

in an attempt to estimate the true asymptotic behaviour of the contract. Substituting the above into (6.11) and (6.20) yields two ordinary differential equations (ODEs) in which C_i and $\overline{C}_i$ are the dependent variables. These are solved numerically alongside the rest of the domain. Errors introduced by the above approximations are small in the region of interest, as verified by numerical experiments. At $t = T$, (6.1) and (6.16) suggest

$$V_i(S,W,D,T) = \overline{V}_i(S,W,D,T) = 0 \ \forall i \in \mathscr{S}.$$

We use Crank-Nicolson time-stepping with Rannacher smoothing [28]. We discretize the diffusive term using a second-order centered difference, while the convective term is discretized using a centered difference only when the corresponding backward Euler scheme is monotone. Otherwise, an upwind discretization is employed. Variable-size timestepping is used (see Johnson [15] for an expository treatment). The resulting linear system is solved using fixed-point iteration. The details of this approach are described by d'Halluin et al. [6] and Kennedy [16].

6.4.3 Determining the Hedging Cost Fee

At contract inception, the withdrawal and death benefits are set to the initial value of the investment account, $S(0)$. That is, $W(0) = S(0)$ and $D(0) = S(0)$. If we overload our previous definition of V_i as parameterized by the fee, α_R, the problem becomes one of determining α_R such that

$$V_I(S(0), W(0), D(0), 0; \alpha_R) - \underbrace{\mathscr{R}(0)}_{1} S(0) = 0, \tag{6.26}$$

where I is the regime observed at time zero. This is a requirement stating that α_R must be selected so as to compensate the writer for the hedging costs. We term such a value of α_R the *hedging cost fee*. Equation (6.26) is solved numerically using Newton's method.

6.5 Results

We begin by performing experiments under the assumptions (i) that the policyholder behaves so as to maximize the writer's losses and (ii) that the policyholder always withdraws at the contract rate. We consider a handful of numerical tests based on perturbations to the base case data in Table 6.1. We subsequently move to considering consumption-optimal strategies, in which we use the base case data in Tables 6.1 and 6.3. Throughout this section, various rates are presented in basis points (bps).

6.5.1 Loss-Maximizing and Contract Rate Withdrawal

All tests in this section are performed on perturbations to the base case data in Table 6.1. Table 6.2 documents wide variation in the hedging cost fee across different volatility and interest rate parameters for the two regimes considered, and for the cases with a ratcheting death benefit, with a nonratcheting death benefit, and without a death benefit. Of course, in any otherwise identical scenario, the loss-maximizing withdrawal assumption results in a higher fee since this represents the worst case scenario for the insurer. As we might expect, higher volatility is associated with an increase in the cost of hedging and thus a higher fee. The fee is also quite sensitive to the levels of the risk-free interest rate across the two regimes. The presence of a death benefit results in a notably increased fee, particularly if this feature is ratcheting.

Table 6.1 Pricing system base case data with regime-dependent parameters obtained from O'Sullivan and Moloney [25] by calibration to FTSE 100 options in January 2007

Parameter			Value	
Volatility	σ_1	σ_2	0.0832	0.2141
Risk-free rate	r_1	r_2	0.0521	0.0521
Rate of transition	$q^Q_{1\to 2}$	$q^Q_{2\to 1}$	0.0525	0.1364
Jump magnitude	$J_{1\to 2}$	$J_{2\to 1}$	1	1
Initial regime	I		1	
Initial investment	$S(0)$		100	
Management rate	α_M		100 bps	
Contract rate	G		0.05	
Bonus rate	B		0.05	
Initial age	x_0		65	
Expiry time	T		57	
Mortality data			[26]	
Ratchets			Triennial	
Withdrawals			Annual	

Time t	Penalty $\kappa(t)$
1	0.03
2	0.02
3	0.01
≥ 4	0

Table 6.2 The value of the hedging cost fee for perturbations to the data in Table 6.1. For each perturbation, fees are calculated under the loss-maximizing (left) and contract rate withdrawal (right) strategies. Values are reported to the nearest basis point

Parameters	Hedging cost fee α_R (bps)					
	Ratcheting death benefit		Nonratcheting death benefit		No death benefit	
Base case (Table 6.1)	54		48	37	24	27	19
Initial regime = 2	158		113	139	75	86	52
$(r_1, r_2) = (0.04, 0.06)$	79		72	62	43	44	33
$(r_1, r_2) = (0.03, 0.07)$	124		114	106	76	73	57
$(r_1, r_2) = (0.02, 0.08)$	239		212	224	156	129	104
$(\sigma_1, \sigma_2) = (0.10, 0.20)$	62		56	45	29	31	22
$(\sigma_1, \sigma_2) = (0.15, 0.25)$	133		123	107	69	70	51

6.5.1.1 Withdrawal Analysis

We now turn to a brief exploration of loss-maximizing withdrawal strategies by the policyholder. Figures 6.2 and 6.3 show these strategies under each regime (Table 6.1) at $t = 1, 2, \ldots, 6$ assuming that the corresponding hedging cost fee is charged for hedging the contract and that $D = 100$. In either regime, if W is much bigger than S, the strategy always involves withdrawing at the contract rate, but the strategy in other regions can be quite complex. We note that in the less volatile regime (Fig. 6.2), the withdrawal strategy does not involve surrender for $t \leq 3$, prior to the vanishing of surrender charges at $t > 3$ (Table 6.1). However, in the more volatile regime (Fig. 6.3), the policyholder is more willing to surrender the contract, despite the large penalties at times $t = 1$ and $t = 2$. Also note that in this regime, the policyholder's willingness to surrender (for large values of S) vanishes at $t = 3$ in anticipation of the triennial ratchet. The complexity of these loss-maximizing strategies provides some further motivation for our consumption-based approach, since it may seem implausible that individual policyholders would actually implement such strategies.

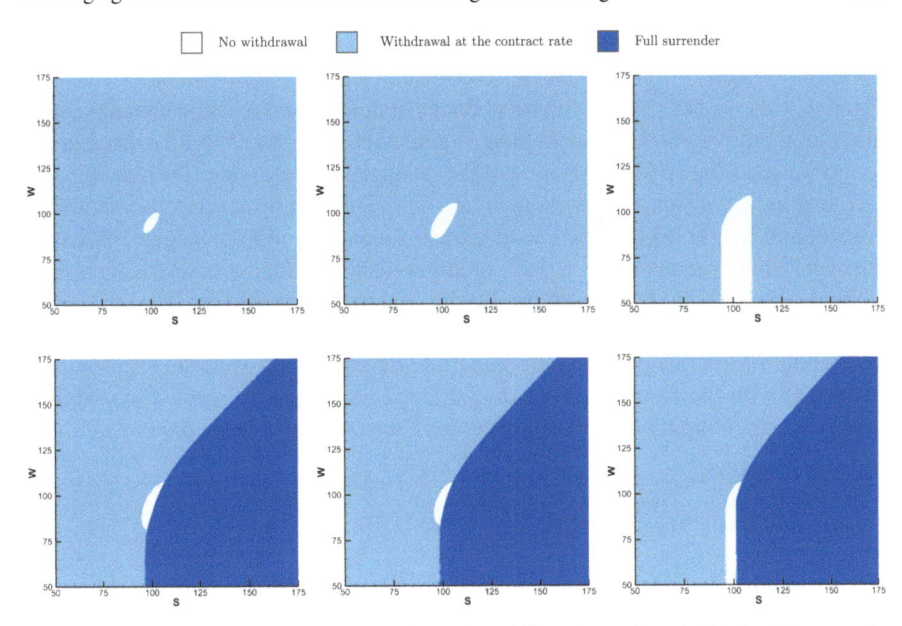

6 Hedging Costs for Variable Annuities Under Regime-Switching   153

Fig. 6.2 Observed loss-maximizing strategies at $D = 100$ under regime 1. The hedging cost fee $\alpha_R \approx 37$ bps is used (Table 6.2). The subfigures, from top-left to bottom-right, correspond to $t = 1, 2, \ldots, 6$

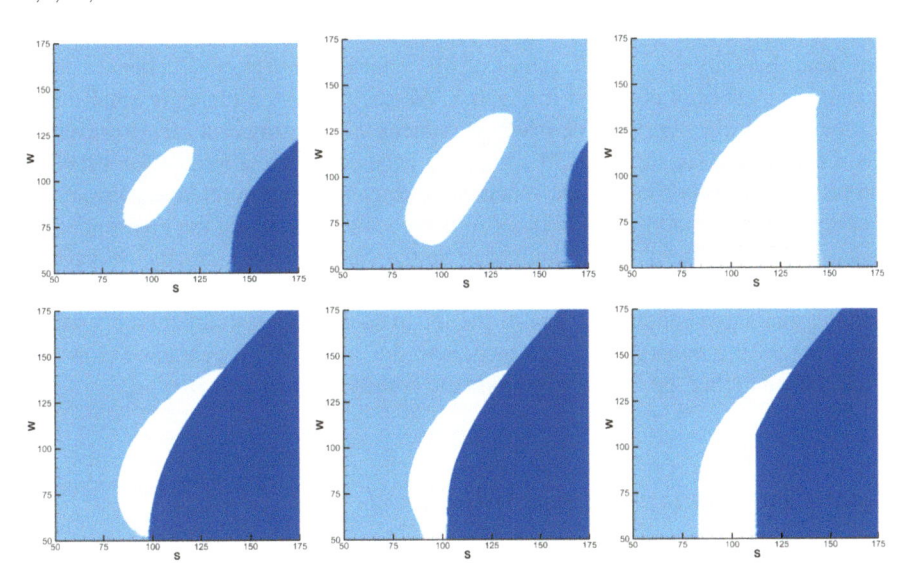

Fig. 6.3 Observed loss-maximizing strategies at $D = 100$ under regime 2. The hedging cost fee $\alpha_R \approx 139$ bps is used (Table 6.2). The subfigures, from top-left to bottom-right, correspond to $t = 1, 2, \ldots, 6$

6.5.1.2 Management Rate

Figure 6.4 shows the relationship between the hedging cost fee and the management rate (i.e. the proportional management expense fee α_M). As is to be expected, the fee grows superlinearly as a function of the management rate, since the management rate acts as a drag on the investment account. This confirms the observation by Forsyth and Vetzal [8] that the use of mutual funds with high management fees as the underlying investment for variable annuities results in higher costs for the insurer compared to a policy written on funds with low management fees (e.g. exchange-traded index funds). We also see that for both the loss-maximizing and contract rate withdrawal strategy, the death benefit adds significant value to the contract, consistent with the results reported in Table 6.2. Again, the disparity between the ratcheting and nonratcheting death benefit is even more pronounced.

6.5.1.3 Alternate Fee Structure

Some insurers have adopted alternate fee structures that are functions of the auxiliary accounts. In general, the risky account evolves according to

$$dS = (\mu S - \alpha F(S,W,D,t)) dt + \sigma S dZ.$$

A comparison of the usual fee structure $F = S$ with $F = S \vee W$ on a contract without death benefits for various values of the management rate α_M under the loss-maximizing strategy is shown in Fig. 6.5. We see that for sufficiently small management rates, the alternate fee structure reduces the hedging cost fee. However, as the management fee increases, the fee calculated under the alternate fee structure surpasses its vanilla counterpart. When the management rate is relatively low, it has a comparatively small impact in terms of decreasing the value of the investment account and hence exerts limited influence on the value of the guarantee. Moreover, since the total rate $\alpha = \alpha_M + \alpha_R$ applies to the greater of the investment account and the guarantee benefit, the size of the fee in such cases is comparatively small. However, as the management rate increases, the value of the guarantee rises and eventually a higher fee is needed to fund the cost of hedging.

6.5.2 Consumption-Optimal Withdrawal

All tests in this section are performed on perturbations to the base case data in Tables 6.1 and 6.3.

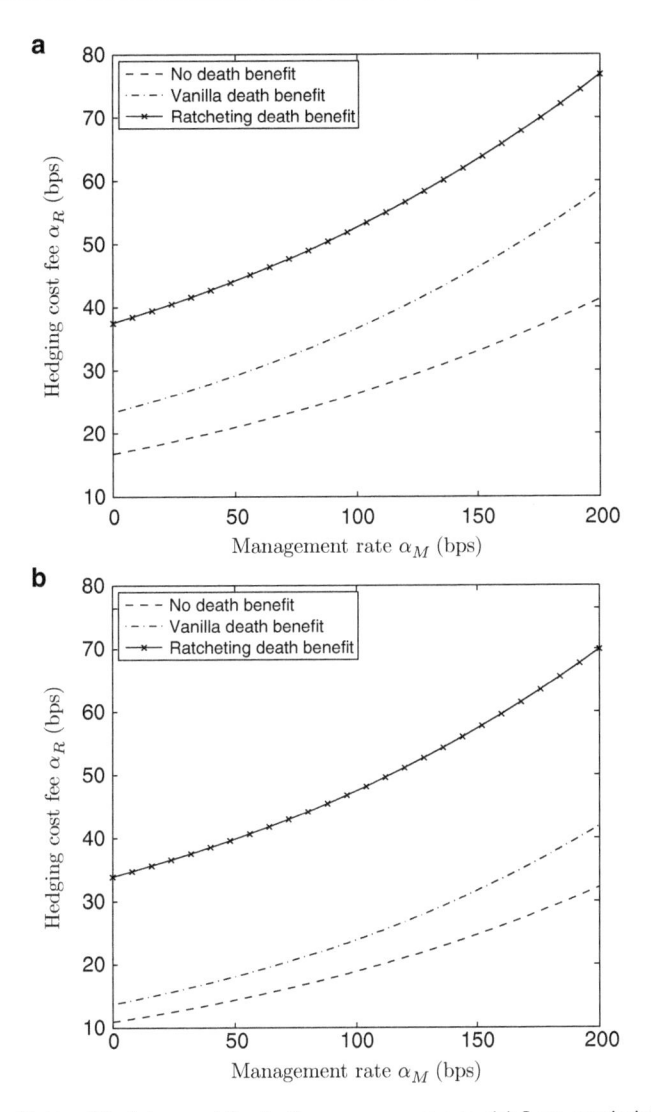

Fig. 6.4 Sensitivity of hedging cost fee to the management rate. (**a**) Loss-maximizing strategy. (**b**) Contract-rate strategy

6.5.2.1 Risk-Aversion

Suppose the management rate, α_M, is zero. If for all regimes $i \in \mathscr{S}$ we take the parameterization shown in Table 6.4, the consumption-optimal strategy reduces to the loss-maximizing strategy (this can be verified by direct substitution). Reflecting this, we refer to this parameterization as the *degeneracy parameterization*. Since the degeneracy parameterization corresponds to the loss-maximizing strategy, it is

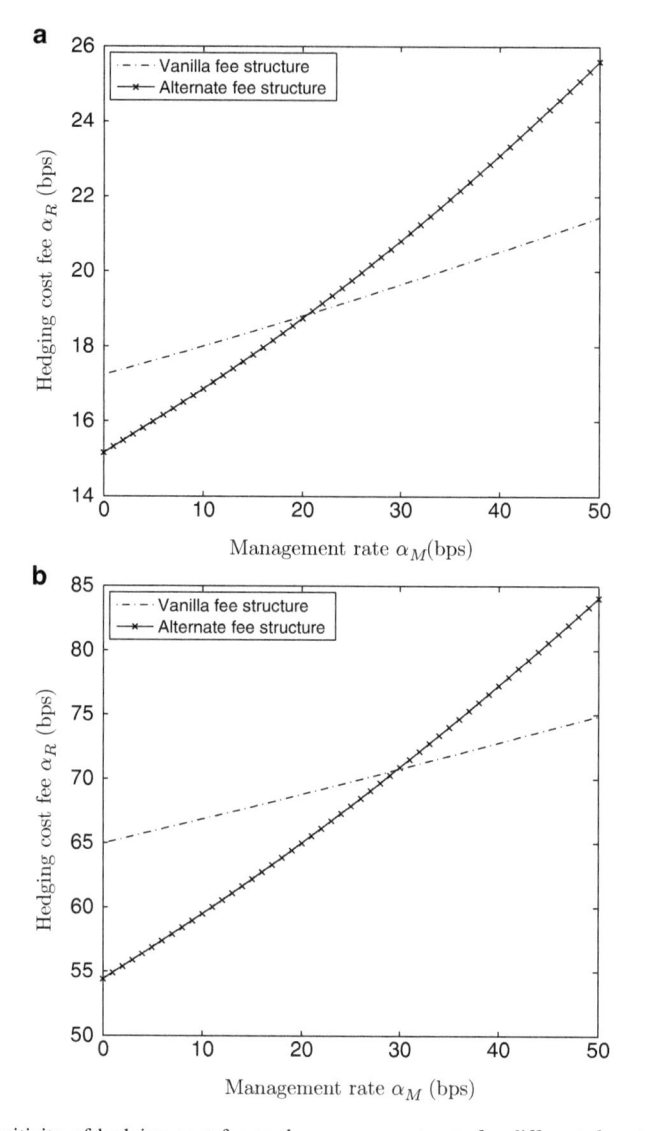

Fig. 6.5 Sensitivity of hedging cost fee to the management rate for different fee structures. (**a**) Initial regime $I = 1$; no death benefit. (**b**) Initial regime $I = 2$; no death benefit

guaranteed to yield the highest possible hedging cost fee. We stress that this holds only when the management rate is zero, as in Table 6.4. The utility parameters under this parameterization $u_i^B(x) = h_i u_i^C(x; a_i = 1, b_i = 0, p_i = 1)$ correspond to the case of risk-neutral utility: $u_i^B(x) = u_i^C(x) = x$.

Although the above only holds under the degeneracy parameterization, we expect to see large hedging cost fees under parameterizations that are close to the degeneracy parameterization. Figure 6.6 shows the effect of simultaneously varying the

Table 6.3 Consumption system base case data with rate of time preference obtained from Nishiyama and Smetters [24]

Parameter			Value	
Drift rate	μ_1	μ_2	0.1	0.1
Time preference	β_1	β_2	0.032	0.032
HARA scaling	a_1	a_2	1	1
HARA offset	b_1	b_2	0	0
Risk-aversion	p_1	p_2	0.5	0.5
Bequest motive	h_1	h_2	1	1
Rate of transition	$q_{1\to 2}$	$q_{2\to 1}$	0.0525	0.1364

regime-dependent drifts μ_1 and μ_2 and risk-aversion parameters p_1 and p_2 on the hedging cost fee for the base case data in Tables 6.1 and 6.3 for a contract without death benefits. When $\mu_1 = \mu_2 = 0.0521$ and $p_1 = p_2 = 1$, a global maximum appears on each surface. As expected, the parameterization $\mu_1 = \mu_2 = 0.0521$ and $p_1 = p_2 = 1$ is close to the degeneracy parameterization (Tables 6.1 and 6.3 specify $\alpha = 100\,\text{bps} \approx 0$ and $\beta_i = 0.032 \approx 0.0521 = r_i$), and hence these maxima (27 and 84 bps, rounded to the nearest basis point) are very close to the hedging cost fees for each regime calculated under the loss-maximizing strategy (27 and 86 bps, rounded to the nearest basis point; see Table 6.2). Realistically, these maxima are not of great interest to the insurer as they occur where the drift of the investment account is equal to the risk-free rate of return. More interestingly, both surfaces exhibit a large "plateau" region (i.e. where the gradient is approximately zero) for which the consumption-optimal hedging cost fee is close to that calculated under the contract rate withdrawal strategy. This suggests that for a large family of parameters, the policyholder withdraws at nearly the contract rate. This can be verified by comparing the hedging cost fee here for the two regimes with those shown in Table 6.2 (19 and 52 bps, rounded to the nearest basis point).

Table 6.4 Degeneracy parameterization

Parameter	α_M	μ_i	β_i	a_i	b_i	p_i	h_i
Value	0	$r_i - \rho_i^Q$	r_i	1	0	1	1

6.5.2.2 Taxation

It has been suggested by Moenig and Bauer [22] that a policyholder's strategy depends on the taxation of his withdrawals. We assume that withdrawals are taxed on the American *last-in first-out* (LIFO) basis and that earnings in the underlying investment account grow on a tax-deferred basis.

This requires the addition of another path dependent variable $Q(t)$, which is referred to as the *tax base* at time t. The tax base denotes what amount of the

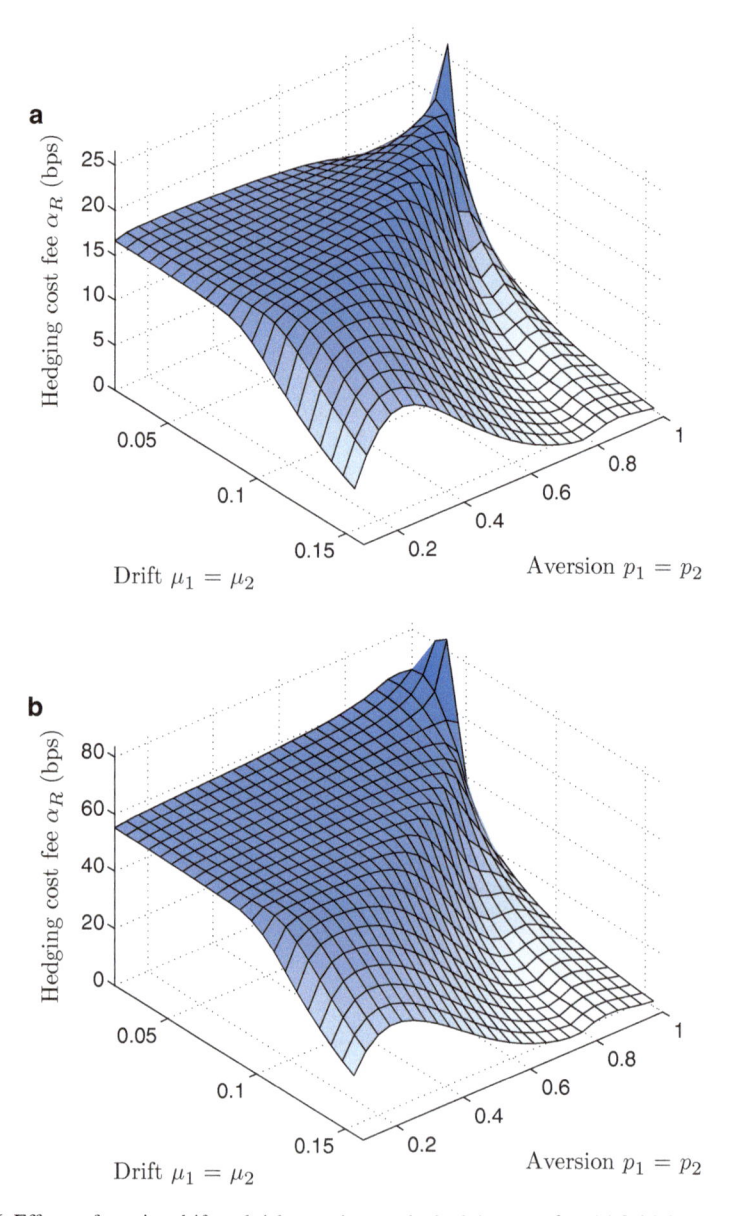

Fig. 6.6 Effects of varying drift and risk-aversion on the hedging cost fee. (**a**) Initial regime $I = 1$; no death benefit. (**b**) Initial regime $I = 2$; no death benefit

Table 6.5 Sensitivity of the hedging cost fee to the tax rate. Values are reported to the nearest tenth of a basis point

	0 %	10 %	20 %	30 %	40 %	50 %
Initial regime $I = 1$	18.0	18.9	19.2	18.7	17.7	16.3
Initial regime $I = 2$	54.7	55.8	56.3	56.7	57.0	57.2

underlying investment account is nontaxable. Initially, $Q(0) = S(0)$. Q is piecewise constant between withdrawals. When a withdrawal of size w is made at time t,

$$Q(t^+) = Q(t^-) - \underbrace{(w - [S(t^-) - Q(t^-)] \vee 0) \vee 0}_{\text{Nontaxable portion of the withdrawal}}.$$

The introduction of the tax base variable introduces an additional dimension for which the PDEs must be solved. We assume that policyholders optimize their after-tax consumption. Table 6.5 shows the effect of the tax rate on the hedging cost fee for the base case contract without death benefits. We find that for typical levels of risk-aversion, taxation has a small effect on the fee. Even for extreme tax rates of 50 %, the fee changes by at most several basis points.

6.6 Conclusion

We have introduced a general methodology that allows for the decoupling of policyholder behaviour from the pricing (i.e. determining the cost of hedging) of a variable annuity. Assuming that the underlying investment follows a regime-switching process, this yields two weakly coupled systems of PDEs: the pricing and utility systems. When considering strategies contingent on the policyholder's level of consumption, the utility system is used to generate policyholder withdrawal behaviour, which is in turn fed into the pricing system as a means to determine the cost of hedging the contract. Our methodology is general enough to allow us to consider any withdrawal strategy contingent on either the cost of hedging the contract or the policyholder's level of consumption.

We have adopted the GLWDB as a case study. A similarity reduction transforms our systems of three-dimensional PDEs to systems of two-dimensional PDEs, allowing us to generate numerical solutions with speed. In the absence of a death benefit, these systems further simplify into systems involving one-dimensional PDEs, which (for a reasonable number of regimes) can be solved with minimal computational effort.

Since GLWDB contracts are held over long periods of time, regime-switching serves as a natural model for the process followed by the underlying asset. This process can incorporate stochastic interest rates and volatility in a simple and intuitive manner. It is also possible to have policyholder preferences which differ between regimes. Results obtained under various regime-switching processes indicate that the hedging cost fee is extremely sensitive to the regime-dependent parameters.

We show that the inclusion of a death-benefit yields large fees for typical contract values under both the loss-maximizing strategy and the static strategy of always withdrawing at the contract rate. We observe an even more pronounced disparity between the no-arbitrage fee generated by a contract with nonratcheting death benefits compared to a contract with ratcheting death benefits. These findings are consistent with the phasing out of products including ratcheting death benefits from the Canadian market.

We find that for a large family of utility functions, the consumption-optimal strategy yields a hedging cost fee that is very close to the hedging cost fee calculated by assuming that the policyholder withdraws at the contract rate. This can be understood as substantiating the otherwise seemingly naïve assumption that the policyholder "generally" withdraws at the contract rate. Adopting the contract rate withdrawal strategy renders the pricing problem computationally simple, as this strategy is deterministic and can easily be implemented in either the PDE or an equivalent Monte Carlo formulation.

Acknowledgements This work was supported by the Natural Sciences and Engineering Research Council of Canada (NSERC) and the Global Risk Institute in Financial Services (Toronto).

Appendix

In this Appendix, we derive the no-arbitrage regime-switching PDEs for general contingent claims. Following along the lines of this Appendix, the reader should have no difficulty combining these arguments with those in Sect. 6.2 to obtain the final equation (6.11).

Regime-Switching Model

Regime-Switching PDEs

Consider the M-regime process S evolving according to

$$dS(t) = a_i(S(t),t)dt + b_i(S(t),t)dZ(t) + \sum_{j=1}^{M} S(t)(J_{i \to j} - 1)dX_{i \to j}(t)$$

in which dS describes the increment of S assuming that the regime at time t is i. We restrict $J_{i \to i} = 1$ for all i so that jumps in the underlying are not experienced unless there is a change in regime.

6 Hedging Costs for Variable Annuities Under Regime-Switching

In the relevant literature, it is often mentioned that the introduction of the regime-switching underlying S yields an incomplete market [34, 7], if the hedging portfolio contains only the underlying asset and the risk-free account. We consider instead a complete market consisting of the bond and M independent hedging instruments. Note that the assumption of the availability of M instruments is not farfetched; we need only find M instruments written on the regime-switching underlying S. Often, it is possible to take S itself as one of these instruments (this scenario is detailed elsewhere in this Appendix).

We follow the formulation of a regime-switching framework as derived by Kennedy [16]. Consider a portfolio Π short an option V and with positions in instruments $F^{(1)}, F^{(2)}, \ldots, F^{(M)}$. We assume that the trading instruments depend only on $S(t)$ and t. Let B represent the money market process with risk-free rate r (i.e. $dB = rBdt$). Denote by V_i and $F_i^{(k)}$ the values of the option and kth instrument in regime i, and $\omega^{(k)}$ is the number of units of instrument i. Assuming that regime i is observed at time t,

$$\Pi(S(t),t) = -V_i(S(t),t) + \sum_{k=1}^{M} \left[\omega^{(k)} F_i^{(k)}(S(t),t)\right] + B(t). \tag{6.27}$$

The increment of the above portfolio can be written as

$$d\Pi(S(t),t) = -dV_i(S(t),t) + \sum_{k=1}^{M} \left[\omega^{(k)} dF_i^{(k)}(S(t),t)\right] + dB(t). \tag{6.28}$$

where

$$dV_i = \hat{\mu}_i dt + \hat{\sigma}_i dZ + \sum_{j=1}^{M} \Delta V_{i \to j} dX_{i \to j}$$

$$\hat{\mu}_i = \frac{1}{2} b_i^2 \frac{\partial^2 V_i}{\partial S^2} + a_i \frac{\partial V_i}{\partial S} + \frac{\partial V_i}{\partial t}$$

$$\hat{\sigma}_i = b_i \frac{\partial V_i}{\partial S}$$

$$\Delta V_{i \to j} = V_j(J_{i \to j} S, t) - V_i(S, t)$$

and

$$dF_i^{(k)} = \bar{\mu}_i^{(k)} dt + \bar{\sigma}_i^{(k)} dZ + \sum_{j=1}^{M} \Delta F_{i \to j}^{(k)} dX_{i \to j}$$

$$\bar{\mu}_i^{(k)} = \frac{1}{2} b_i^2 \frac{\partial^2 F_i^{(k)}}{\partial S^2} + a_i \frac{\partial F_i^{(k)}}{\partial S} + \frac{\partial F_i^{(k)}}{\partial t}$$

$$\bar{\sigma}_i^{(k)} = b_i \frac{\partial F_i^{(k)}}{\partial S}$$

$$\Delta F_{i \to j}^{(k)} = F_j^{(k)}(J_{i \to j} S, t) - F_i^{(k)}(S, t).$$

Substituting these expressions into (6.28) yields

$$d\Pi(t) = \left[\sum_{k=1}^{M}\left[\omega^{(k)}\bar{\mu}_i^{(k)}\right] + rB - \hat{\mu}_i\right]dt + \left[\sum_{k=1}^{M}\left[\omega^{(k)}\bar{\sigma}_i^{(k)}\right] - \hat{\sigma}_i\right]dZ$$
$$+ \sum_{j=1}^{M}\left[\sum_{k=1}^{M}\left[\omega^{(k)}\Delta F_{i\to j}^{(k)}\right] - \Delta V_{i\to j}\right]dX_{i\to j}. \quad (6.29)$$

To make the portfolio deterministic, we eliminate Brownian risk by

$$\sum_{k=1}^{M}\omega^{(k)}\bar{\sigma}_i^{(k)} = \hat{\sigma}_i \quad (6.30)$$

and jump risk by

$$\sum_{k=1}^{M}\omega^{(k)}\Delta F_{i\to j}^{(k)} = \Delta V_{i\to j} \ \forall j \in \mathscr{S}. \quad (6.31)$$

Note that the jump risk equation corresponding to $j = i$ relates a zero change in the hedging instruments to zero change in the option, so that to eliminate jump risk, we need only satisfy $M - 1$ equations.

Given that the portfolio is deterministic, the principle of no-arbitrage requires $r\Pi dt = d\Pi$. Using the expressions (6.27) and (6.29), we write this as

$$\sum_{k=1}^{M}\omega^{(k)}\left(\bar{\mu}_i^{(k)} - rF_i^{(k)}\right) = \hat{\mu}_i - rV_i. \quad (6.32)$$

Equations (6.30)–(6.32) make for a total of $M + 1$ equations in M unknowns. This system has a solution if and only if one of the equations is a linear combination of the others. We denote by $\xi_i, q_{i\to 1}^{\mathbb{Q}}, q_{i\to 2}^{\mathbb{Q}}, \ldots, q_{i\to M}^{\mathbb{Q}}$ the weights under which the linear dependence requirement

$$\xi_i\left(\sum_{k=1}^{M}\left[\omega^{(k)}\bar{\sigma}_i^{(k)}\right] - \hat{\sigma}_i\right) = \sum_{\substack{j=1 \\ j\neq i}}^{M} q_{i\to j}^{\mathbb{Q}}\left(\sum_{k=1}^{M}\left[\omega^{(k)}\Delta F_{i\to j}^{(k)}\right] - dV_{i\to j}\right)$$
$$+ \sum_{k=1}^{M}\left[\omega^{(k)}\left(\bar{\mu}^{(k)} - rF_i^{(k)}\right)\right] - (\hat{\mu}_i - rV_i)$$

holds true. Rearranging this expression,

$$\sum_{k=1}^{M}\left[\omega^{(k)}\left(\xi_i\bar{\sigma}_i^{(k)} - \sum_{\substack{j=1 \\ j\neq i}}^{M}\left[q_{i,j}^{\mathbb{Q}}\Delta F_{i\to j}^{(k)}\right] - \left(\bar{\mu}_i - rF_i^{(k)}\right)\right)\right]$$
$$- \xi_i\hat{\sigma}_i + \sum_{\substack{j=1 \\ j\neq i}}^{M}\left[q_{i\to j}^{\mathbb{Q}}\Delta V_{i\to j}\right] + \hat{\mu}_i - rV_i = 0.$$

Since this must hold for any position $\omega^{(1)}, \omega^{(2)}, \ldots, \omega^{(M)}$, we write the above as

$$\xi_i \bar{\sigma}_i^{(k)} - \sum_{\substack{j=1 \\ j \neq i}}^{M} q_{i \to j}^{\mathbb{Q}} \Delta F_{i \to j}^{(k)} = \left(\bar{\mu}_i^{(k)} - r F_i^{(k)} \right) \quad \forall k \in \mathscr{S} \tag{6.33}$$

and

$$\xi_i \hat{\sigma}_i - \sum_{\substack{j=1 \\ j \neq i}}^{M} q_{i \to j}^{\mathbb{Q}} \Delta V_{i \to j} = \hat{\mu}_i - r V_i. \tag{6.34}$$

This procedure effectively decouples the hedging instruments from the option V. Resolving the symbols $\hat{\mu}_i$ and $\hat{\sigma}_i$ in (6.34) yields

$$\frac{1}{2} b_i^2 \frac{\partial^2 V_i}{\partial S^2} + (a_i - \xi_i b_i) \frac{\partial V_i}{\partial S} - r V_i + \sum_{\substack{j=1 \\ j \neq i}}^{M} \left[q_{i \to j}^{\mathbb{Q}} \Delta V_{i \to j} \right] + \frac{\partial V_i}{\partial t} = 0, \tag{6.35}$$

which describes a system of M PDEs: one for each regime. The more familiar form above reveals $a_i - \xi_i b_i$ as the risk-neutral drift and the $q_{i \to j}^{\mathbb{Q}}$ terms as the risk-neutral transition intensities.

We express this more compactly by defining

$$q_{i \to i}^{\mathbb{Q}} = - \sum_{\substack{j=1 \\ j \neq i}}^{M} q_{i \to j}^{\mathbb{Q}}$$

and noting that

$$\sum_{\substack{j=1 \\ j \neq i}}^{M} q_{i \to j}^{\mathbb{Q}} \Delta V_{i \to j} = \sum_{\substack{j=1 \\ j \neq i}}^{M} q_{i \to j}^{\mathbb{Q}} V_j (J_{i \to j} S, t) - V_i \sum_{\substack{j=1 \\ j \neq i}}^{M} q_{i \to j}^{\mathbb{Q}} = \sum_{\substack{j=1 \\ j \neq i}}^{M} q_{i \to j}^{\mathbb{Q}} V_j (J_{i \to j} S, t) + q_{i \to i}^{\mathbb{Q}} V_i$$

so that (6.35) becomes

$$\frac{1}{2} b_i^2 \frac{\partial^2 V_i}{\partial S^2} + (a_i - \xi_i b_i) \frac{\partial V_i}{\partial S} - \left(r - q_{i \to i}^{\mathbb{Q}} \right) V_i + \sum_{\substack{j=1 \\ j \neq i}}^{M} \left[q_{i \to j}^{\mathbb{Q}} V_j (J_{i \to j} S, t) \right] + \frac{\partial V_i}{\partial t} = 0. \tag{6.36}$$

Eliminating the Market Price of Risk

It is often possible to eliminate the market price of risk $\xi_i b_i$ from (6.36) [16]. For example, let

$$a_i (S(t), t) = (\mu_i - \alpha) S(t)$$

and
$$b_i(S(t),t) = \sigma_i S(t).$$

Under these parameters, (6.36) becomes

$$\frac{1}{2}\sigma_i^2 S^2 \frac{\partial^2 V_i}{\partial S^2} + (\mu_i - \alpha - \xi_i \sigma_i) S \frac{\partial V_i}{\partial S} - \left(r - q_{i \to i}^{\mathbb{Q}}\right) V_i + \sum_{\substack{j=1 \\ j \neq i}}^{M} \left[q_{i \to j}^{\mathbb{Q}} V_j \left(J_{i \to j} S, t\right)\right] + \frac{\partial V_i}{\partial t} = 0. \quad (6.37)$$

Suppose further that S itself is not tradeable but tracks the tradeable index $\hat{S}$ with

$$d\hat{S}(t) = \mu_i \hat{S}(t) dt + \sigma_i \hat{S}(t) dZ(t).$$

Take the 1st instrument, $F^{(1)}$, to be $\hat{S}$ so that

$$\bar{\mu}_i^{(1)} = \mu_i \hat{S}$$
$$\bar{\sigma}_i^{(1)} = \sigma_i \hat{S}$$
$$\Delta F_{i \to j}^{(1)} = \hat{S}(J_{i \to j} - 1).$$

Substituting this into (6.33) for $k = 1$ yields

$$\xi_i \sigma_i \hat{S} - \sum_{\substack{j=1 \\ j \neq i}}^{M} q_{i \to j}^{\mathbb{Q}} \hat{S}(J_{i \to j} - 1) = \xi_i \sigma_i \hat{S} - \rho_i^{\mathbb{Q}} \hat{S} = \mu_i \hat{S} - r\hat{S}.$$

More compactly, we write this as

$$\xi_i \sigma_i \hat{S} = \left(\rho_i^{\mathbb{Q}} + \mu_i - r\right) \hat{S} \quad (6.38)$$

where

$$\rho_i^{\mathbb{Q}} = \sum_{\substack{j=1 \\ j \neq i}}^{M} q_{i \to j}^{\mathbb{Q}} (J_{i \to j} - 1) = \sum_{j=1}^{M} q_{i \to j}^{\mathbb{Q}} J_{i \to j}.$$

Whenever $\hat{S}$ is equal to zero, S is necessarily zero so that the term associated with the market price of risk in (6.37) also vanishes. We are thus only interested in the case in which $\hat{S} \neq 0$, under which (6.38) states that

$$\xi_i \sigma_i = \rho_i^{\mathbb{Q}} + \mu_i - r.$$

Substituting the above into (6.37),

$$\frac{1}{2}\sigma_i^2 S^2 \frac{\partial^2 V_i}{\partial S^2} + \left(r - \alpha - \rho_i^{\mathbb{Q}}\right) S \frac{\partial V_i}{\partial S} - \left(r - q_{i \to i}^{\mathbb{Q}}\right) V_i + \sum_{\substack{j=1 \\ j \neq i}}^{M} \left[q_{i \to j}^{\mathbb{Q}} V_j \left(J_{i \to j} S, t\right)\right] + \frac{\partial V_i}{\partial t} = 0.$$

References

1. Azimzadeh, P.: Hedging costs for variable annuities. MMath thesis, Cheriton School of Computer Science, University of Waterloo (2013)
2. Bauer, D., Kling, A., Russ, J.: A universal pricing framework for guaranteed minimum benefits in variable annuities. ASTIN Bull. Actuar. Stud. Non Life Insur. **38**(2), 621 (2008)
3. Bélanger, A., Forsyth, P.A., Labahn, G.: Valuing the guaranteed minimum death benefit clause with partial withdrawals. Appl. Math. Financ. **16**(6), 451–496 (2009)
4. Butrica, B.A., Iams, H.M., Smith, K.E., Toder, E.J.: The disappearing defined benefit pension and its potential impact on the retirement incomes of baby boomers. Soc. Secur. Bull. **69**(3), 1–27 (2009)
5. Chen, Z., Vetzal, K.R., Forsyth, P.A.: The effect of modelling parameters on the value of GMWB guarantees. Insur.: Math. Econ. **43**(1), 165–173 (2008)
6. d'Halluin, Y., Forsyth, P., Vetzal, K.R.: Robust numerical methods for contingent claims under jump diffusion processes. IMA J. Numer. Anal. **25**(1), 87–112 (2005)
7. Elliott, R.J., Chan, L., Siu, T.K.: Option pricing and Esscher transform under regime switching. Ann. Financ. **1**(4), 423–432 (2005)
8. Forsyth, P.A., Vetzal, K.R.: An optimal stochastic control framework for determining the cost of hedging of variable annuities. Working paper, University of Waterloo (2013)
9. Friedman, A.: Partial differential equations of parabolic type, vol. 196, 1983 edn. Prentice-Hall, Englewood Cliffs (1964)
10. Hamilton, J.D.: A new approach to the economic analysis of nonstationary time series and the business cycle. Econometrica **57**, 357–384 (1989)
11. Hardy, M.R.: A regime-switching model of long-term stock returns. N. Am. Actuar. J. **5**(2), 41–53 (2001)
12. Holz, D., Kling, A., Russ, J.: GMWB for life an analysis of lifelong withdrawal guarantees. Zeitschrift für die gesamte Versicherungswissenschaft **101**, 305–325 (2012)
13. Hull, J., White, A.: The pricing of options on assets with stochastic volatilities. J. Financ. **42**(2), 281–300 (1987)
14. Jin, Z., Wang, Y., Yin, G.: Numerical solutions of quantile hedging for guaranteed minimum death benefits under a regime-switching jump-diffusion formulation. J. Comput. Appl. Math. **235**(8), 2842–2860 (2011)
15. Johnson, C.: Numerical Solution of Partial Differential Equations by the Finite Element Method. Dover, Mineola (2009)
16. Kennedy, J.: Hedging contingent claims in markets with jumps. Ph.D. thesis, Applied Mathematics, University of Waterloo (2007)
17. Kling, A., Ruez, F., RUß, J.: The impact of stochastic volatility on pricing, hedging, and hedge efficiency of withdrawal benefit guarantees in variable annuities. ASTIN Bull. **41**(2), 511–545 (2011)

18. Levi, E.E.: Sulle equazioni lineari totalmente ellittiche alle derivate parziali. Rendiconti del circolo Matematico di Palermo **24**(1), 275–317 (1907)
19. Lin, X.S., Tan, K.S., Yang, H.: Pricing annuity guarantees under a regime-switching model. N. Am. Actuar. J. **13**(3), 316 (2009)
20. Merton, R.: Optimum consumption and portfolio rules in a continuous-time model. J. Econ. Theory **3**, 373–413 (1970)
21. Milevsky, M., Salisbury, T.: Financial valuation of guaranteed minimum withdrawal benefits. Insur.: Math. Econ. **38**(1), 21–38 (2006)
22. Moenig, T., Bauer, D.: Revisiting the risk-neutral approach to optimal policyholder behavior: a study of withdrawal guarantees in variable annuities. Working paper, Georgia State University (2011)
23. Ngai, A., Sherris, M.: Longevity risk management for life and variable annuities: the effectiveness of static hedging using longevity bonds and derivatives. Insur.: Math. Econ. **49**(1), 100–114 (2011)
24. Nishiyama, S., Smetters, K.: Consumption taxes and economic efficiency with idiosyncratic wage shocks. Technical report, National Bureau of Economic Research (2005)
25. O'Sullivan, C., Moloney, M.: The variance gamma scaled self-decomposable process in actuarial modelling. J. Bus. **75**, 305–332 (2010)
26. Pasdika, U., Wolff, J., Gen Re, MARC Life: coping with longevity: the new German Annuity Valuation table DAV 2004 R. In: The Living to 100 and Beyond Symposium, Orlando (2005)
27. Piscopo, G., Haberman, S.: The valuation of guaranteed lifelong withdrawal benefit options in variable annuity contracts and the impact of mortality risk. N. Am. Actuar. J. **15**(1), 59 (2011)
28. Rannacher, R.: Finite element solution of diffusion problems with irregular data. Numer. Math. **43**(2), 309–327 (1984)
29. Shah, P., Bertsimas, D.: An analysis of the guaranteed withdrawal benefits for life option. Working paper, MIT (2008)
30. Siu, T.K.: Fair valuation of participating policies with surrender options and regime switching. Insur.: Math. Econ. **37**(3), 533–552 (2005)
31. Windcliff, H., Forsyth, P.A., Vetzal, K.R.: Valuation of segregated funds: shout options with maturity extensions. Insur.: Math. Econ. **29**(1), 1–21 (2001)
32. Windcliff, H., Forsyth, P.A., Vetzal, K.R.: Analysis of the stability of the linear boundary condition for the Black-Scholes equation. J. Comput. Financ. **8**, 65–92 (2004)
33. Yuen, F.L., Yang, H.: Pricing Asian options and equity-indexed annuities with regime switching by the trinomial tree method. N. Am. Actuar. J. **14**(2), 256–277 (2010)
34. Zhou, X.Y., Yin, G.: Markowitz's mean-variance portfolio selection with regime switching: a continuous-time model. SIAM J. Control Optim. **42**(4), 1466–1482 (2003)

Chapter 7
A Stochastic Approximation Approach for Trend-Following Trading

Duy Nguyen, George Yin, and Qing Zhang

Abstract This work develops a feasible computation procedure for trend-following trading under a bull-bear switching market model. In the asset model, the drift of the stock price switches between two parameters corresponding to an uptrend (bull market) and a downtrend (bear market) according to a partially observable Markov chain. The objective is to buy and sell the underlying stock to maximize an expected return. It is shown in Dai et al. (SIAM J Financ Math 1:780–810, 2010; Optimal trend following trading rules. Working paper) that an optimal trading strategy can be obtained in terms of two threshold levels. Finding the threshold levels turns out to be a difficult task. In this paper, we develop a stochastic approximation algorithm to approximate the threshold levels. One of the main advantages of this approach is that one need not solve the associated HJB equations. We also establish the convergence of the algorithm and provide numerical examples to illustrate the results.

7.1 Introduction

This paper develops a numerical procedure to approximate the trend-following trading strategy. The idea of trend following is to go long at the beginning of a bull market and exit when the trend reverses. A trend-following trader purchases shares when the prices go up to a certain level from the bottom and sells when the prices start to fall.

There is an extensive literature devoted to trading and portfolio management strategies. For instance, Merton [11] pioneered the continuous-time portfolio

---

D. Nguyen • Q. Zhang (✉)
Department of Mathematics, The University of Georgia, Athens, GA 30602, USA
e-mail: dnguyen@math.uga.edu; qingz@math.uga.edu

G. Yin
Department of Mathematics, Wayne State University, Detroit, MI 48202, USA
e-mail: gyin@math.wayne.edu

selection with utility maximization, which was subsequently extended to incorporate transaction costs by Magill and Constantinides [10]; see also Davis and Norman [5], Shreve and Soner [12], Liu and Loewenstein [9], Dai and Yi [2], and references therein. Zhang and Zhang [17] obtained an optimal trading strategy in a mean reverting market. Other work relevant to mean reverting strategies can be found in Dai et al. [1], Song et al. [13], Zervos et al. [16], among others. In connection with trend-following trading, Iwarere and Barmish [7] developed an confidence interval approach to identify trading opportunities. Dai et al. [3] provided a theoretical justification of a trend-following strategy in a bull-bear switching market. They employed the conditional probability in the bull market to generate trade signals. However, the work imposed a somewhat unrealistic assumption: Only one share of stock is allowed to be traded. Such restriction was removed in Dai et al. [4]. It is shown in these papers that the optimal trading rule can be determined by two threshold levels. Buy if the conditional probability of bull market given up to date stock price is higher than the upper level and sell if it falls below the lower level. Note that in both [3, 4] partial differential equations are extensively used to character the optimal trading rules. Then a finite difference method is used to solve the associated HJB equations.

While the recent progress provided some useful results, the computational issue remains a real challenge. In this paper, we focus on computational aspect and aim to provide an easily implementable alternative. Instead of solving the complicated partial differential equations, we use a stochastic approximation approach to find the optimal threshold levels. We model the market trends using a geometric Brownian motion with regime switching. Two regimes are considered: the uptrend (bull market) and downtrend (bear market). The switching process is modeled as a two-state Markov chain which is not directly observable. In addition, we assume all funds are available to trade. Moreover, to keep things simple, we consider only one round trip trading (buying the stock and then selling it). The objective is to maximize the expected percentage gain. We construct a stochastic approximation algorithm and establish its convergence. In addition, we also carry out extensive Monte Carlo simulations and real market tests.

The rest of the paper is arranged as follows. We present the problem formulation in the next section. Section 7.3 is devoted to convergence analysis. Then in Sect. 7.4, we report numerical experiment results.

7.2 Problem Formulation

Let S_t be the asset price at time t satisfying the equation

$$\frac{dS_t}{S_t} = \mu(\alpha_t)dt + \sigma dW_t, \ S_0 = x, \qquad (7.1)$$

where $\mu(i) = \mu_i, i = 1, 2$, is the expected rate of return, $\sigma > 0$ the volatility, W_t a standard Brownian motion, and α_t a Markov chain with state space $\mathscr{M} = \{1, 2\}$. The process α_t represents the market mode at each time t. That is, $\alpha_t = 1$ indicates a bull market and $\alpha_t = 2$ a bear market. Naturally, we assume that $\mu_1 > 0 > \mu_2$. Let $Q = \begin{pmatrix} -\lambda_1 & \lambda_1 \\ \lambda_2 & -\lambda_2 \end{pmatrix}$ be the generator of α_t for some $\lambda_1 > 0$ and $\lambda_2 > 0$. Throughout the paper, we assume that $\{\alpha_t\}$ and $\{W_t\}$ are independent.

Let $\mathscr{T}$ denote a finite time and let $0 \leq \tau \leq \nu \leq \mathscr{T}$ be stopping times so that one buys the stock at τ and sells at ν. In this paper, we consider only one round-trip trading (one buying and selling cycle). In addition, we trade all available funds. Our objective is to choose τ and ν to maximize the reward function

$$J(\tau, \nu) = E\left[\log\left(\frac{S_\nu}{S_\tau}\right)\right]. \tag{7.2}$$

In this paper, we assume that only the stock price S_t is observable at time t. The market trend α_t is not directly observable, which leads to a control problem with incomplete information. It is necessary to convert the problem into a completely observable one. One way to achieve this is to use the Wonham filter [14]; see also Elliott et al. [6].

Let $p_t = P(\alpha_t = 1 | S_s : 0 \leq s \leq t)$ denote the conditional probability of $\alpha_t = 1$ (bull market) given the stock price up to time t. Then the Wonham filter in terms of p_t satisfies the following stochastic differential equation (SDE)

$$dp_t = [-(\lambda_1 + \lambda_2)p_t + \lambda_2]dt + \frac{(\mu_1 - \mu_2)p_t(1-p_t)}{\sigma}d\hat{W}_t, \tag{7.3}$$

where $d\hat{W}_t$ is an innovation process given by

$$d\hat{W}_t = \frac{d\log(S_t) - [(\mu_1 - \mu_2)p_t + \mu_2 - \sigma^2/2]dt}{\sigma}. \tag{7.4}$$

To have a quick view of the dependence of p_t on (S_t, α_t), we take $\lambda_1 = 0.36$, $\lambda_2 = 2.53$, $\mu_1 = 0.18$, $\mu_2 = -0.77$, and $\sigma = 0.184$ and plot the sample path of randomly generated (S_t, α_t) and the corresponding p_t in Fig. 7.1.

In view of (7.4), we rewrite (7.1) in terms of p_t and $\hat{W}_t$:

$$dS_t = S_t[(\mu_1 - \mu_2)p_t + \mu_2]dt + S_t\sigma d\hat{W}_t.$$

It follows that

$$S_\nu = S_\tau \exp\left(\int_\tau^\nu [(\mu_1 - \mu_2)p_r + \mu_2 - \sigma^2/2]dr + \sigma(\hat{W}_\nu - \hat{W}_\tau)\right).$$

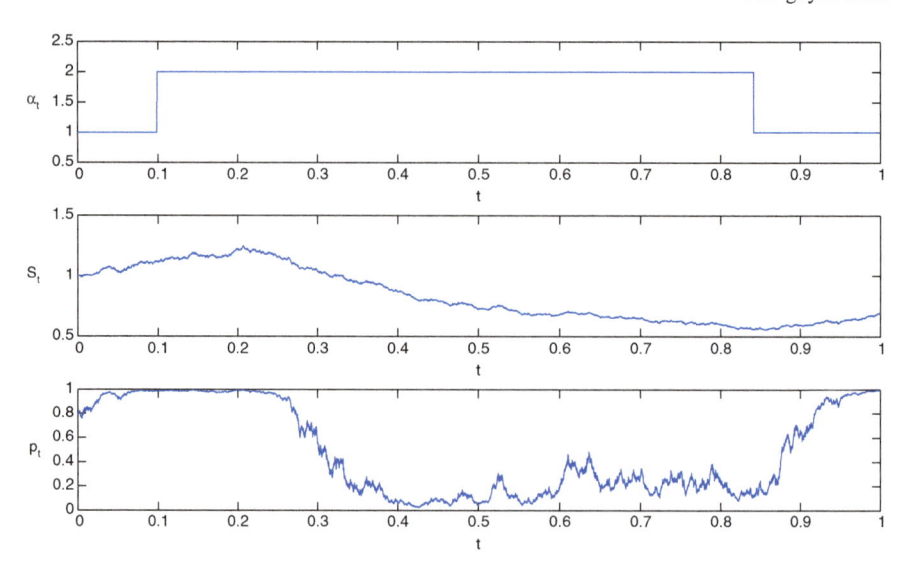

Fig. 7.1 A sample path of (S_t, α_t) and the corresponding p_t

Therefore, the reward in (7.2) can be written in terms of p_t and $\hat{W}_t$ as

$$\begin{aligned} J(\tau, v) &= E\left[\int_\tau^v [(\mu_1 - \mu_2)p_r + \mu_2 - \sigma^2/2]dr + \sigma(\hat{W}_v - \hat{W}_\tau)\right] \\ &= E\left[\int_\tau^v [(\mu_1 - \mu_2)p_r + \mu_2 - \sigma^2/2]dr\right]. \end{aligned} \quad (7.5)$$

In [3], it is shown that the optimal strategy is determined by threshold levels θ^b (buy level) and θ^s (sell level) with $\theta^s < \theta^b$. In this paper, we develop an alternative approach and use a stochastic approximation approach to find the optimal threshold level (θ^s, θ^b). Let

$$\phi(\theta) = \phi(\theta^b, \theta^s) = E\left(\int_{\tau(\theta)}^{v(\theta)} \left[(\mu_1 - \mu_2)p_r + \mu_2 - \frac{\sigma^2}{2}\right]dr\right), \quad (7.6)$$

where

$$\tau(\theta) = \inf\{t : p_t \geq \theta^b\},$$
$$v(\theta) = \inf\{t > \tau(\theta) : p_t \leq \theta^s\}.$$

The idea can be explained as follows. Although $\phi(\theta)$ is not observable due to the involvement of expectation, we can measure $\phi(\theta)$ with some noise. The simplest of such measurement may be written as $\phi(\theta) + \chi(\xi, \theta)$ where $\chi(\xi, \theta)$ is the noise. In lieu of such a simple form, we consider a much more general form in that the noise corrupted observation or measurement of $\phi(\theta)$ is $\tilde{\phi}(\theta, \xi)$. That is, nonadditive noise is allowed and complex nonlinear function form can be incorporated in our formulation. We also assume that $E_\xi[\tilde{\phi}(\theta,\xi)] = \phi(\theta)$ and for the moment let us assume that $\tilde{\phi}(\theta,\xi)$ depends smoothly on θ.

To proceed, we describe the stochastic approximation procedure and illustrate the use of $\tilde{\phi}(\theta,\xi)$.

0. Initialization: Choose an initial threshold estimate $\theta_0 = (\theta_0^b, \theta_0^s)$.
1. Simulation: Generate $\{p_t : t \leq \mathcal{T}\}$ using (7.3)
2. Iteration: With $n > 0$ and $\theta_n = (\theta_n^b, \theta_n^s)$ computed, carry out one step stochastic approximation to find updated threshold $\theta_{n+1} = (\theta_{n+1}^b, \theta_{n+1}^s)$. Let $e_1 = (1,0)$, $e_2 = (0,1)$, $c_n = 1/n^{1/6}$, and $\xi_{n,i}^\pm$ are noise sequences.

   (a) Find $\tau(\theta_n + c_n e_1) < v(\theta_n)$, compute $\tilde{\phi}(\theta_n + c_n e_1, \xi_{n,1}^+)$.
   (b) Find $\tau(\theta_n - c_n e_1) < v(\theta_n)$, compute $\tilde{\phi}(\theta_n - c_n e_1, \xi_{n,1}^-)$.
   (c) Find $\tau(\theta_n) < v(\theta_n + c_n e_2)$, compute $\tilde{\phi}(\theta_n + c_n e_2, \xi_{n,2}^+)$.
   (d) Find $\tau(\theta_n) < v(\theta_n - c_n e_2)$, compute $\tilde{\phi}(\theta_n - c_n e_2, \xi_{n,2}^-)$.
   (e) Use (a)–(d), find the gradient estimate $\nabla\tilde{\phi}(\theta_n, \xi_n) = (\nabla^i\tilde{\phi}(\theta_n, \xi_n))$ of $\phi(\theta)$ by

   $$\nabla^i\tilde{\phi}(\theta_n, \xi_n) = \frac{1}{2c_n}[\tilde{\phi}(\theta_n + c_n e_i, \xi_{n,i}^+) - \tilde{\phi}(\theta_n - c_n e_i, \xi_{n,i}^-)], \text{ for } i = 1,2.$$

   (f) Update one step the parameter estimate by using stochastic approximation method (SA) given by

   $$\theta_{n+1} = \theta_n + \frac{1}{n}\nabla\tilde{\phi}(\theta_n, \xi_n). \tag{7.7}$$

3. Repeat step 2 with $n \leftarrow n+1$ until $\|\theta_n - \theta_{n+1}\|_2 < \varepsilon$ with a prescribed tolerance level ε or with $n = N$ for some large N.

To proceed, we use the techniques developed in [8] to analyze the algorithm.

7.3 Asymptotic Properties

In the convergence analysis, we use the idea that on each "small" interval, the noise ξ varies much faster than the "state" θ. Thus with θ "fixed", the noise will be eventually averaged out resulting in an averaged system that can be characterize by a system of ordinary differential equations. Define

$$t_n = \sum_{j=1}^n \frac{1}{j},$$
$$m(t) = \max\{n : t_n \leq t\},$$
$$\theta^0 = \theta_n, \text{ for } t \in [t_n, t_{n+1}) \text{ and } \theta^n(t) = \theta^0(t_n + t).$$

Note that $\theta^0(\cdot)$ is a piecewise constant interpolation of θ_n on the interval $[t_n, t_{n+1})$ and $\theta^n(\cdot)$ is its shift. Next, for $i = 1, 2$, we define

$$b_n^i = \frac{\phi(\theta_n + c_n e_i) - \phi(\theta_n - c_n e_i)}{2c_n} - \frac{\partial \phi(\theta_n)}{\partial \theta^i},$$

$$\rho_n^i = [\tilde{\phi}(\theta_n + c_n e_i, \xi_{n,i}^+) - \tilde{\phi}(\theta_n - c_n e_i, \xi_{n,i}^-)] \\ - E_n[\tilde{\phi}(\theta_n + c_n e_i, \xi_{n,i}^+) - \tilde{\phi}(\theta_n - c_n e_i, \xi_{n,i}^-)],$$

$$\psi_n^i = [E_n \tilde{\phi}(\theta_n + c_n e_i, \xi_{n,i}^+) - \phi(\theta_n + c_n e_i)] \\ - [E_n \tilde{\phi}(\theta_n - c_n e_i, \xi_{n,i}^-) - \phi(\theta_n - c_n e_i)],$$

where E_n denotes the condition expectation with respect to $\mathscr{F}_n$ - the σ-algebra generated by $\{\xi_j^\pm : j < n\}$. Write $b_n = (b_n^1, b_n^2)'$. In the above, ψ_n^i and b_n^i for $i = 1, 2$ represent the noise and bias, and $\{\rho_n = (\rho_n^1, \rho_n^2)'\}$ is a martingale difference sequence. It is also reasonable to assume that after taking the conditional expectation, the resulting function is smooth. Thus we have separate the noise into two part, uncorrelated noise $\{\rho_n\}$ and correlated noise $\{\psi_n = (\psi_n^1, \psi_n^2)'\}$. With these notations, the algorithm in (7.7) can be written as following

$$\theta_{n+1} = \theta_n + \frac{1}{n}\nabla \phi(\theta_n) + \frac{1}{n}\frac{\psi_n}{2c_n} + \frac{1}{n}\frac{\rho_n}{2c_n} + \frac{b_n}{n}. \tag{7.8}$$

Note that in fact, both ψ_n and b_n are θ dependent. In what follow, when it is needed, we write $\psi_n = \psi(\theta_n, \xi_n)$ where ξ_n includes $\xi_n^\pm$.

To proceed with the analysis of the algorithm (7.7), we assume the following conditions hold.

(A1) For each ξ, $\tilde{\phi}(\cdot, \xi)$ is a continuous function.
(A2) For each $0 < N < \infty$ and each $0 < T < \infty$, the set $\{\sup_{|\theta| \leq N} \tilde{\phi}(\theta, \xi_n) : n \leq m(T)\}$ is uniformly integrable.
(A3) The sequences $\{\xi_n^\pm\}$ are bounded. For each θ in a bounded set and for each $0 < T < \infty$

$$\sup_n \sum_{j=n}^{m(T+t_n)-1} \frac{1}{j}\sqrt{E\left|E_j\frac{\psi_j(\theta, \xi_j)}{2c_j}\right|} < \infty, \quad \lim_n \sup_{0 \leq i \leq m(T+t_n)} E|\psi_i^n| = 0,$$

where for $i \leq m(T + t_n)$,

$$\psi_i^n = (n+i)\sum_{j=n+i}^{m(T+t_n)+i-1} \frac{1}{2jc_j}E_{n+i}[\psi_j(\theta_{n+i+1}, \xi_j) - \psi_j(\theta_{n+i}, \xi_j)].$$

(A4) The second derivative of $\phi(\cdot)$ is continuous.

Before we proceed to the analysis of the algorithm, let us recall the definition of weak convergence. We say a sequence of $\mathbb{R}^2$-valued random variables Z_n converges weakly to a random variable Z iff for any bounded and continuous function $h(\cdot)$,

$$Eh(Z_n) \to Eh(Z) \text{ as } n \to \infty,$$

7 A Stochastic Approximation Approach for Trend-Following Trading

and Z_n is said to be tight iff for each $\eta > 0$, there exist a compact set $K_\eta \subset \mathbb{R}^2$ such that
$$P(Z_n \in K_\eta) \geq 1 - \eta \text{ for all } n.$$

Let $S_v = \{x \in \mathbb{R}^r : |x| < v\}$ be the v-sphere. We say that the process $\zeta^{n,v}$ is an v-truncation of the process ζ^n if $\zeta^{n,v}(t) = \zeta(t)$ up until the first exist from S_v and
$$\lim_{m \to \infty} \limsup_n P\left(\sup_{t \leq T} |\zeta^{n,v}| \geq m\right) = 0 \text{ for each } T < \infty.$$

Next let q^v be a smooth function such that $q^v(\theta) = 1$ when $|\theta| \leq v$ and $q^v(\theta) = 0$ when $|\theta| \geq v+1$. We define $\{\theta_n^v\}$ recursively by $\theta_1^v = \theta_1$ and
$$\theta_{n+1}^v = \theta_n^v + \left[\frac{1}{n}\nabla\phi(\theta_n) + \frac{1}{n}\frac{\psi_n}{2c_n} + \frac{1}{n}\frac{\rho_n}{2c_n} + \frac{b_n}{n}\right]q^v(\theta_n^v), \quad n \geq 1. \quad (7.9)$$

Define the interpolation of θ_n^v as $\theta^{0,v} = \theta_n^v$ for $t \in [t_n, t_{n+1})$ and $\theta^{n,v}(t) = \theta^{0,v}(t + t_n)$. Thus $\theta^{n,v}$ is a v-truncation of θ^n (see [8, p. 278]).

In what follows, we first show that the truncated process $\{\theta^{n,v}(\cdot)\}$ is tight in $D^2[0,\infty)$ – the space of $\mathbb{R}^2$-valued functions that are right continuous, have left-hand limits, and endowed with the Skorohod topology. We then obtain the weak convergence of $\theta^{n,v}$ and characterize the limit as a solution of an ODE. Finally, by letting $v \to \infty$, we conclude that the untruncated process $\{\theta^n(\cdot)\}$ also converges.

Lemma 7.1. *Under conditions* (A1)–(A4),
$$\sum_{k=m(t+t_n)}^{m(t+s+t_n)-1} \frac{1}{k}\left[\frac{\psi_k + \rho_k}{2c_k}\right]q^v(\theta_k^v) \text{ converges in probability as } n \to \infty.$$

The convergence is uniform in t.

Proof. First recall that $\{\rho_n\}$ is a martingale difference sequence. The orthogonality implies that
$$E_{m(t+t_n)}\left|\sum_{k=m(t+t_n)}^{m(t+s+t_n)-1} \frac{1}{k}\frac{\rho_k}{2c_k}q^v(\theta_k^v)\right|^2 \\ \leq K \sum_{k=m(t+t_n)}^{m(t+s+t_n)-1} \frac{1}{k^{10/6}}E_{m(t+t_n)}|\rho_k|^2 \to 0 \text{ as } n \to \infty. \quad (7.10)$$

Using the same technique as in [15, Lemma 3.1], we can show that
$$E_{m(t+t_n)}\left|\sum_{k=m(t+t_n)}^{m(t+s+t_n)-1} \frac{1}{k}\frac{\psi_k}{2c_k}q^v(\theta_k^v)\right|^2 \to 0 \text{ in mean as } n \to \infty. \quad (7.11)$$

Thus the lemma is proved. $\square$

Theorem 7.1. *Under conditions* (A1)–(A4), $\theta^n(\cdot)$ *converges weakly to* $\theta(\cdot)$, *which is a solution of*

$$\dot{\theta} = \nabla \phi(\theta)$$

provided that the ordinary differential equation above has a unique solution for each initial condition.

Proof. The proof of the weak convergence requires that we first verify the sequence $\{\theta^n(\cdot)\}$ is tight. By virtue of the Prohorov's theorem which states that on a complete separable metric space, tightness is equivalent to sequentially compact, can then extract a convergent subsequence. We next characterize the limit by showing that it is the solution of a martingale problem with a desired operator. The proof is divided into three steps.

(i) Tightness of $\{\theta^{n,\nu}(\cdot)\}$. Note that

$$\theta^{n,\nu}(t) = \tilde{\theta}^{n,\nu}(t) + \sum_{k=1}^{m(t+t_n)-1} \frac{1}{k} \left[\frac{\psi_k + \rho_k}{2c_k} \right] q^\nu(\theta_k^\nu).$$

For any $\eta > 0$, let $0 \leq t$, and $0 \leq s \leq \eta$. We have

$$E_{m(t+t_n)} |\tilde{\theta}^{n,\nu}(t+s) - \tilde{\theta}^{n,\nu}(t)|^2$$
$$\leq K E_{m(t+t_n)} \left| \sum_{k=m(t+t_n)}^{m(t+s+t_n)-1} \frac{1}{k} [\nabla \phi(\theta_k^\nu) + b_k] q^\nu(\theta_k^\nu) \right|^2,$$

where $E_{m(t+t_n)}$ denotes the conditional expectation on the σ-algebra generated by $\mathscr{F}_{m(t+t_n)}$. By the mean of the boundedness of $\{\theta_k^\nu\}$, we have

$$E_{m(t+t_n)} \left| \sum_{k=m(t+t_n)}^{m(t+s+t_n)-1} \frac{1}{k} \nabla \phi(\theta_k^\nu) q^\nu(\theta_k^\nu) \right|^2 \leq K[(t+s+t_n) - (t+t_n)]^2 \leq K\eta^2. \tag{7.12}$$

Similarly

$$E_{m(t+t_n)} \left| \sum_{k=m(t+t_n)}^{m(t+s+t_n)-1} \frac{1}{k} b_k q^\nu(\theta_k^\nu) \right|^2 \leq K\eta^2. \tag{7.13}$$

Using (7.11)–(7.13), taking $\limsup_n$ and then followed by letting $\eta \to 0$, we obtain

$$\lim_{\eta \to 0} \limsup_{n \to \infty} \sup_{0 \leq s \leq \eta} E[E_{m(t+t_n)} |\theta^{n,\nu}(t+s) - \theta^{n,\nu}(t)|^2] = 0 \tag{7.14}$$

Combining (7.14) with the result of Lemma 7.1, the tightness criterion in [8, p. 47] yields the tightness of $\{\theta^{n,\nu}(\cdot)\}$.

(ii) Characterization of the limit process. Sine $\{\theta^{n,\nu}(\cdot)\}$ is tight by the Prohorov's theorem, we can extract a convergent subsequence which is sill denoted by $\{\theta^{n,\nu}(\cdot)\}$ for notational simplicity. Let $\theta^\nu(\cdot)$ be its limits. We characterize the limit process. For each $t, s \geq 0$, Lemma 7.1 implies that

$$\theta^{n,v}(t+s) - \theta^{n,v}(t) = \sum_{k=m(t+t_n)}^{m(t+s+t_n)-1} \frac{1}{k}[\nabla \phi(\theta_k^{n,v}) + b_k]q^v(\theta_k^v) + o(1), \quad (7.15)$$

where $o(1) \to 0$ in probability uniformly in t. By a truncated Taylor expansion and (A4), it can be verified that

$$b_n q^v(\theta_n^{n,v}) = O(c_n^2/c_n) = O(c_n).$$

Thus

$$E \left| \sum_{k=m(t+t_n)}^{m(t+s+t_n)-1} \frac{1}{k} b_k q^v(\theta_k^v) \right| \to 0 \text{ as } n \to \infty \text{ uniformly in } t.$$

Therefore

$$\sum_{k=m(t+t_n)}^{m(t+s+t_n)-1} \frac{1}{k} b_k q^v(\theta_k^v) \to 0 \text{ as } n \to \infty \text{ in probability uniformly in } t \quad (7.16)$$

Note that there is an increasing sequence of positive integers $\{m_l\}$ and a deceasing sequence of positive real numbers $\{\delta_l\}$ such that $m(t+t_n) \leq m_l \leq m_{l+1} - 1 \leq m(t+t_n+s) - 1$ for any $t, s > 0$ and that

$$\frac{1}{\delta_l} \sum_{j=m_l}^{m_{l+1}-1} \frac{1}{j} \to 1 \text{ as } n \to \infty.$$

Also note that in the choice above, δ_l depends on n so we could write it as δ_l^n. However, to keep the notation simple, we write it as δ_l in what follows. Then we can write the right-hand side of (7.15) as follow

$$\sum_{k=m(t+t_n)}^{m(t+s+t_n)-1} \frac{1}{k} \nabla \phi(\theta_k^{n,v}) q^v(\theta_k^v)$$

$$= \sum_{l: m(t+t_n) \leq m_l \leq m_{l+1}-1 \leq m(t+t_n+s)-1} \sum_{k=m_l}^{m_{l+1}-1} \frac{1}{k} \nabla \phi(\theta_k^{n,v}) q^v(\theta_k^v)$$

$$= \sum_{l: m(t+t_n) \leq m_l \leq m_{l+1}-1 \leq m(t+t_n+s)-1} \sum_{k=m_l}^{m_{l+1}-1} \frac{1}{k} \nabla \phi(\theta_{m_l}^{n,v}) q^v(\theta_{m_l}^v) + o(1)$$

$$= \sum_{l: m(t+t_n) \leq m_l \leq m_{l+1}-1 \leq m(t+t_n+s)-1} \delta_l \frac{1}{\delta_l} \sum_{k=m_l}^{m_{l+1}-1} \frac{1}{k} \nabla \phi(\theta_k^{n,v}) q^v(\theta_k^v) + o(1)$$

$$= \sum_{l: m(t+t_n) \leq m_l \leq m_{l+1}-1 \leq m(t+t_n+s)-1} \delta_l \nabla \phi(\theta_{m_l}^{n,v}) q^v(\theta_{m_l}^v) + o(1),$$

$$(7.17)$$

where $o(1) \to 0$ in probability uniformly in t. Using (7.15)–(7.17) for any bounded and continuous function $h(\cdot)$, continuous differentiable function $f(\cdot)$, any positive integer κ, any $t_i \leq t$ with $i \leq \kappa$, we can show that there is a sequence $\{\tilde{e}_n\}$ of real numbers such that $\tilde{e}_n \to 0$ as $n \to \infty$, and that

$$\begin{aligned}
& Eh(\theta^{n,v}(t_i) : i \leq \kappa)[f(\theta^{n,v}(t+s)) - f(\theta^{n,v}(t))] \\
&= Eh(\theta^{n,v}(t_i) : i \leq \kappa) \\
&\quad \times \left[\sum_{l:m(t+t_n) \leq m_l \leq m_{l+1}-1 \leq m(t+t_n+s)-1} \delta_l \nabla f'(\theta^{n,v}_{m_l}) \nabla \phi(\theta^{n,v}_{m_l}) q^v(\theta^v_{m_l}) \right] + \tilde{e}_n \\
&\to Eh(\theta^{n,v}(t_i) : i \leq \kappa) \int_t^{t+s} \nabla f'(\theta^n(u)) \nabla \phi(\theta^n(u)) q^v(\theta^v(u)) du \text{ as } n \to \infty.
\end{aligned} \tag{7.18}$$

In the above, we have used the weak convergence of $\theta^{n,v}(\cdot)$ to $\theta^v(\cdot)$ and Skorohod representation. On the other hand, the weak convergence and the Skorohod representation yield that

$$\begin{aligned}
& Eh(\theta^{n,v}(t_i) : i \leq \kappa)[f(\theta^{n,v}(t+s)) - f(\theta^{n,v}(t))] \\
&\to Eh(\theta^v(t_i) : i \leq \kappa)[f(\theta^v(t+s)) - f(\theta^v(t))] \text{ as } n \to \infty.
\end{aligned} \tag{7.19}$$

Then (7.17) and (7.18) lead to the fact that $\theta^v(\cdot)$ is solution of the martingale problem with the operator given by

$$L^v f(\theta^v) = \nabla f'(\theta^v) \nabla \phi(\theta^v) q^v(\theta),$$

or equivalently, $\theta^v(\cdot)$ is the solution of the truncated differential equation

$$\dot{\theta}^v = \nabla \phi(\theta^v) q^v(\theta^v).$$

(iii) The convergence of the untruncated sequence $\{\theta^n(\cdot)\}$. We so far have worked with a fixed but arbitrary integer v. In this step, we examine the asymptotic properties as $v \to \infty$. The details are similar to that of [8, p. 284] and thus omitted for brevity. □

7.4 Numerical Examples

In this section, we demonstrate the performance of our algorithm. First, we carry out Monte Carlo simulations to illustrate the method. Then we test how the algorithm works in real market.

Using the SDE in (7.3), we rewrite p_t in term of the stock price S_t by

$$dp_t = f(p_t)dt + \frac{(\mu_1 - \mu_2)p_t(1 - p_t)}{\sigma^2} d\log(S_t), \tag{7.20}$$

where f is a third-order polynomial of p_t given by

$$f(p) = -(\lambda_1 + \lambda_2)p + \lambda_2 - \frac{(\mu_1 - \mu_2)p(1-p)((\mu_1 - \mu_2)p + \mu_2 - \sigma^2/2)}{\sigma^2}.$$

It is easy to check that $f(0) = \lambda_2 > 0$ and $f(1) = -\lambda_1 < 0$. Furthermore, it is readily seen that $f(p)$ approaches $\pm\infty$ as $|p| \to \infty$. Therefore, f has exactly one root $\xi \in (0,1)$. When the stock prices stay as a constant, p_t is attracted to ξ. This attractor is unbiased choice for p_0. Moreover, since $(\mu_1 - \mu_2)p_t(1 - p_t)/\sigma^2 \geq 0$, p_t moves in the same direction as the stock prices. This is also intuitive since the movement of stock price forms trends. We will estimate p_t simply by replacing the differential equation in (7.20) with a difference using the trading day as the step size on the finite horizon $[0, Ndt]$

$$p_{t+1} = p_t + f(p_t)dt + \frac{(\mu_1 - \mu_2)p_t(1-p_t)}{\sigma^2} \log\left(\frac{S_{t+1}}{S_t}\right),$$

where $t = 0, dt, 2dt, \ldots, Ndt$.
Note that it is possible that $p_t > 1$ or $p_t < 0$ for some t. To keep $p_t \in [0,1]$, we truncate the process and follow the truncated equation instead:

$$p_{t+1} = \min\left(\max\left(p_t + f(p_t)dt + \frac{(\mu_1 - \mu_2)p_t(1-p_t)}{\sigma^2} \log(\frac{S_{t+1}}{S_t}), 0\right), 1\right).$$

Example 7.1 (Monte Carlo simulations). In this example, we choose the parameters with the following specifications:

Table 7.1 Monte Carlo parameter values

λ_1	λ_2	μ_1	μ_2	σ	K	ρ
0.36	2.53	0.18	−0.77	0.184	0.001	0.0679

These parameters were used in [4] for DJIA index. We generate sample paths of S_t and p_t using Eqs. (7.1) and (7.3) with $N = 50{,}000$ steps, step size $dt = 1/N$, and $T = 1$. One sample path of S_t is given in Fig. 7.2. The corresponding p_t is given in Fig. 7.3 and (θ_n^b, θ_n^s) in Fig. 7.4. We start with initial guess $(\theta_0^b, \theta_0^s) = (0.9, 0.7)$ and perform stochastic approximation (7.7) with 1,000 iterations. We then use different seeds to perform the algorithm using 500 replications. After taking the sample mean we obtain $(\theta^b, \theta^s) = (0.916, 0.725)$, and $\phi = 0.0401$ which is equivalent to 4.01% gain obtained in one round-trip transaction.

We next perturb the parameters to see the dependence of the threshold level. Our tests show that the threshold (θ^b, θ^a) is not sensitive to changes in these parameters. These are summarized in Table 7.2.

Next we simulate this trend-following strategy and compare it with the buy-and-hold strategy by using a large number of sample paths. Again, we use the parameter values given in Table 7.1. The average returns of the trend-following (TF) strategy on one unit invested on simulated paths are listed on Table 7.3 together with the average number of trades on each path. We also list the average return of the buy and hold (BH) strategy for comparison. These results are given in Table 7.3.

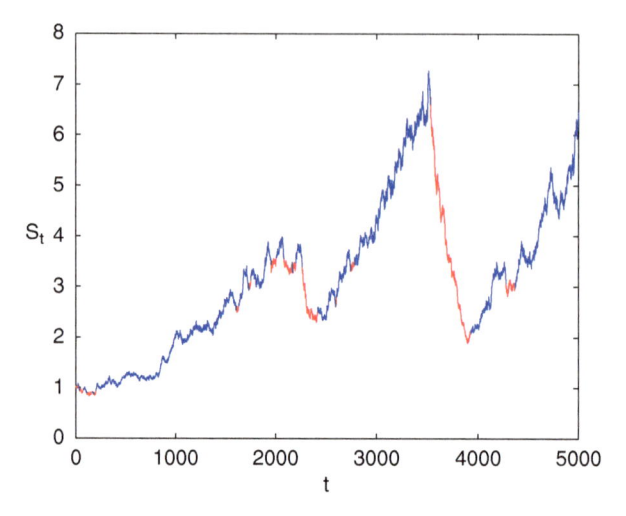

Fig. 7.2 A random sample path of stock prices

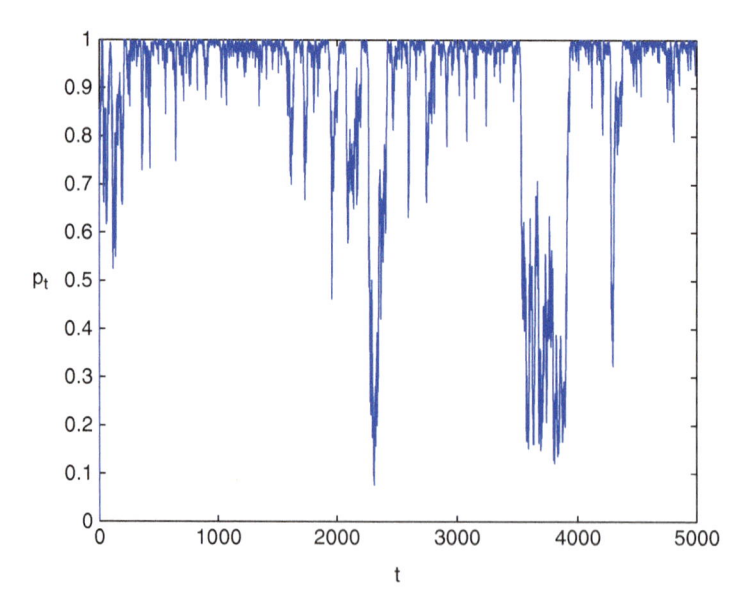

Fig. 7.3 A sample path of p_t

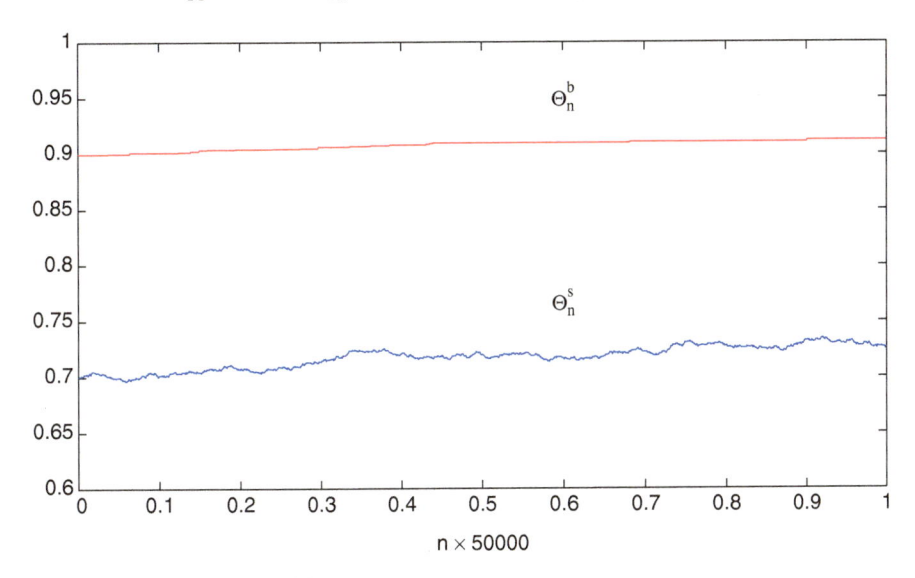

Fig. 7.4 Buy and sell thresholds

Table 7.2 Thresholds with different parameters

λ_1	λ_2	μ_1	μ_2	σ	θ^b	θ^s
0.36	2.53	0.18	−0.77	0.184	0.916	0.725
0.36	2.53	0.18	−0.77	**0.174**	0.917	0.725
0.36	2.53	0.18	−0.77	**0.194**	0.916	0.727
0.36	2.53	**0.17**	**−0.71**	0.184	0.915	0.730
0.36	2.53	**0.19**	**−0.83**	0.184	0.917	0.731
0.36	**2**	0.18	−0.77	0.184	0.915	0.724
0.36	**3**	0.18	−0.77	0.184	0.916	0.725
0.3	2.53	0.18	−0.77	0.184	0.915	0.733
0.42	2.53	0.18	−0.77	0.184	0.916	0.718

Table 7.3 Monte Carlo simulation: 20 years

No. of sample paths	TF (%)	BH (%)	No. of trades
10,000	11.51	4.64	36.53
50,000	11.55	4.71	36.58
100,000	11.51	4.67	36.64

We next investigate the performance of our strategy on each sample path. We use the parameters in Table 7.1 and the buy-sell threshold $(\theta^b, \theta^s) = (0.916, 0.725)$. The result of simulation is collected in Table 7.4. We can see that the result is very sensitive to each individual sample path, but our strategy clearly outperforms the buy and hold strategy.

Table 7.4 Ten single-path simulations

Trend following (%)	Buy and hold (%)	No. of trades
11.61	4.42	22
7.10	2.40	40
6.70	2.92	46
5.50	2.11	50
13.89	6.31	32
14.72	9.32	30
13.50	10.70	40
12.78	5.35	32
9.96	6.56	40
12.33	3.70	32

Next, we show that the trend-following strategy is not sensitive to small changes in these thresholds by examining the dependence of the performance on the threshold level (θ^b, θ^s). The results are summarized in Table 7.5. We can see from Table 7.5 that shifting the thresholds has little impact on the performance.

Table 7.5 Shifting the thresholds

θ^b	θ^s	TF (%)	BH (%)	No. of trades
0.906	0.718	11.55	4.71	37.16
0.915	0.724	11.54	4.69	36.58
0.916	0.727	11.59	4.70	36.78
0.917	0.731	11.44	4.58	37.39
0.925	0.733	11.51	4.72	36.25

Example 7.2 (Real Market Tests). In this example, we examine how our algorithm works in the real markets. We test on the historical data of SSE (01/03/2000–12/30/2011) and SP500 (01/03/1972–12/30/2011). To run the SA algorithm, we need to determine the parameters. We regard a decline of more than 19 % as a bear market and a rally of 24 % or more as a bull market. Statistic of bull and bear markets for SSE index in 12 years from 2000 to 2011 are shown in Table 7.6. Similar for SP 500, see [3] for more details.

Table 7.6 Statistics of bull and bear markets

Index	λ_1	λ_2	μ_1	μ_2	σ
SSE(00-11)	1	1	1	−1	0.253
SP500(72-11)	0.353	2.208	0.196	−0.616	0.173

The testing results for trading SSE and SP 500 are summarized in Table 7.7. The annual return with trend following, and buy and hold strategy is collected along with annual Sharpe ratio. Using our trend-following trading strategy, e.g. SSE, one dollar invested in the beginning of 2000 returns $5.30 at the end of 2011 while buy and hold strategy returns $1.61 in the same period. The annual return for trend following and buy and hold are 14.90 and 4.05 %, respectively. The story for SP500 is similar: one dollar invested in the beginning of 1972 with trend following returns $15.68 which corresponds to the annual return 7.12% at the end of 2011 while buy hold returns $12.36 which corresponds to the annual return 6.49 % in the same period. The equity curves for SSE and SP 500 tests are given in Figs. 7.5 and 7.7, respectively. The natural logarithms of the corresponding indices (setting the initial to 1) are given in Figs. 7.6 and 7.8, respectively. For example, in Fig. 7.5, the equity curve is obtained by following our trend-following rule and invest $1 to begin with. Then in Fig. 7.7, we plot the equity curve beginning with $1 in the SP500 index. It is clear that the trend-following strategy not only outperforms the buy and hold strategy in total return, but also has a smoother equity curve which means higher Sharpe ratio. In addition, when sitting on cash, 3 % interest rate is added to the equity curve for the SSE and 6.79 % for the SP500.

Table 7.7 Testing results for trend following with real stock data

Index(time frame)	TF (%)	BH (%)	TF Sharpe ratio	BH Sharpe ratio	10 year bond (%)
SSE(00-11)	14.90	4.05	1.159	0.461	3.00
SP500(72-11)	7.12	6.49	0.1172	0.1052	6.79

Acknowledgements We thank Professor Qiji Zhu for fruitful discussions that led to the improvement of the computational results. This research was supported in part by the Army Research Office under W911NF-12-1-0223.

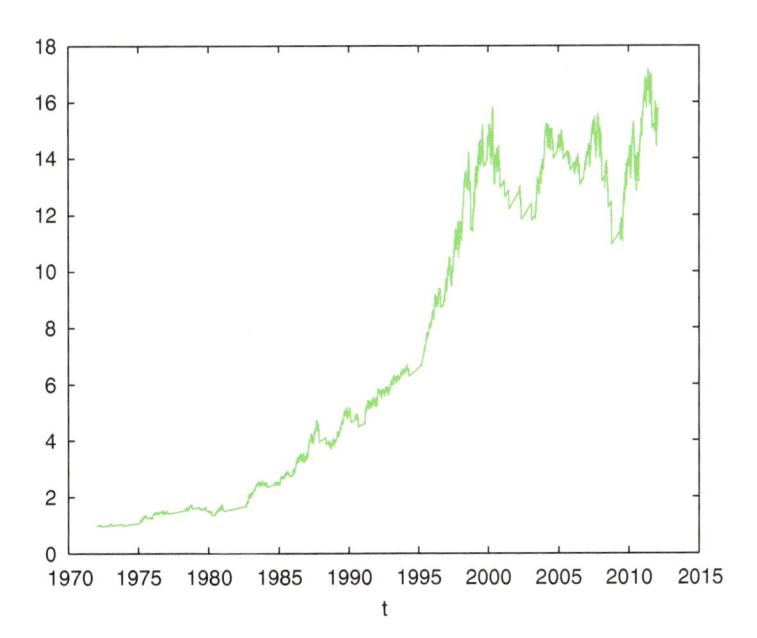

Fig. 7.5 The equity curve with trend following of SP500 between 1972 and 2011

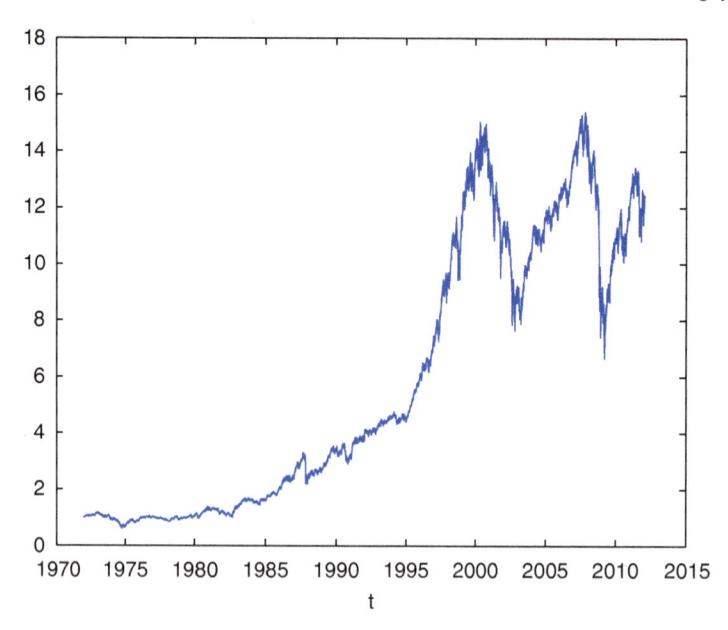

Fig. 7.6 The equity curve (buy and hold) of SP 500 between 1972 and 2011

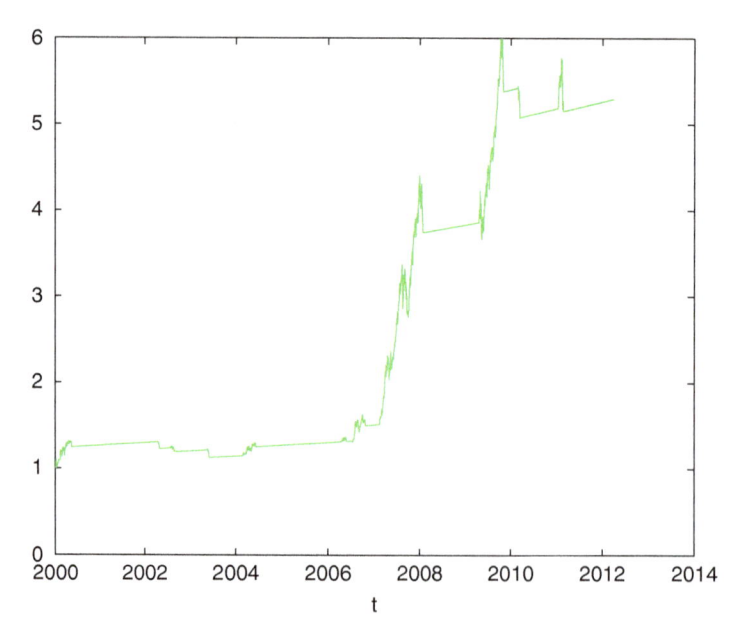

Fig. 7.7 The equity curve with trend following of SSE between 2000 and 2011

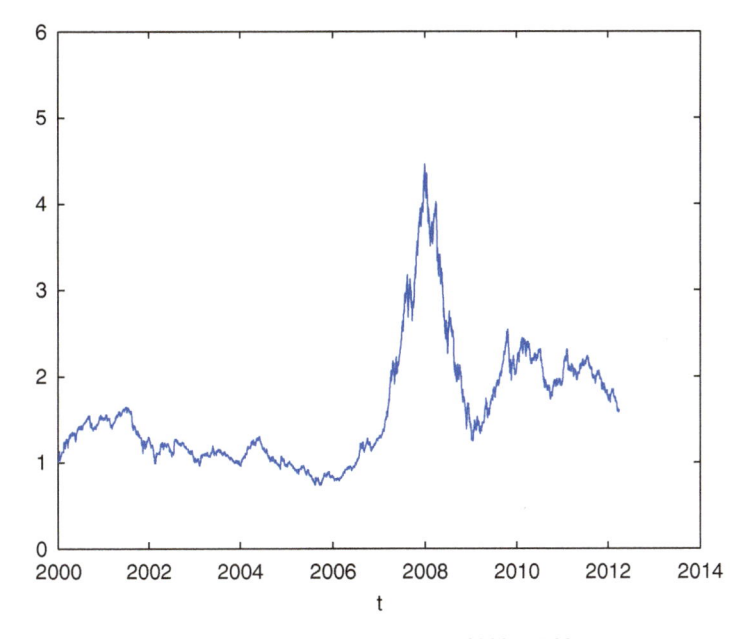

Fig. 7.8 The equity curve (buy and hold) of SSE between 2000 and 2011

References

1. Dai, M., Jin, H., Zhong, Y., Zhou, X.Y.: Buy low and sell high. In: Chiarella, C., Novikov, A. (eds.) Contemporary Quantitative Finance: Essays in Honor of Eckhard Platen, pp. 317–334. Springer, Berlin/London (2010)
2. Dai, M., Yi, F.: Finite horizontal optimal investment with transaction costs: a parabolic double obstacle problem. J. Differ. Equ. **246**, 1445–1469 (2009)
3. Dai, M., Zhang, Q., Zhu, Q.: Trend following trading under a regime switching model. SIAM J. Financ. Math. **1**, 780–810 (2010)
4. Dai, M., Zhang, Q., Zhu, Q.: Optimal trend following trading rules. Working paper
5. Davis, M.H.A., Norman, A.R.: Portfolio selection with transaction costs. Math. Oper. Res. **15**, 676–713 (1990)
6. Elliott, R.J., Aggoun, L., Moore, J.B.: Hidden Markov Models. Springer, New York (1995)
7. Iwarere, S., Barmish, B.R.: A confidence interval triggering method for stock trading via feedback control. In: Proceedings of American Control Conference, Baltimore (2010)
8. Kushner, H.J., Yin, G.: Stochastic Approximation and Recursive Algorithms and Applications, 2nd edn. Springer, New York (2003)
9. Liu, H., Loewenstein, M.: Optimal portfolio selection with transaction costs and finite horizons. Rev. Financ. Stud. **15**, 805–835 (2002)

10. Magill, M.J.P., Constantinides, G.M.: Portfolio selection with transaction costs. J. Econ. Theory **13**, 264–271 (1976)
11. Merton, R.C.: Optimal consumption and portfolio rules in a continuous time model. J. Econ. Theory **3**, 373–413 (1971)
12. Shreve, S.E., Soner, H.M.: Optimal investment and consumption with transaction costs. Ann. Appl. Probab. **4**, 609–692 (1994)
13. Song, Q.S., Yin, G., Zhang, Q.: Stochastic optimization methods for buyinglow and-selling-high strategies. Stoch. Anal. Appl. **27**, 523–542 (2009)
14. Woham, W.M.: Some applications of stochastic differential equations to optimal nonlinear filtering. SIAM J. Control **2**, 347–369 (1965)
15. Yin, G., Liu, R.H., Zhang, Q.: Recursive algorithms for stock liquidation: a stochastic optimization approach. SIAM J. Optim. **13**, 240–263 (2002)
16. Zervos, M., Johnsony, T.C., Alazemi, F.: Buy-low and sell-high investment strategies. Working paper (2011)
17. Zhang, H., Zhang, Q.: Trading a mean-reverting asset: buy low and sell high. Automatica **44**, 1511–1518 (2008)

Chapter 8
A Hidden Markov-Modulated Jump Diffusion Model for European Option Pricing

Tak Kuen Siu

Abstract The valuation of a European-style contingent claim is discussed in a hidden Markov regime-switching jump-diffusion market, where the evolution of a hidden economic state process over time is described by a continuous-time, finite-state, hidden Markov chain. A two-stage procedure is used to discuss the option valuation problem. Firstly filtering theory is employed to transform the original market with hidden quantities into a filtered market with complete observations. Then a generalized version of the Esscher transform based on a Doléan-Dade stochastic exponential is employed to select a pricing kernel in the filtered market. A partial-differential-integral equation for the price of a European-style option is presented.

8.1 Introduction

The valuation of contingent claims has long been a very important issue in the theory and practice of finance. The seminal works of Black and Scholes [2] and Merton [27] pioneered the development of option valuation theory and significantly advanced the practice of option valuation in the finance industry. The Black-Scholes-Merton option valuation is deeply immersed in the practice in the finance industry to the extent that it is rather uneasy to find a market practitioner in the City who has never heard of the Black-Scholes-Merton option pricing model. There may be two major reasons why the Black-Scholes-Merton option pricing model is so popular in the finance industry. Firstly, the pricing model is preference-free which means that the price of an option does not depend on the subjective view or risk preference of a market agent. Secondly, there is a closed-form expression for the price of a standard

---

T.K. Siu (✉)
Cass Business School, City University London, 106 Bunhill Row, EC1Y 8TZ London, UK

Faculty of Business and Economics, Department of Applied Finance and Actuarial Studies, Macquarie University, Sydney, NSW 2109, Australia
e-mail: Ken.Siu.1@city.ac.uk; ktksiu2005@gmail.com

European call option which is easy to implement in practice. Despite the popularity of the Black-Scholes-Merton option pricing model, its use has been constantly challenged by both academic researchers and market practitioners. Particularly, the geometric Brownian motion assumption underlying the model cannot explain some important empirical features of asset price dynamics, such as the heavy-tails of the return's distribution, the time-varying volatility, jumps and regime-switchings. Furthermore, the model cannot explain some systematic empirical features of option prices data, such as the implied volatility smile or smirk. There is a large amount of literature which extend the Black-Scholes-Merton model with a view to providing more realistic modeling frameworks for option valuation.

Markovian regime-switching models are one of the major classes of econometric models which can incorporate some stylized facts of asset price dynamics, such as the heavy-tails of the return's distribution, the time-varying volatility and regime-switchings. Though Markovian regime-switching models have a long history in engineering, their general philosophy and principle appeared in some pioneering works in statistics and econometrics. Quandt [31] and Goldfeld and Quandt [18] described nonlinearity in economic data using regime-switching regression models. Tong [37, 38] pioneered the fundamental principle of probability switching in nonlinear time series analysis. Hamilton [20] pioneered and popularized the use of Markov-switching autoregressive time series models in economics and econometrics. Recently much effort has been devoted to the use of Markovian regime-switching models for option valuation. A general belief is that Markovian regime-switching models can incorporate the impact of structural changes in economic conditions on asset prices which is particularly relevant for pricing long-dated options. Some works on option valuation in Markovian regime-switching models include Naik [30], Guo [19], Buffington and Elliott [3], Elliott et al. [9, 13], Siu [33, 34], Siu et al. [35], Elliott and Siu [10, 12], amongst others.

Jump-diffusion models are an important extension of the geometric Brownian motion for modeling asset price dynamics. This class of models captures jumps, or spikes, in returns due to extraordinary market events or news via jump components described by compound Poisson processes. There is a main difference between Markovian regime-switching models and jump-diffusion models. In a Markovian regime-switching model, there are jumps in the model coefficients corresponding to regime switches, but no jumps in the return process. In a jump-diffusion model, there are jumps in the return process, but no jumps in the model coefficients. Merton [28] pioneered the use of a jump-diffusion model for option valuation, where a compound Poisson model with lognormally distributed jump sizes was used to describe the jump component. Kou [25] pioneered option valuation under another jump-diffusion model for option valuation, where the jump amplitudes were exponentially distributed. It seems a general belief that jump-diffusion option valuation models may be suitable for pricing short-lived options by capturing the impact of sudden jumps in the return processes on option prices. Furthermore it is known that jump-diffusion option valuation models can incorporate some empirical features of asset price dynamics, such as jumps, heavy-tails of the return's distribution, and of option prices, such as implied volatility smiles. Bakshi et al. [1] provided a comprehensive empirical study on various option valuation models and found that incorporating both jumps and stochastic volatility is vital for pricing and internal consistency. Pan [32] and Liu et al. [26] provided theoretical and empirical supports on the use of jump-risk premia in explaining systematic empirical behavior of option prices data, respectively.

Both jump-diffusion models and Markovian regime-switching models play an important role in modeling asset price dynamics and option valuation. It may be of interest to combine the two classes of models and establish a class of "second-generation" models, namely, Markovian regime-switching jump-diffusion models. The rationale behind this initiative is to fuse the empirical advantages of the two classes of models so that a generalized option valuation model based on the wider class of "second-generation" models may be suitable for pricing short-lived and long-dated options traded in the finance and insurance industries, respectively. Indeed, this initiative was undertaken by some researchers, for example, Elliott et al. [13] and Siu et al. [35], where Markovian regime-switching jump-diffusion models were used to price financial options and participating life insurance policies, respectively. In both Elliott et al. [13] and Siu et al. [35], the modulating Markov chain governing the evolution of the "true" state of an underlying economy over time was assumed observable. However, in practice, it is difficult, if not impossible, to directly observe the "true" state of the underlying economy. Consequently it is of practical interest to consider a general situation where the modulating Markov chain is hidden or unobservable. In a recent paper, Elliott and Siu [12] considered a hidden Markovian regime-switching pure jump model for option valuation and addressed the corresponding filtering issue.

In this paper, the valuation of a European-style contingent claim in a hidden Markov regime-switching jump-diffusion market is discussed. In such market, the price process of an underlying risky security is described by a generalized jump-diffusion process with stochastic drift and jump intensity being modulated by a continuous-time, finite-state, hidden Markov chain whose states represent different states of a hidden economic environment. A two-stage procedure is used to discuss the option valuation problem. Firstly filtering theory is employed to transform the original market with hidden quantities into a filtered market where the hidden quantities in the original market are replaced by their filtered estimates. Consequently, the filtered market is one with complete observations. Then the option valuation problem is considered in the filtered market which is deemed to be incomplete due to the presence of price jumps in the market. We employ a generalized version of the Esscher transform based on a Doléan-Dade stochastic exponential to select a pricing kernel in the filtered market. A partial-differential-integral equation (PDIE) for the price of a European-style option is presented. This work is different from that in Elliott and Siu [12] in at least two aspects. Firstly, the price process of the risky share we consider here has a diffusion component. Secondly, in Elliott and Siu [12], the selection of a pricing kernel using a generalized version of the Esscher transform was first considered in a market with hidden observations. Filtering theory was then applied to transform the market into one with complete observations. This paper may partly serve as a brief review for some mathematical techniques which are hopefully relevant to pricing European-style options under a hidden regime-switching jump-diffusion model. In Siu [36], an American option pricing problem is considered under the hidden regime-switching jump-diffusion model.

The paper is structured as follows. The next section presents the price dynamics in the hidden Markov regime-switching jump-diffusion market. In Sect. 8.3, we discuss the use of filtering theory to turn the original market into the filtered market and give the filtering equations for the hidden Markov chain. The use of the generalized Esscher transform to select a pricing kernel and the derivation of the PDIE for the price of the European-style option are discussed in Sect. 8.4. The final section gives concluding remarks.

8.2 Hidden Regime-Switching Jump-Diffusion Market

A continuous-time financial market with two primitive investment securities, namely a bond and a share, is considered, where these securities can be traded continuously over time in a finite time horizon $\mathscr{T} := [0,T]$, where $T < \infty$. As usual, uncertainty is described by a complete probability space $(\Omega, \mathscr{F}, \mathbb{P})$ where $\mathbb{P}$ is a real-world probability measure. The following standard institutional assumptions for the continuous-time financial market are imposed:

1. The market is frictionless, (i.e., there are no transaction costs and taxes in trading the investment securities);
2. Securities are perfectly divisible, (i.e., any fractional units of the securities can be traded);
3. There is a single market interest rate for borrowing and lending;

To describe the evolution of the hidden economic state over time, we consider a continuous-time, finite-state, hidden Markov chain $\mathbf{X} := \{\mathbf{X}(t) | t \in \mathscr{T}\}$ on $(\Omega, \mathscr{F}, \mathbb{P})$. In practice, the "true" state of an underlying economy is not observable. Consequently, it makes practical sense to use a hidden Markov chain to represent different modes of the underlying economic environment. Using the convention in Elliott et al. [8], we identify the state space of the chain $\mathbf{X}$ with a finite set of standard unit vectors $\mathscr{E} := \{\mathbf{e}_1, \mathbf{e}_2, \cdots, \mathbf{e}_N\}$ in $\mathfrak{R}^N$, where the jth-component of $\mathbf{e}_i$ is the Kronecker product δ_{ij} for each $i, j = 1, 2, \cdots, N$. The space $\mathscr{E}$ is called the canonical state space of the chain $\mathbf{X}$. The statistical laws of the chain $\mathbf{X}$ are described by a family of rate matrices $\{\mathbf{A}(t) | t \in \mathscr{T}\}$, where $\mathbf{A}(t) := [a_{ij}(t)]_{i,j=1,2,\cdots,N}$ and $a_{ij}(t)$ is the instantaneous transition rate of the chain $\mathbf{X}$ from state $\mathbf{e}_i$ to state $\mathbf{e}_j$ at time t. So if $p^i(t) := \mathbb{P}(\mathbf{X}(t) = \mathbf{e}_i)$ and $\mathbf{p}(t) := (p^1(t), p^2(t), \cdots, p^N(t))' \in \mathfrak{R}^N$, then $\mathbf{p}(t)$ satisfies the following Kolmogorov forward equation:

$$\frac{d\mathbf{p}(t)}{dt} = \mathbf{A}(t)\mathbf{p}(t), \quad \mathbf{p}(0) = \mathrm{E}[\mathbf{X}(0)].$$

Here E is an expectation under $\mathbb{P}$.

Let $\mathbb{F}^{\mathbf{X}} := \{\mathscr{F}^{\mathbf{X}}(t)|t \in \mathscr{T}\}$ be the right-continuous, $\mathbb{P}$-complete, natural filtration generated by the chain $\mathbf{X}$. Then Elliott et al. [8] obtained the following semimartingale dynamics for the chain $\mathbf{X}$ under $\mathbb{P}$:

$$\mathbf{X}(t) = \mathbf{X}(0) + \int_0^t \mathbf{A}(u)\mathbf{X}(u-)du + \mathbf{M}(t), \quad t \in \mathscr{T}. \tag{8.1}$$

Here $\mathbf{M} := \{\mathbf{M}(t)|t \in \mathscr{T}\}$ is an $\mathfrak{R}^N$-valued, square-integrable, $(\mathbb{F}^{\mathbf{X}}, P)$-martingale. Since the process $\{\int_0^t \mathbf{A}(u)\mathbf{X}(u-)du|t \in \mathscr{T}\}$ is $\mathbb{F}^{\mathbf{X}}$-predictable, $\mathbf{X}$ is a special semimartingale, and so the above decomposition is unique. This is called the canonical decomposition (see, for example, Elliott [6], Chapter 12 therein).

For each $t \in \mathscr{T}$, let $r(t)$ be the instantaneous interest rate of the bond B at time t, where $r(t) > 0$. Then the price process of the bond $\{B(t)|t \in \mathscr{T}\}$ evolves over time as follows:

$$B(t) = \exp\left(\int_0^t r(u)du\right), \quad t \in \mathscr{T},$$

$$B(0) = 1.$$

To simplify our discussion, we assume that the interest rate process $\{r(t)|t \in \mathscr{T}\}$ is a deterministic function of time t. In general, one may consider the situation where the interest rate depends on the hidden Markov chain $\mathbf{X}$. However, in this situation, it may be difficult, if not impossible, to use filtering theory, (or in particular the separation principle), to turn the hidden Markovian regime-switching market into one with complete observations. This is one of the main focuses in the paper.

As in Elliott and Siu (2013), we now describe the jump component in the price process of the risky share. Let $Z := \{Z(t)|t \in \mathscr{T}\}$ be a real-valued pure jump process on $(\Omega, \mathscr{F}, \mathbb{P})$ with $Z(0) = 0$, $\mathbb{P}$-a.s. It is clear that

$$Z(t) = \sum_{0 < s \leq t} (Z(s) - Z(s-)).$$

Let $\mathscr{B}(\mathscr{T})$ and $\mathscr{B}(\mathfrak{R}_0)$ be the Borel σ-fields generated by open subsets of $\mathscr{T}$ and $\mathfrak{R}_0 := \mathfrak{R} \setminus \{0\}$, respectively. Suppose $\{\gamma(\cdot,\cdot,\omega)|\omega \in \Omega\}$ is the random measure which selects finite jump times T_k and the corresponding non-zero random jump sizes $\Delta Z(T_k) := Z(T_k) - Z(T_k-)$, $k = 1, 2, \cdots$, of the pure jump process Z. Then

$$\gamma(dt, dz, \omega) = \sum_{k \geq 0} \delta_{(T_k, \Delta Z(T_k))}(dt \times dz) I_{\{T_k < \infty, \Delta Z(T_k) \neq 0\}},$$

where $\delta_{(T_k, \Delta Z(T_k))}(dt \times dz)$ is the random Dirac measure, or point mass, at the point $(T_k, \Delta Z(T_k))$ and I_E is the indicator function of the event E. To simplify the notation, we write $\gamma(du, dz)$ for $\gamma(dt, dz, \omega)$ unless otherwise stated.

So, for each $t \in \mathscr{T}$,

$$Z(t) = \int_0^t \int_{\mathfrak{R}_0} z\gamma(du, dz).$$

To specify the statistical laws of the Poisson random measure $\gamma(dt,dz)$, we consider the hidden Markov regime-switching compensator:

$$v_{\mathbf{X}(t-)}(dt,dz) := \sum_{i=1}^{N} \langle \mathbf{X}(t-), \mathbf{e}_i \rangle \lambda_i(t) \eta_i(dz|t) dt ,$$

where for each $i = 1, 2, \cdots, N$

1. $\{\lambda_i(t)|t \in \mathscr{T}\}$ is the jump intensity process of Z when the economy is in the ith state; we suppose that $\{\lambda_i(t)|t \in \mathscr{T}\}$ is $\mathbb{F}^Z$-predictable, where $\mathbb{F}^Z := \{\mathscr{F}^Z(t)|t \in \mathscr{T}\}$ is the right-continuous, $\mathbb{P}$-complete, natural filtration generated by the pure jump process Z;
2. For each $t \in \mathscr{T}$ and each $i = 1, 2, \cdots, N$, $\eta_i(dz|t)$ is the conditional Lévy measure of the random jump size of $\gamma(dt,dz)$ given that there is a jump at time t and that the economy is in the ith state; we assume that $\{\eta_i(dz|t)|t \in \mathscr{T}\}$ is an $\mathbb{F}^Z$-predictable measure-valued process on $(\Omega, \mathscr{F}, \mathbb{P})$.
3. The subscript "$\mathbf{X}(t-)$" is used here to emphasize the dependence of $v_{\mathbf{X}(t-)}(dt,dz)$ on $\mathbf{X}(t-)$.

Then the random measure $\tilde{\gamma}(\cdot,\cdot)$ defined by putting:

$$\tilde{\gamma}(dt,dz) := \gamma(dt,dz) - v_{\mathbf{X}(t-)}(dt,dz) ,$$

is a martingale random measure under $\mathbb{P}$, and hence, it is called the compensated random measure of $\gamma(\cdot,\cdot)$. For discussions on random measures, one may refer to Elliott (1982), Chapter 15 therein.

Let $W := \{W(t)|t \in \mathscr{T}\}$ be a standard Brownian motion on $(\Omega, \mathscr{F}, \mathbb{P})$ with respect to the $\mathbb{P}$-augmentation of its natural filtration $\mathbb{F}^W := \{\mathscr{F}^W(t)|t \in \mathscr{T}\}$. To simplify our discussion, we assume that W is stochastically independent of $\mathbf{X}$ and Z under $\mathbb{P}$. For each $t \in \mathscr{T}$, let $\mu^{\mathbf{X}}(t)$ and $\sigma(t)$ be the appreciation rate and the volatility of the risky share at time t, respectively. We suppose that $\mu^{\mathbf{X}}(t)$ is modulated by the chain $\mathbf{X}$ as:

$$\mu^{\mathbf{X}}(t) := \langle \mu(t), \mathbf{X}(t) \rangle .$$

Here $\mu(t) := (\mu_1(t), \mu_2(t), \cdots, \mu_N(t))' \in \Re^N$ such that for each $i = 1, 2, \cdots, N$ and each $t \in \mathscr{T}$, $\mu_i(t) > r(t)$, $\mathbb{P}$-a.s., and $\{\mu(t)|t \in \mathscr{T}\}$ is an $\mathbb{F}^W$-predictable process; $\mu_i(t)$ represents the appreciation rate of the risky share at time t when the hidden economy is in the ith-state at that time; the scalar product $\langle \cdot, \cdot \rangle$ selects the component in the vector of the appreciation rates that is in force at a particular time according to the state of the hidden economy at that time; the superscript $\mathbf{X}$ in $\mu^{\mathbf{X}}$ is used to emphasize the dependence of the appreciation rate $\mu^{\mathbf{X}}$ on the chain $\mathbf{X}$.

Furthermore, we assume that the volatility process $\{\sigma(t)|t \in \mathscr{T}\}$ is $\mathbb{F}^W$-predictable and that for each $t \in \mathscr{T}$, $\sigma(t) > 0$, $\mathbb{P}$-a.s. In general, one may consider a situation where the volatility depends the hidden Markov chain $\mathbf{X}$. However, there may be two potential concerns about this generalization. Firstly, it complicates the filtering issue and it is difficult, if not impossible, to derive an exact, finite-dimensional filter of the chain **X** in this general situation. Secondly, some difficulties may arise in the interpretation of the information structure of the asset price model. Particularly, it was noted in Guo [19] and Gerber and Shiu [17] that in the case of a Markovian regime-switching geometric Brownian motion, the volatility parameter can be completely determined from a given price path of the risky share. More specifically, it can be identified by means of the predictable quadratic variation. Thirdly, it was noted in Merton [29] that appreciation rates of risky securities are a lot harder to estimate than their volatilities. It may not be unreasonable to assume that the volatility does not depend on the hidden Markov chain **X**. Lastly, if the volatility is assumed to be modulated by the chain **X**, filtering theory (or in particular the separation principle) may be difficult to apply to turn the hidden Markovian regime-switching market into one with complete observations. One may also refer to Elliott and Siu [11] for related discussions.

We suppose that under the real-world measure $\mathbb{P}$ the price process of the risky share is governed by the following hidden Markovian regime-switching, jump-diffusion model:

$$\frac{dS(t)}{S(t-)} = \left(\mu^{\mathbf{X}}(t) + \sum_{i=1}^{N}(e^z - 1) \langle \mathbf{X}(t), \mathbf{e}_i \rangle \lambda_i(t)\eta_i(dz|t) \right) dt$$
$$+ \sigma(t)dW(t) + \int_{\Re_0} (e^z - 1)\tilde{\gamma}(dt,dz) .$$

Write, for each $t \in \mathcal{T}$,

$$Y(t) := \ln(S(t)/S(0)) .$$

This is the logarithmic return from the risky share over the time interval $[0,t]$.

Applying Itô's differentiation rule to $Y(t)$ then gives:

$$dY(t) = \left(\mu^{\mathbf{X}}(t) - \frac{1}{2}\sigma^2(t) + \sum_{i=1}^{N} z\langle \mathbf{X}(t), \mathbf{e}_i \rangle \lambda_i(t)\eta_i(dz|t) \right) dt$$
$$+ \sigma(t)dW(t) + \int_{\Re_0} z\tilde{\gamma}(dt,dz) .$$

Since the coefficients in the price process, or the return process, of the risky share depends on the hidden Markov chain **X**, the hidden Markovian regime-switching jump-diffusion market is one with partial observations. In the next section we shall use filtering theory to transform this market into one with complete observations.

We end this section by specifying the information structure of our market model. Let $\mathbb{F}^Y := \{\mathscr{F}^Y(t)|t \in \mathcal{T}\}$ be the $\mathbb{P}$-augmentation of the natural filtration generated by the return process $Y := \{Y(t)|t \in \mathcal{T}\}$. This is the observable filtration in our market model. For each $t \in \mathcal{T}$, let $\mathscr{G}(t) := \mathscr{F}^{\mathbf{X}}(t) \vee \mathscr{F}(t)$. Write $\mathbb{G} := \{\mathscr{G}(t)|t \in \mathcal{T}\}$ representing the full information structure of the model.

8.3 Filtering Theory and Filtered Market

Filtering theory has been widely used by the electrical engineering community to decompose observations from stochastic dynamical systems into signal and noise. Particularly, it has been widely used in signal processing, system and control engineering, radio and telecommunication engineering. In this section we shall first discuss the use of filtering theory to transform the original market with partial observations to a filtered market with complete observations. The general philosophy of this idea is in the spirit of that of the separation principle used in stochastic optimal control theory for partially observed stochastic dynamical systems, see, for example, Fleming and Rishel [15], Kallianpur [22] and Elliott [6]. Then we shall outline the basic idea of a reference probability approach, whose history can be traced back to the work of Zakai [39], to derive a stochastic differential equation for the unnormalized filter of the hidden Markov chain **X** given observations about the return process of the risky asset. This filtering equation is called the Zakai equation in the filtering literature. The derivation of the filtering equation resembles to that in Wu and Elliott [11] and Elliott and Siu [11], so only key steps are presented and the results are stated without giving the proofs. Due to the presence of stochastic integrals in the Zakai equation, its numerical computation may be rather uneasy. From the numerical perspective, it may be more convenient to consider ordinary differential equations than stochastic differential equations. Using the gauge transformation technique in Clark [5], we shall give a (pathwise) linear ordinary differential equation governing the evolution of a "transformed" unnormalized filter of the chain **X** over time. This filter is robust with respect to the observation process in the Skorohod topology and has an advantage from the numerical perspective. Using a version of the Bayes' rule, the normalized filter can be recovered from the (transformed) unnormalized one.

8.3.1 The Separation Principle

The use of the filtering theory to transform the original market to the filtered market involves the use of the innovations approach which is also called the separation principle. This approach has two steps. The first step introduces innovations processes which are adapted to the observable filtration. The second step expresses the price processes with hidden quantities in terms of these innovations processes and filtered estimates of the hidden quantities.

For any integrable, $\mathbb{G}$-adapted process $\{\phi(t)|t \in \mathscr{T}\}$, let $\{\hat{\phi}(t)|t \in \mathscr{T}\}$ be the $\mathbb{F}^Y$-optional projection of $\{\phi(t)|t \in \mathscr{T}\}$ under the measure $\mathbb{P}$. Then, for each $t \in \mathscr{T}$,

$$\hat{\phi}(t) = \mathrm{E}[\phi(t)|\mathscr{F}^Y(t)] \text{ , } \mathscr{P}\text{-a.s.}$$

The optional projection takes into account the measurability in $(t, \omega) \in \mathscr{T} \times \Omega$.

Define, for each $t \in \mathcal{T}$,

$$h(t) := \mu(t) - \frac{1}{2}\sigma^2(t) .$$

Then the return process Y of the risky share is written as:

$$Y(t) = \int_0^t h(u)du + \int_0^t \sigma(u)dW(u) + \int_0^t \int_{\Re_0} z\gamma(du,dz) .$$

Write, for each $t \in \mathcal{T}$,

$$Y_1(t) = \int_0^t h(u)du + \int_0^t \sigma(u)dW(u) , \quad Y_2(t) := \int_0^t \int_{\Re_0} z\gamma(du,dz) ,$$

so that

$$Y(t) = Y_1(t) + Y_2(t) .$$

Consider then the following $\mathbb{F}^Y$-adapted process $\hat{W} := \{\hat{W}(t) | t \in \mathcal{T}\}$:

$$\hat{W}(t) := W(t) + \int_0^t \left(\frac{h(u) - \hat{h}(u)}{\sigma(u)} \right) du , \quad t \in \mathcal{T} .$$

Then following standard filtering theory, (see, for example, [15, 22, 6]), $\hat{W}$ is an $(\mathbb{F}^Y, \mathbb{P})$-standard Brownian motion. The process $\hat{W}$ is called the innovation process for the diffusion part Y_1 of the return process Y of the risky share.

We now define the innovation process for the jump part Y_2 of the return process Y of the risky share. Consider the $\mathbb{G}$-adapted process $Q := \{Q(t) | t \in \mathcal{T}\}$ on $(\Omega, \mathcal{F}, \mathcal{P})$ which is defined by putting:

$$Q(t) := \int_0^t \int_0^\infty z(\gamma(du,dz) - \nu_{\mathbf{X}(u-)}(du,dz)) , \quad t \in \mathcal{T} ,$$

so that Q is a $(\mathbb{G}, \mathbb{P})$-martingale.

Define

$$\hat{\nu}(dt,dz) := \sum_{i=1}^N \langle \hat{\mathbf{X}}(t), \mathbf{e}_i \rangle \lambda_i(t) \eta_i(dz|t) dt ,$$

and

$$\hat{\gamma}(dt,dz) := \gamma(dt,dz) - \hat{\nu}(dt,dz) .$$

The following lemma was due to Elliott [7]. We state the result here without giving the proof.

Lemma 8.1. *Let $\hat{Q} := \{\hat{Q}(t) | t \in \mathscr{T}\}$ be the $\mathbb{F}^Z$-adapted process defined by setting:*

$$\hat{Q}(t) := \int_0^t \int_{\Re_0} z \hat{\gamma}(du, dz) .$$

Then $\hat{Q}$ is an $(\mathbb{F}^Y, \mathbb{P})$-martingale.

The process $\hat{Q}$ is then used here as the innovations process of the jump part Y_2 of the return process of the risky share.

The following lemma then gives a representation for the price process of the risky share in terms of the two innovations processes $\hat{W}$ and $\hat{Q}$ under the real-world measure $\mathbb{P}$. Since the result is rather standard, we just state it without giving the proof.

Lemma 8.2. *Under the real-world measure $\mathbb{P}$, the return process of the risky share is:*

$$dY(t) = \left(\hat{h}(t) + \sum_{i=1}^{N} z \langle \hat{\mathbf{X}}(t), \mathbf{e}_i \rangle \lambda_i(t) \eta_i(dz|t) \right) dt + \sigma(t) d\hat{W}(t)$$
$$+ \int_{\Re_0} z \hat{\gamma}(dt, dz) ,$$

and the price process of the risky share is:

$$\frac{dS(t)}{S(t-)} = \left(\hat{\mu}^{\mathbf{X}}(t) + \sum_{i=1}^{N} (e^z - 1) \langle \hat{\mathbf{X}}(t), \mathbf{e}_i \rangle \lambda_i(t) \eta_i(dz|t) \right) dt + \sigma(t) d\hat{W}(t)$$
$$+ \int_{\Re_0} (e^z - 1) \hat{\gamma}(dt, dz) .$$

It is obvious that the return and price processes of the risky share in Lemma 8.2 only involve observable quantities. Consequently, we adopt these dynamics as the return and price processes of the risky share in the filtered market.

8.3.2 Filtering Equations

The reference probability approach to derive a filtering equation for the chain $\mathbf{X}$ is now discussed. We start with a reference probability measure $\mathbb{P}^\dagger$ on $(\Omega, \mathscr{F})$ under which the return process Y of the risky share becomes simpler and does not depend on the chain $\mathbf{X}$. That is, under $\mathbb{P}^\dagger$,

1. Y_1 is a Brownian motion with $\langle Y_1, Y_1 \rangle (t) = \int_0^t \sigma^2(u) du$, where $\{\langle Y_1, Y_1 \rangle (t) | t \in \mathscr{T}\}$ is the predictable quadratic variation of Y_1;

2. The Poisson random measure $\gamma(dt,dy)$ has an unit intensity for random jump times and the conditional Lèvy measure $\eta(dz|t)$ for random jump sizes so that

$$\tilde{\gamma}^\dagger(dt,dz) := \gamma(dt,dz) - \eta(dz|t)dt ,$$

is an $(\mathbb{F}^Y, \mathbb{P}^\dagger)$-martingale random measure;
3. Y_1 and $\gamma(\cdot,\cdot)$ are stochastically independent;
4. The chain $\mathbf{X}$ has the family of rate matrices, $\{\mathbf{A}(t)|t \in \mathscr{T}\}$.

Define

$$\Lambda^1(t) := \exp\left(\int_0^t \sigma^{-2}(u)h(u)dY_1(u) - \frac{1}{2}\int_0^t \sigma^{-2}(u)h^2(u)du\right) ,$$

$$\Lambda^2(t) := \exp\left[-\int_0^t \left(\sum_{i=1}^N \langle \mathbf{X}(u-), \mathbf{e}_i \rangle \int_{\Re_0} (g_i(z|u)-1)\eta(z|u)dz\right)du \right.$$
$$\left. + \int_0^t \int_{\Re_0} \left(\sum_{i=1}^N \langle \mathbf{X}(u-), \mathbf{e}_i \rangle \log(g_i(z|u))\right) \gamma(du,dz)\right] ,$$

where $g_i(z|t) := \frac{\lambda_i(t)\eta_i(dz|t)}{\eta(dz|t)}$, for each $i = 1, 2, \cdots, N$.

Consider the $\mathbb{G}$-adapted process $\Lambda := \{\Lambda(t)|t \in \mathscr{T}\}$ defined by putting:

$$\Lambda(t) := \Lambda^1(t) \cdot \Lambda^2(t) .$$

It is not difficult to check that Λ is a $(\mathbb{G}, \mathbb{P})$-martingale, and hence, $E[\Lambda(T)] = 1$. Consequently the real-world measure $\mathbb{P}$ equivalent to $\mathbb{P}^\dagger$ on $\mathscr{G}(T)$ can be reconstructed using the Radon-Nikodym derivative $\Lambda(T)$ as follows:

$$\left.\frac{d\mathbb{P}}{d\mathbb{P}^\dagger}\right|_{\mathscr{G}(T)} := \Lambda(T) .$$

Using a version of Girsanov's theorem for jump-diffusion processes, it can be shown that under $\mathbb{P}$,

1. $\Lambda := \{\Lambda(t)|t \in \mathscr{T}\}$ is the unique solution of the following stochastic differential-integral equation:

$$\Lambda(t) = 1 + \int_0^t \Lambda(u)h(u)\sigma^{-2}(u)dY_1(u)$$
$$+ \sum_{i=1}^N \int_0^t \langle \mathbf{X}(u), \mathbf{e}_i \rangle \int_{\Re_0} \Lambda(u-)(g_i(z|u)-1)\tilde{\gamma}^\dagger(du,dz) .$$

2. The process

$$W(t) := \int_0^t \sigma^{-1}(u)dY_1(u) - \int_0^t \sigma^{-1}(u)h(u)du, \quad t \in \mathcal{T},$$

is a standard Brownian motion.
3. $\gamma(dt,dz)$ is the Poisson random measure with the compensator:

$$v_{\mathbf{X}(t-)}(dt,dz) := \sum_{i=1}^N \langle \mathbf{X}(t-), \mathbf{e}_i \rangle \lambda_i(t) \eta_i(dz|t)dt,$$

so that

$$\tilde{\gamma}(dt,dz) := \gamma(dt,dz) - v_{\mathbf{X}(t-)}(dt,dz),$$

is a martingale random measure.

Consequently, under $\mathbb{P}$, the return process Y of the risky share is:

$$Y(t) = Y_1(t) + Y_2(t) = \int_0^t h(u)du + \int_0^t \sigma(u)dW(u) + \int_0^t \int_{\Re_0} z\gamma(du,dz).$$

The ultimate goal of filtering is to evaluate $\hat{\mathbf{X}}$ which is an $\mathbb{F}^Y$-optional projection of $\mathbf{X}$ under $\mathbb{P}$. Then, for each $t \in \mathcal{T}$,

$$\hat{\mathbf{X}}(t) = \mathrm{E}[\mathbf{X}(t)|\mathscr{F}^Y(t)], \quad \mathbb{P}\text{-a.s.}$$

Indeed, $\hat{\mathbf{X}}(t)$ is an optimal estimate of $\mathbf{X}(t)$ in the least square sense.

Write, for each $t \in \mathcal{T}$,

$$\mathbf{q}(t) := \mathrm{E}^\dagger[\Lambda(t)\mathbf{X}(t)|\mathscr{F}(t)],$$

where $\mathrm{E}^\dagger$ is an expectation under the reference probability measure $\mathbb{P}^\dagger$ and $\mathbf{q}(t)$ is called an unnormalized filter of $\mathbf{X}(t)$. Instead of evaluating $\hat{\mathbf{X}}(t)$ directly, a filtering equation governing the evolution of the unnormalized filter $\mathbf{q}(t)$ over time is first derived. Before presenting the filtering equation, we need to define some notation.

For each $t \in \mathcal{T}$, let $Z(t,\cdot) : \Omega \to \Re_0$ be a random variable with a strictly positive conditional Lèvy measure $\eta(dz|t)$ under the reference measure $\mathbb{P}^\dagger$. Then, for each $t \in \mathcal{T}$ and each $i = 1, 2, \cdots, N$, the random variable $G_i(t,\cdot) : \Omega \to \Re_+$, where $\Re_+$ is the positive real line, is defined as:

$$G_i(t,\omega) := \frac{\lambda_i(t)\eta_i(Z(t,\omega)|t)}{\eta(Z(t,\omega)|t)} := g_i(Z(t,\omega)|t),$$

for some measurable function $g_i(\cdot|t)$ on $(\Re_0, \mathscr{B}(\Re_0))$.

Note that $G_i(t,\omega)$ is well-defined since $\eta(dz|t) > 0$. Again, to simplify the notation, we suppress the notation "ω" unless otherwise stated. We now define the following diagonal matrices:

$$\mathbf{diag}(\mathbf{G}(t) - \mathbf{1}) := \mathbf{diag}(G_1(t) - 1, G_2(t) - 1, \cdots, G_N(t) - 1),$$
$$\mathbf{diag}(\lambda(t) - \mathbf{1}) := \mathbf{diag}(\lambda_1(t) - 1, \lambda_2(t) - 1, \cdots, \lambda_N(t) - 1).$$

Here $\mathbf{diag}(\mathbf{y})$ denotes the diagonal matrix with diagonal elements being given by the components in a vector $\mathbf{y}$; $\mathbf{1} := (1,1,\cdots,1)' \in \Re^N$.

For each $i = 1, 2, \cdots, N$ and each $t \in \mathcal{T}$, let $h_i(t) := \mu_i(t) - \frac{1}{2}\sigma^2(t)$ and $\mathbf{h}(t) := (h_1(t), h_2(t), \cdots, h_N(t))' \in \Re^N$. Write, for each $t \in \mathcal{T}$,

$$J(t) := \int_0^t \int_{\Re_0} \gamma(du, dz).$$

Then the following theorem is standard and gives the Zakai stochastic differential equation for the unnormalized filter $\mathbf{q}(t)$ (see, for example, Elliott and Siu [11], Theorem 4.1 therein). We state the result without giving the proof.

Theorem 8.1. *For each $t \in \mathcal{T}$, let*

$$\mathbf{B}(t) := \mathbf{diag}(\mathbf{h}(t)).$$

Then under $\mathbb{P}^\dagger$, the unnormalized filter $\mathbf{q}(t)$ satisfies the Zakai stochastic differential equation:

$$\mathbf{q}(t) = \mathbf{q}(0) + \int_0^t \mathbf{A}(u)\mathbf{q}(u)du + \int_0^t \mathbf{B}(u)\mathbf{q}(u)\sigma^{-2}(u)dY_1(u)$$
$$+ \int_0^t \mathbf{diag}(\mathbf{G}(u) - \mathbf{1})\mathbf{q}(u)dJ(u) - \int_0^t \mathbf{diag}(\lambda(u) - \mathbf{1})\mathbf{q}(u)du.$$
(8.2)

The filtering equation in Theorem 8.1 involves stochastic integrals. This may be a disadvantage for numerical implementation. Using the gauge transformation technique of Clark [5], the filtering equation can be simplified as a (pathwise) linear ordinary differential equation. The key steps are presented in the sequel.

Define, for each $i = 1, 2, \cdots, N$ and each $t \in \mathcal{T}$,

$$\gamma_i(t) := \exp\left(\int_0^t h_i(u)\sigma^{-2}(u)dY_1(u) - \frac{1}{2}\int_0^t h_i^2(u)\sigma^{-4}(u)du \right.$$
$$\left. + \int_0^t (1 - \lambda_i(u))du + \int_0^t \log G_i(u)dJ(u)\right).$$
(8.3)

Then the gauge transformation matrix $\Gamma(t)$ is defined as:

$$\Gamma(t) := \mathbf{diag}(\gamma_1(t), \gamma_2(t), \ldots, \gamma_N(t)).$$

Write, for each $t \in \mathscr{T}$, $\Gamma^{-1}(t)$ for the inverse of $\Gamma(t)$. The existence of $\Gamma^{-1}(t)$ is guaranteed by the definition of $\Gamma(t)$ and the positivity of $\gamma_i(t)$ for each $i = 1, 2, \cdots, N$.

Take, for each $t \in \mathscr{T}$,

$$\bar{\mathbf{q}}(t) := \Gamma^{-1}(t)\mathbf{q}(t) .$$

This is called a transformed unnormalized filter of $\mathbf{X}(t)$.

Then the following theorem gives a (pathwise) linear ordinary differential equation governing the transformed process $\{\bar{\mathbf{q}}(t) | t \in \mathscr{T}\}$. Again we state the result without giving the proof which is rather standard (see, for example, Elliott and Siu [11], Theorem 4.2 therein).

Theorem 8.2. $\bar{\mathbf{q}}$ *satisfies the following first-order linear ordinary differential equation:*

$$\frac{d\bar{\mathbf{q}}(t)}{dt} := \Gamma^{-1}(t)\mathbf{A}(t)\Gamma(t)\bar{\mathbf{q}}(t) , \quad \bar{\mathbf{q}}(0) = \mathbf{q}(0) .$$

Finally, using a version of the Bayes' rule,

$$\begin{aligned}
\hat{\mathbf{X}}(t) &:= \mathrm{E}[\mathbf{X}(t)|\mathscr{F}(t)] \\
&= \frac{\mathrm{E}^\dagger[\Lambda(t)\mathbf{X}(t)|\mathscr{F}(t)]}{\mathrm{E}^\dagger[\Lambda(t)|\mathscr{F}(t)]} \\
&= \frac{\mathbf{q}(t)}{\langle \mathbf{q}(t), \mathbf{1} \rangle} \\
&:= \frac{\Gamma(t)\bar{\mathbf{q}}(t)}{\langle \Gamma(t)\bar{\mathbf{q}}(t), \mathbf{1} \rangle} ,
\end{aligned}$$

so the normalized filter $\hat{\mathbf{X}}(t)$ can be "recovered" from the (transformed) unnormalized one $\bar{\mathbf{q}}(t)$.

8.4 Generalized Esscher Transform in the Filtered Market

The main theme of this section is to determine a pricing kernel in the filtered market described in Sect. 8.3.1 using a generalized version of the Esscher transform based on a Doléan-Dade stochastic exponential. Firstly, let us recall that in the filtered market, the price process of the risky share under the real-world measure $\mathbb{P}$ is governed by the following stochastic differential equation with jumps:

$$\begin{aligned}
\frac{dS(t)}{S(t-)} &= \left(\hat{\mu}(t) + \sum_{i=1}^{N}(e^z - 1)\langle \hat{\mathbf{X}}(t), \mathbf{e}_i \rangle \lambda_i(t)\eta_i(dz|t) \right)dt + \sigma(t)d\hat{W}(t) \\
&\quad + \int_{\mathfrak{R}_0}(e^z - 1)\hat{\gamma}(dt, dz) .
\end{aligned}$$

The presence of jumps renders the filtered market incomplete. Consequently, there is more than one equivalent martingale measure, or pricing kernel, in the market. Though there are different approaches to select a pricing kernel in an incomplete market, we focus here on the Esscher transformation approach which was pioneered by the seminal work of Gerber and Shiu [16]. Note that the local characteristics of the price process of the risky share in the filtered market are $\mathbb{F}^Y$-predictable processes, so the price process is a semimartingale beyond the class of Lévy processes. Consequently, the original version of the Esscher transform in Esscher [14] and Gerber and Shiu [16] cannot be applied in this situation. Bühlmann et al. [4], Kallsen and Shiryaev [23] and Jacod and Shiryaev [21] considered a generalized version of the Esscher transform for measure changes for general semimartingales and discussed its application for option valuation. This version of the Esscher transform is defined using the concepts of Doléan Dade stochastic exponential and the Laplace cumulant process. It was used in Elliott and Siu [12] to select a pricing kernel in a hidden regime-switching pure jump process. In the sequel, we shall first define the generalized Esscher transform and give the local condition in terms of the local characteristics of the price process of the risky share in the filtered market. Then we present the price dynamics of the risky share under an equivalent (local)-martingale measure selected by the generalized Esscher transform.

Let $\mathscr{L}(Y)$ be the space of processes $\theta := \{\theta(t) | t \in \mathscr{T}\}$ satisfying the following conditions:

1. θ is $\mathbb{F}^Y$-predictable;
2. θ is integrable with respect to the return process Y; that is, the (stochastic) integral process $\{(\theta \cdot Y)(t) | t \in \mathscr{T}\}$, where $(\theta \cdot Y)(t) := \int_0^t \theta(u) dY(u)$ is well-defined.

Consider, for each $\theta \in \mathscr{L}(Y)$, the following exponential process:

$$\mathscr{D}^\theta(t) := \exp((\theta \cdot Y)(t)), \quad t \in \mathscr{T}.$$

Note that $\{(\theta \cdot Y)(t) | t \in \mathscr{T}\}$ is a semimartingale, so $\mathscr{D}^\theta := \{\mathscr{D}^\theta(t) | t \in \mathscr{T}\}$ is also called an exponential semimartingale.

Define, for each $\theta \in \mathscr{L}(Y)$, the following semimartingale:

$$\begin{aligned}\mathscr{H}^\theta(t) := & \int_0^t \left(\hat{h}(u)\theta(u) + \frac{1}{2}\sigma^2(u)\theta^2(u) \right. \\ & \left. + \sum_{i=1}^N \int_{\Re_0} (e^z - 1 - \theta(u)z) \langle \hat{\mathbf{X}}(u), \mathbf{e}_i \rangle \lambda_i(u) \eta_i(dz|u) \right) du \\ & + \int_0^t \sigma(u)\theta(u) d\hat{W}(u) + \int_0^t \int_{\Re_0} (e^{\theta(u)z} - 1) \hat{\gamma}(du, dz).\end{aligned}$$

Using Itô's differentiation rule, it can be shown that

$$\mathscr{D}^\theta(t) = 1 + \int_0^t \mathscr{D}^\theta(u-) d\mathscr{H}^\theta(u).$$

From Theorem 13.5 in Elliott [6], $\mathscr{D}^\theta$ is the unique solution of the above integral equation. It is the Doléans-Dade exponential of $\mathscr{H}^\theta := \{\mathscr{H}^\theta(t)|t \in \mathscr{T}\}$. Symbolically, it is typified as:

$$\mathscr{D}^\theta(t) = \mathscr{E}(\mathscr{H}^\theta)(t), \quad t \in \mathscr{T}.$$

Again from Theorem 13.5 in Elliott [6],

$$\mathscr{D}^\theta(t) = \exp\left(\mathscr{H}^\theta(t) - \frac{1}{2} < (\mathscr{H}^\theta)^c, (\mathscr{H}^\theta)^c > (t)\right) \prod_{0<u\leq t} (1+\Delta\mathscr{H}^\theta(u))e^{-\Delta\mathscr{H}^\theta(u)},$$

where

1. $(\mathscr{H}^\theta)^c := \{(\mathscr{H}^\theta)^c(t)|t \in \mathscr{T}\}$ is the continuous part of $\mathscr{H}^\theta$;
2. $\{< (\mathscr{H}^\theta)^c, (\mathscr{H}^\theta)^c > (t)|t \in \mathscr{T}\}$ is the predictable quadratic variation of $(\mathscr{H}^\theta)^c$.

Note that $\mathscr{H}^\theta$ is called the stochastic logarithm of $\mathscr{D}^\theta$ or the exponential transform of $\{(\theta \cdot Y)(t)|t \in \mathscr{T}\}$. Since $\mathscr{H}^\theta$ is a special semimartingale, its predictable part of finite variation, denoted as $\mathscr{K}^\theta := \{\mathscr{K}^\theta(t)|t \in \mathscr{T}\}$, is uniquely determined as:

$$\mathscr{K}^\theta(t) = \int_0^t \left(\hat{h}(u)\theta(u) + \frac{1}{2}\sigma^2(u)\theta^2(u) \right.$$
$$\left. + \sum_{i=1}^N \int_{\Re_0} (e^z - 1 - \theta(u)z)\langle\hat{\mathbf{X}}(u), \mathbf{e}_i\rangle \lambda_i(u)\eta_i(dz|u)\right)du.$$

It was noted in Kallsen and Shiryaev [23] that the Laplace cumulant process of $\{(\theta \cdot Y)(t)|t \in \mathscr{T}\}$ is the predictable part of finite variation of $\mathscr{H}^\theta$. This is also called the Laplace cumulant process of Y at θ. The Doléan-Dade stochastic exponential $\mathscr{E}(\mathscr{K}^\theta) := \{\mathscr{E}(\mathscr{K}^\theta)(t)|t \in \mathscr{T}\}$ of $\mathscr{K}^\theta$ is the unique solution of the following linear stochastic differential equation:

$$\mathscr{E}(\mathscr{K}^\theta)(t) = 1 + \int_0^t \mathscr{E}(\mathscr{K}^\theta)(u)d\mathscr{K}^\theta(u).$$

As in Kallsen and Shiryaev [23], the modified Laplace cumulant process of Y at θ is defined by the process $\tilde{\mathscr{K}}^\theta := \{\tilde{\mathscr{K}}^\theta(t)|t \in \mathscr{T}\}$ such that

$$\exp(\tilde{\mathscr{K}}^\theta(t)) = \mathscr{E}(\mathscr{K}^\theta)(t).$$

By differentiation, $\tilde{\mathscr{K}}^\theta(t) = \mathscr{K}^\theta(t)$, $\mathscr{P}$-a.s., for each $t \in \mathscr{T}$.

Then the density process of the generalized Esscher transform associated with $\theta \in \mathscr{L}(Y)$ is defined as the process $\Lambda^\theta := \{\Lambda^\theta(t)|t \in \mathscr{T}\}$, where

$$\Lambda^\theta(t) := \frac{\exp((\theta \cdot Y)(t))}{\mathscr{E}(\mathscr{K}^\theta)(t)}.$$

Note that the Laplace cumulant process $\mathscr{E}(\mathscr{K}^\theta)$ is the normalization constant and may be thought of as a generalization of the moment generation function in the "first-generation" of the Esscher transform in Gerber and Shiu [16].

From the definition of $\tilde{\mathscr{K}}^\theta$,

$$\begin{aligned}
\Lambda^\theta(t) &= \exp\left((\theta \cdot Y)(t) - \tilde{\mathscr{K}}^\theta(t)\right) \\
&= \exp\left((\theta \cdot Y)(t) - \mathscr{K}^\theta(t)\right) \\
&= \exp\Bigg(-\sum_{i=1}^{N}\int_0^t\int_{\Re_0}(e^z-1-\theta(u)z)\langle\hat{\mathbf{X}}(u),\mathbf{e}_i\rangle\lambda_i(u)\eta_i(dz|u) \\
&\quad + \int_0^t\int_{\Re_0}\theta(u)z\hat{\gamma}(du,dz)\Bigg) \times \exp\left(\int_0^t\theta(u)\sigma(u)d\hat{W}(u)\right. \\
&\quad \left. - \frac{1}{2}\int_0^t\theta^2(u)\sigma^2(u)du\right),\quad \mathbb{P}\text{-a.s.}
\end{aligned}$$

Using Itô's differentiation rule,

$$\Lambda^\theta(t) = 1 + \int_0^t \Lambda^\theta(u)\theta(u)\sigma(u)d\hat{W}(u) + \int_0^t\int_{\Re_0}\Lambda^\theta(u-)(e^{\theta(u)z}-1)\hat{\gamma}(du,dz).$$

This is an $(\mathbb{F}^Y,\mathbb{P})$-local martingale. We suppose here that $\theta \in \mathscr{L}(Y)$ is such that Λ^θ is an $(\mathbb{F}^Y,\mathbb{P})$-martingale, so $\mathrm{E}[\Lambda^\theta(T)] = 1$.

Consequently, for each $\theta \in \mathscr{L}(Y)$, a new probability measure $\mathbb{P}^\theta \sim \mathbb{P}$ on $\mathscr{F}^Y(T)$ by putting:

$$\left.\frac{d\mathbb{P}^\theta}{d\mathbb{P}}\right|_{\mathscr{F}^Y(T)} := \Lambda^\theta(T).$$

To preclude arbitrage opportunities, we must determine $\theta \in \mathscr{L}(Y)$ such that the discounted price process of the risky share $\{\tilde{S}(t)|t\in\mathscr{T}\}$, where $\tilde{S}(t) := \exp(-\int_0^t r(u)du)S(t)$, is an $(\mathbb{F}^Y,\mathbb{P}^\theta)$-(local)-martingale, (i.e., $\mathbb{P}^\theta$ is an equivalent (local)-martingale measure). This is called the (local)-martingale condition. A necessary and sufficient condition for the (local)-martingale condition is given in the following theorem.

Theorem 8.3. *The (local)-martingale condition holds if and only if for each $t \in \mathscr{T}$, there exists an $\mathbb{F}^Y$-progressively measurable process $\{\theta(t)|t\in\mathscr{T}\}$ such that*

$$\hat{\mu}(t) - r(t) + \theta(t)\sigma^2(t) + \sum_{i=1}^{N}\int_{\Re_0}\langle\hat{\mathbf{X}}(t),\mathbf{e}_i\rangle e^{\theta(t)z}(e^z-1)\lambda_i(t)\eta_i(dz|t) = 0,\quad \mathscr{P}\text{-a.s.} \tag{8.4}$$

Proof. The proof is standard, so only key steps are given. Note that $\{\tilde{S}(t)|t\in\mathscr{T}\}$ is an $(\mathbb{F}^Y,\mathbb{P}^\theta)$-(local)-martingale if and only if $\{\Lambda^\theta(t)\tilde{S}(t)|t\in\mathscr{T}\}$ is an $(\mathbb{F}^Y,\mathbb{P})$-(local)-martingale, for some $\mathbb{F}^Y$-progressively measurable process $\{\theta(t)|t\in\mathscr{T}\}$.

The result then follows by applying Itô's product rule to $\Lambda^\theta(t)\tilde{S}(t)$ and equating the finite variation term of $\Lambda^\theta(t)\tilde{S}(t)$ to zero. □

From Theorem 8.3, the risk-neutral Esscher parameter $\{\theta(t)|t \in \mathscr{T}\}$ of the generalized Esscher transform Λ^θ is obtained by solving Equation (8.4). In the particular case where the jump component is absent, the risk-neutral Esscher parameter is given by:

$$\theta(t) = \frac{r(t) - \hat{\mu}(t)}{\sigma^2(t)} = -\frac{\beta(t)}{\sigma(t)},$$

where $\beta(t)$ is the market price of risk of the particular case of the filtered market where jumps are absent and is defined as:

$$\beta(t) := \frac{\hat{\mu}(t) - r(t)}{\sigma(t)}.$$

The following lemma gives the probability laws of the return process Y of the risky share under $\dot{\mathbb{P}}^\theta$. It is a direct consequence of a Girsanov transform for a measure change, so we only state the result.

Lemma 8.3. *The process defined by:*

$$\hat{W}^\theta(t) := \hat{W}(t) - \int_0^t \theta(u)\sigma(u)du, \quad t \in \mathscr{T},$$

is an $(\mathbb{F}^Y, \mathbb{P}^\theta)$-*standard Brownian motion. Furthermore, the process defined by:*

$$\hat{\gamma}^\theta(dt, dz) := \gamma(dt, dz) - \sum_{i=1}^N \langle \hat{\mathbf{X}}(t), \mathbf{e}_i \rangle \lambda_i(t) e^{\theta(t)z} \eta_i(dz|t) dt,$$

is an $(\mathbb{F}^Y, \mathbb{P}^\theta)$-*martingale.*

Under $\mathbb{P}^\theta$, *the price process of the risky share is given by:*

$$dS(t) = S(t)r(t)dt + S(t)\sigma(t)d\hat{W}^\theta(t) + S(t-)\int_{\Re_0}(e^z - 1)\hat{\gamma}^\theta(dt, dz).$$

To simplify our analysis, we suppose here that the probability law of the chain $\mathbf{X}$ remains unchanged after the measure change from $\mathbb{P}$ to $\mathbb{P}^\theta$. Consequently, under $\mathbb{P}^\theta$, the semimartingale dynamics for the chain $\mathbf{X}$ are:

$$\mathbf{X}(t) = \mathbf{X}(0) + \int_0^t \mathbf{A}(u)\mathbf{X}(u)du + \mathbf{M}(t), \quad t \in \mathscr{T}.$$

Furthermore, we assume that $\hat{W}^\theta$, $\hat{\gamma}^\theta$ and $\mathbf{X}$ are stochastically independent under $\mathbb{P}^\theta$.

8.5 European-Style Option

In this section, we shall consider the valuation of a standard European-style option in the filtered market and derive a partial differential integral equation (PDIE) for the option price. Consider a European-style option written on the risky share S whose payoff at the maturity time T is $H(S(T)) \in L^2(\Omega, \mathscr{F}^Y(T), \mathbb{P})$, where $L^2(\Omega, \mathscr{F}^Y(T), \mathbb{P})$ is the space of square-integrable random variables on $(\Omega, \mathscr{F}^Y(T), \mathbb{P})$ and $H : \mathfrak{R} \to \mathfrak{R}$ is a measurable function.

Conditional on the observed information $\mathscr{F}^Y(t)$ at the current time t, the price of the European option at time t is given by:

$$V(t) = \mathrm{E}^\theta \left[\exp\left(-\int_t^T r(u)du\right) H(S(T)) | \mathscr{F}^Y(t) \right].$$

Here E^θ is an expectation under the measure $\mathbb{P}^\theta$.

Note that $\{(S(t), \mathbf{q}(t)) | t \in \mathscr{T}\}$ is jointly Markovian with respect to the observed filtration $\mathbb{F}^Y$. Consequently if $S(t) = s \in (0, \infty)$ and $\mathbf{q}(t) = \mathbf{q} \in \mathfrak{R}^N$,

$$\begin{aligned} V(t) &= \mathrm{E}^\theta \left[\exp\left(-\int_t^T r(u)du\right) H(S(T)) | \mathscr{F}^Y(t) \right] \\ &= \mathrm{E}^\theta \left[\exp\left(-\int_t^T r(u)du\right) H(S(T)) | (S(t), \mathbf{q}(t)) = (s, \mathbf{q}) \right] \\ &:= V^\dagger(t, s, \mathbf{q}), \end{aligned}$$

for some function $V^\dagger : \mathscr{T} \times (0, \infty) \times \mathfrak{R}^N \to \mathfrak{R}$.

For each $t \in \mathscr{T}$, let $V(t, s, \mathbf{q}) := \exp(-\int_0^t r(u)du) V^\dagger(t, s, \mathbf{q})$. Then

$$V(t, s, \mathbf{q}) = \mathrm{E}^\theta \left[\exp\left(-\int_0^T r(u)du\right) H(S(T)) | \mathscr{F}^Y(t) \right].$$

This is an $(\mathbb{F}^Y, \mathbb{P}^\theta)$-martingale.

For each $t \in \mathscr{T}$, let

$$\hat{f}^\theta(t) := \int_0^t \int_{\mathfrak{R}_0} \hat{\gamma}^\theta(du, dz).$$

Then using Lemma 8.3 and Theorem 8.1, the unnormalized filter process under the risk-neutral measure $\mathbb{P}^\theta$ is given by:

$$\begin{aligned} d\mathbf{q}(t) = & \left(\mathbf{A}(t) + \mathbf{B}(t)\sigma^{-2}(t)(\hat{h}(t) + \sigma^2(t)\theta(t)) \right) \\ & + \sum_{i=1}^N \int_{\mathfrak{R}_0} \mathbf{diag}(\mathbf{G}(t) - \mathbf{1}) \langle \hat{\mathbf{X}}(t), \mathbf{e}_i \rangle e^{\theta(t)z} \lambda_i(t) \eta_i(dz|t) \end{aligned}$$

$$-\mathbf{diag}(\lambda(t)-1)\Big)\mathbf{q}(t)dt+\mathbf{B}(t)\sigma^{-1}(t)\mathbf{q}(t)d\hat{W}^\theta(t)$$
$$+\mathbf{diag}(\mathbf{G}(t)-1)\mathbf{q}(t)d\hat{J}^\theta(t).$$

To simplify the notation, we write

$$\alpha^\theta(t) := \mathbf{A}(t)+\mathbf{B}(t)\sigma^{-2}(t)(\hat{h}(t)+\sigma^2(t)\theta(t))$$
$$+\sum_{i=1}^{N}\int_{\Re_0}\mathbf{diag}(\mathbf{G}(t)-1)\langle \hat{\mathbf{X}}(t),\mathbf{e}_i\rangle e^{\theta(t)z}\lambda_i(t)\eta_i(dz|t)$$
$$-\mathbf{diag}(\lambda(t)-1) \in \Re^N \otimes \Re^N,$$

so

$$d\mathbf{q}(t) = \alpha^\theta(t)\mathbf{q}(t)dt + \mathbf{B}(t)\sigma^{-1}(t)\mathbf{q}(t)d\hat{W}^\theta(t) + \mathbf{diag}(\mathbf{G}(t)-1)\mathbf{q}(t)d\hat{J}^\theta(t).$$

Under $\mathbb{P}^\theta$, the price process of the risky asset is:

$$dS(t) = S(t)r(t)dt + S(t)\sigma(t)d\hat{W}^\theta(t) + S(t-)\int_{\Re_0}(e^z-1)\hat{\gamma}^\theta(dt,dz).$$

Suppose $V: \mathscr{T} \times (0,\infty) \times (0,\infty)^N \to \Re$ is a function in $\mathscr{C}^{1,2}(\mathscr{T} \times (0,\infty) \times (0,\infty)^N)$, where $\mathscr{C}^{1,2}(\mathscr{T} \times (0,\infty) \times (0,\infty)^N)$ is the space of functions $V(t,s,\mathbf{q})$ which are continuously differentiable in $t \in \mathscr{T}$ and twice continuously differentiable in $(s,\mathbf{q}) \in (0,\infty) \times (0,\infty)^N$.

The following theorem gives the partial differential integral equation (P.D.I.E.) for the price of the European-style option V.

Theorem 8.4. *Let $\mathbf{q}_- := \mathbf{q}(t-)$ and $s_- := S(t-)$, for each $t \in \mathscr{T}$. Write $\mathbf{y}'$ for the transpose of a vector, or matrix, $\mathbf{y}$. Define, for each $t \in \mathscr{T}$,*

$$\beta^\theta(t) := \sum_{i=1}^{N}\mathbf{diag}(\mathbf{G}(t)-1)\langle \hat{\mathbf{X}}(t),\mathbf{e}_i\rangle \lambda_i(t)\int_{\Re_0}e^{\theta(t)z}(e^z-1)\eta_i(dz|t) \in \Re^N \otimes \Re^N.$$

Then the option price $V^\dagger(t,s,\mathbf{q})$ at time t satisfies:

$$\frac{\partial V^\dagger}{\partial t} + \frac{\partial V^\dagger}{\partial s}s(\hat{\mu}(t)+\theta(t)\sigma^2(t)) + \left\langle \frac{\partial V^\dagger}{\partial \mathbf{q}},(\alpha^\theta(t)-\beta^\theta(t))\mathbf{q}(t)\right\rangle$$
$$+\frac{1}{2}\frac{\partial^2 V^\dagger}{\partial s^2}\sigma^2(t)s^2 + s\left\langle \frac{\partial^2 V^\dagger}{\partial \mathbf{q}\partial s},\mathbf{B}(t)\mathbf{q}(t)\right\rangle + \frac{1}{2}(\mathbf{B}(t)\mathbf{q}(t))'\frac{\partial^2 V^\dagger}{\partial \mathbf{q}^2}(\mathbf{B}(t)\mathbf{q}(t))$$
$$+\sum_{i=1}^{N}\langle \hat{\mathbf{X}}(t),\mathbf{e}_i\rangle \int_{\Re_0}\Big(V^\dagger(t,s_-e^z,\mathbf{q}_-\mathbf{diag}(\mathbf{G}(t))) - V^\dagger(t,s_-,\mathbf{q}_-)\Big)$$
$$\times e^{\theta(t)z}\lambda_i(t)\eta_i(dz|t) = r(t)V^\dagger,$$

with the terminal condition $V^\dagger(T,S(T),\mathbf{q}(T)) = H(S(T))$.

8 A Hidden Markov-Modulated Jump Diffusion Model for European Option Pricing

Proof. The proof is standard. For the sake of completeness, we give the proof here. Applying Itô's differentiation rule to $V(t,s,\mathbf{q})$ gives:

$$V(t,s,\mathbf{q}) = V(0,s_0,\mathbf{q}_0) + \int_0^t \frac{\partial V}{\partial u} du + \int_0^t \frac{\partial V}{\partial s} S(u) \left(r(u) - \sum_{i=1}^N \langle \hat{\mathbf{X}}(u), \mathbf{e}_i \rangle \lambda_i(u) \right.$$

$$\left. \times \int_{\Re_0} e^{\theta(u)z}(e^z - 1) \eta_i(dz|u) \right) du - \int_0^t \left\langle \frac{\partial V}{\partial \mathbf{q}}, \mathrm{diag}(\mathbf{G}(u) - \mathbf{1})\mathbf{q}(u) \right\rangle$$

$$\times \sum_{i=1}^N \langle \hat{\mathbf{X}}(u), \mathbf{e}_i \rangle \lambda_i(u) \int_{\Re_0} e^{\theta(u)z}(e^z - 1) \eta_i(dz|u) \bigg) du$$

$$+ \int_0^t \frac{\partial V}{\partial s} S(u) \sigma(u) d\hat{W}^\theta(u) + \int_0^t \left\langle \frac{\partial V}{\partial \mathbf{q}}, \alpha^\theta(u)\mathbf{q}(u) \right\rangle du$$

$$+ \int_0^t \left\langle \frac{\partial V}{\partial \mathbf{q}}, \mathbf{B}(u)\mathbf{q}(u) \right\rangle \sigma^{-1}(u) d\hat{W}^\theta(u) + \int_0^t \frac{\partial^2 V}{\partial s^2} \sigma^2(u) S^2(u) du$$

$$+ \int_0^t S(u) \left\langle \frac{\partial^2 V}{\partial \mathbf{q} \partial s}, \mathbf{B}(u)\mathbf{q}(u) \right\rangle du + \frac{1}{2} \int_0^t (\mathbf{B}(u)\mathbf{q}(u))' \frac{\partial^2 V}{\partial \mathbf{q}^2} (\mathbf{B}(u)\mathbf{q}(u)) du$$

$$+ \sum_{i=1}^N \int_0^t \langle \hat{\mathbf{X}}(u), \mathbf{e}_i \rangle \int_{\Re_0} \Big(V(u, S(u-)e^z, \mathbf{q}(u-)\mathrm{diag}(\mathbf{G}(u)))$$

$$- V(u, S(u-), \mathbf{q}(u-)) \Big) e^{\theta(u)z} \lambda_i(u) \eta_i(dz|u) du$$

$$+ \int_0^t \int_{\Re_0} \Big(V(u, S(u-)e^z, \mathbf{q}(u)) - V(u, S(u-), \mathbf{q}(u-)) \Big) \hat{\gamma}^\theta(du, dz) .$$

Rearranging then gives:

$$V(t,s,\mathbf{q}) = V(0,s_0,\mathbf{q}_0) + \int_0^t \left[\frac{\partial V}{\partial u} + \frac{\partial V}{\partial s} S(u) \left(r(u) - \sum_{i=1}^N \langle \hat{\mathbf{X}}(u), \mathbf{e}_i \rangle \lambda_i(u) \right. \right.$$

$$\left. \times \int_{\Re_0} e^{\theta(u)z}(e^z - 1) \eta_i(dz|u) \right) - \left\langle \frac{\partial V}{\partial \mathbf{q}}, \mathrm{diag}(\mathbf{G}(u) - \mathbf{1})\mathbf{q}(u) \right\rangle$$

$$\times \sum_{i=1}^N \langle \hat{\mathbf{X}}(u), \mathbf{e}_i \rangle \lambda_i(u) \int_{\Re_0} e^{\theta(u)z}(e^z - 1) \eta_i(dz|u) + \left\langle \frac{\partial V}{\partial \mathbf{q}}, \alpha^\theta(u)\mathbf{q}(u) \right\rangle$$

$$+ \frac{1}{2} \frac{\partial^2 V}{\partial s^2} \sigma^2(u) S^2(u) + S(u) \left\langle \frac{\partial^2 V}{\partial \mathbf{q} \partial s}, \mathbf{B}(u)\mathbf{q}(u) \right\rangle + \frac{1}{2} (\mathbf{B}(u)\mathbf{q}(u))' \frac{\partial^2 V}{\partial \mathbf{q}^2} (\mathbf{B}(u)\mathbf{q}(u))$$

$$+ \sum_{i=1}^N \langle \hat{\mathbf{X}}(u), \mathbf{e}_i \rangle \int_{\Re_0} \Big(V(u, S(u-)e^z, \mathbf{q}(u-)\mathrm{diag}(\mathbf{G}(u))) - V(u, S(u-), \mathbf{q}(u-)) \Big)$$

$$\times e^{\theta(u)z} \lambda_i(u) \eta_i(dz|u) \bigg] du + \int_0^t \frac{\partial V}{\partial s} S(u) \sigma(u) d\hat{W}^\theta(u)$$

$$+ \int_0^t \left\langle \frac{\partial V}{\partial \mathbf{q}}, \mathbf{B}(u)\mathbf{q}(u) \right\rangle \sigma^{-1}(u) d\hat{W}^\theta(u)$$

$$+ \int_0^t \int_{\Re_0} \Big(V(u, S(u-)e^z, \mathbf{q}(u)) - V(u, S(u-), \mathbf{q}(u-)) \Big) \hat{\gamma}^\theta(du, dz).$$

Using the martingale condition (8.4) and the definition of $\beta^\theta(t)$,

$$\begin{aligned}
V(t,s,\mathbf{q}) = V(0,s_0,\mathbf{q}_0) &+ \int_0^t \left[\frac{\partial V}{\partial u} + \frac{\partial V}{\partial s} S(u)(\hat{\mu}(u) + \theta(u)\sigma^2(u)) + \left\langle \frac{\partial V}{\partial \mathbf{q}}, (\alpha^\theta(u) - \beta^\theta(u))\mathbf{q}(u) \right\rangle \right. \\
&+ \frac{1}{2} \frac{\partial^2 V}{\partial s^2} \sigma^2(u) S^2(u) + S(u) \left\langle \frac{\partial^2 V}{\partial \mathbf{q} \partial s}, \mathbf{B}(u)\mathbf{q}(u) \right\rangle + \frac{1}{2}(\mathbf{B}(u)\mathbf{q}(u))' \frac{\partial^2 V}{\partial \mathbf{q}^2} (\mathbf{B}(u)\mathbf{q}(u)) \\
&+ \sum_{i=1}^N \langle \hat{\mathbf{X}}(u), \mathbf{e}_i \rangle \int_{\Re_0} \Big(V(u, S(u-)e^z, \mathbf{q}(u-)\mathbf{diag}(\mathbf{G}(u))) - V(u, S(u-), \mathbf{q}(u-)) \Big) \\
&\left. \times e^{\theta(u)z} \lambda_i(u) \eta_i(dz|u) \right] du + \int_0^t \frac{\partial V}{\partial s} S(u) \sigma(u) d\hat{W}^\theta(u) \\
&+ \int_0^t \left\langle \frac{\partial V}{\partial \mathbf{q}}, \mathbf{B}(u)\mathbf{q}(u) \right\rangle \sigma^{-1}(u) d\hat{W}^\theta(u) \\
&+ \int_0^t \int_{\Re_0} \Big(V(u, S(u-)e^z, \mathbf{q}(u)) - V(u, S(u-), \mathbf{q}(u-)) \Big) \tilde{\gamma}^\theta(du, dz).
\end{aligned}$$

Note that the discounted price process $\{V(t, S(t), \mathbf{q}(t)) | t \in \mathcal{T}\}$ is an $(\mathbb{F}^Y, \mathbb{P}^\theta)$-martingale. It must be a special semimartingale. Consequently, the du-integral terms must sum to zero, and hence,

$$\begin{aligned}
&\frac{\partial V}{\partial t} + \frac{\partial V}{\partial s} S(t)(\hat{\mu}(t) + \theta(t)\sigma^2(t)) + \left\langle \frac{\partial V}{\partial \mathbf{q}}, (\alpha^\theta(t) - \beta^\theta(t))\mathbf{q}(t) \right\rangle \\
&+ \frac{1}{2} \frac{\partial^2 V}{\partial s^2} \sigma^2(t) S^2(t) + S(t) \left\langle \frac{\partial^2 V}{\partial \mathbf{q} \partial s}, \mathbf{B}(t)\mathbf{q}(t) \right\rangle + \frac{1}{2}(\mathbf{B}(t)\mathbf{q}(t))' \frac{\partial^2 V}{\partial \mathbf{q}^2} (\mathbf{B}(t)\mathbf{q}(t)) \\
&+ \sum_{i=1}^N \langle \hat{\mathbf{X}}(t), \mathbf{e}_i \rangle \int_{\Re_0} \Big(V(t, S(t-)e^z, \mathbf{q}(t-)\mathbf{diag}(\mathbf{G}(t))) - V(t, S(t-), \mathbf{q}(t-)) \Big) \\
&\times e^{\theta(t)z} \lambda_i(t) \eta_i(dz|t) = 0.
\end{aligned}$$

Therefore, the result follows by applying the differentiation rule to

$$V^\dagger(t,s,\mathbf{q}) = \exp\left(-\int_0^t r(u) du \right) V(t,s,\mathbf{q})$$

again. $\square$

The result in Theorem 8.4 may be extended from the class of smooth functions $\mathscr{C}^{1,2}(\mathcal{T} \times (0, \infty) \times (0, \infty)^N)$ to a wider class of functions in which a generalized Itô's differentiation rule holds. The wider class of functions may include differences of two convex functions, (see, for example, [24]).

8.6 Conclusion

We discussed a two-stage approach for pricing a European-style option in a hidden Markovian regime-switching jump-diffusion model. Filtering theory was first used to turn the original market with partial observations to a filtered market with complete observations. Then the option valuation problem was considered in the filtered market where the hidden quantities in the original market were replaced by their filtered estimates. The generalized Esscher transform for semimartingales was used to select a pricing kernel in the incomplete filtered market. By noticing that the price process and the unnormalized filter process of the hidden Markov chain are jointly Markovian with respect to the observed filtration, a partial differential-integral equation governing the price of the European-style option was derived.

References

1. Bakshi, G., Cao, C., Chen, Z.: Empirical performance of alternative option pricing models. J. Financ. **52**(5), 2003–2049 (1997)
2. Black, F., Scholes, M.: The pricing of options and corporate liabilities. J. Politi. **81**, 637–659 (1973)
3. Buffington, J., Elliott, R.J.: American options with regime switching. Int. J. Theor. Appl. Financ. **5**, 497–514 (2002)
4. Bühlmann, H., Delbaen, F., Embrechts, P., Shiryaev, A.N.: No-arbitrage, change of measure and conditional Esscher transforms. CWI Q. **9**(4), 291–317 (1996)
5. Clark, J.M.C.: The design of robust approximations to the stochastic differential equations for nonlinear filtering. In: Skwirzynski, J.K. (ed.) Communications Systems and Random Process Theory, pp. 721–734. Sijthoff and Noorhoff, Netherlands (1978)
6. Elliott, R.J.: Stochastic Calculus and Applications. Springer, Berlin/Heidelberg/New York (1982)
7. Elliott, R.J.: Filtering and control for point process observations. In: Baras, J.S., Mirelli, V. (eds.) Recent Advances in Stochastic Calculus: Progress in Automation and Information Systems, pp. 1–27. Springer, Berlin/Heidelberg/New York (1990)
8. Elliott, R.J., Aggoun, L., Moore, J.B.: Hidden Markov Models: Estimation and Control, 2nd edn. Springer, Berlin/Heidelberg/New York (1994)
9. Elliott, R.J., Chan, L., Siu, T.K.: Option pricing and Esscher transform under regime switching. Ann. Financ. **1**(4), 423–432 (2005)
10. Elliott, R.J., Siu, T.K.: Pricing and hedging contingent claims with regime switching risk. Commun. Math. Sci. **9**(2), 477–498 (2011)
11. Elliott, R.J., Siu, T.K.: An HMM approach for optimal investment of an insurer. International Journal of Robust and Nonlinear Control **22**(7), 778–807 (2012)
12. Elliott, R.J., Siu, T.K.: Option pricing and filtering with hidden Markov-modulated pure jump processes. Appl. Math. Financ. **20**(1), 1–25 (2013)

13. Elliott, R.J., Siu, T.K., Chan, L., Lau, J.W.: Pricing options under a generalized Markov modulated jump diffusion model. Stoch. Anal. Appl. **25**(4), 821–843 (2007)
14. Esscher, F.: On the probability function in the collective theory of risk. Skandinavisk Aktuarietidskrift **15**, 175–195 (1932)
15. Fleming, W.H., Rishel, R.W.: Deterministic and Stochastic Optimal Control. Springer, Berlin/Heidelberg/New York (1975)
16. Gerber, H.U., Shiu, E.S.W.: Option pricing by Esscher transforms. Trans. Soc. Actuar. **46**, 99–191 (1994)
17. Gerber, H.U., Shiu, E.S.W.: Discussion of Phelim Boyle and Sun Siang Liew's asset allocation with hedge funds on the menu. N. Am. Actuar. J. **12**(1), 89–90 (2008)
18. Goldfeld, S.M., Quandt, R.E.: A Markov model for switching regressions. J. Econom. **1**, 3–16 (1973)
19. Guo, X.: Information and option pricings. Quant. Financ. **1**, 38–44 (2001)
20. Hamilton, J.D.: A new approach to the economic analysis of nonstationary time series and the business cycle. Econometrica **57**, 357–384 (1989)
21. Jacod, J., Shiryaev, A.N.: Limit Theorems for Stochastic Processes. Springer, Berlin/Heidelberg/New York (2003)
22. Kallianpur, G.: Stochastic Filtering Theory. Springer, Berlin/Heidelberg/New York (1980)
23. Kallsen, J., Shiryaev, A.N.: The cumulant process and Esscher's change of measure. Financ. Stoch. **6**, 397–428 (2002)
24. Karatzas, I., Shreve, S.E.: Brownian Motion and Stochastic Calculus, 2nd edn. Springer, Berlin/Heidelberg/New York (1991)
25. Kou, S.G.: A jump diffusion model for option pricing. Manag. Sci. **48**, 1086–1101 (2002)
26. Liu, J., Pan, J., Wang, T.: An equilibrium model of rare-event premia and its implication for option smirks. Rev. Financ. Stud. **18**, 131–164 (2005)
27. Merton, R.C.: The theory of rational option pricing. Bell J. Econ. Manag. Sci. **4**, 141–183 (1973)
28. Merton, R.C.: Option pricing when the underlying stock returns are discontinuous. J. Financ. Econ. **3**, 125–144 (1976)
29. Merton, R.C.: On estimating the expected return on the market. J. Financ. Econ. **8**, 323–361 (1980)
30. Naik, V.: Option valuation and hedging strategies with jumps in the volatility of asset returns. J. Financ. **48**, 1969–1984 (1993)
31. Quandt, R.E.: The estimation of parameters of linear regression system obeying two separate regimes. J. Am. Stat. Assoc. **55**, 873–880 (1958)
32. Pan, J.: The jump-risk premia implicit in options: evidence from an integrated time-series study. J. Financ. Econ. **63**, 3–50 (2002)
33. Siu, T.K.: Fair valuation of participating policies with surrender options and regime switching. Insur.: Math. Econ. **37**(3), 533–552 (2005)
34. Siu, T.K.: A game theoretic approach to option valuation under Markovian regime-switching models. Insur.: Math. Econ. **42**(3), 1146–1158 (2008)

35. Siu, T.K., Lau, J.W., Yang, H.: Pricing participating products under a generalized jump-diffusion. J. Appl. Math. Stoch. Anal. **2008**, 30 p (2008). Article ID: 474623
36. Siu, T.K.: American option pricing and filtering with a hidden regime-switching jump diffusion. Submitted (2013)
37. Tong, H.: On a threshold model. In: Chen, C. (ed.) Pattern Recognition and Signal Processing. NATO ASI Series E: Applied Sc. No. 29, pp. 575–586. Sijthoff & Noordhoff, Netherlands (1978)
38. Tong, H.: Threshold models in non-linear time series analysis. Springer, Berlin/Heidelberg/New York (1983)
39. Zakai, M.: On the optimal filtering of diffusion processes. Zeitschrift fur Wahrscheinlichkeitstheorie und verwandte Gebiete **11**, 230–243 (1969)

Chapter 9
An Exact Formula for Pricing American Exchange Options with Regime Switching

Leunglung Chan

Abstract This paper investigates the pricing of American exchange options when the price dynamics of each underlying risky asset are assumed to follow a Markov-modulated Geometric Brownian motion; that is, the appreciation rate and the volatility of each underlying risky asset depend on unobservable states of the economy described by a continuous-time hidden Markov process. We show that the price of an American exchange option can be reduced to the price of an American option. Then, we modify the result of Zhu and Chan (An analytic formula for pricing American options with regime switching. Submitted for publication, 2012), a closed-form analytical pricing formula for the American exchange option is given.

9.1 Introduction

Option pricing is an important field of research in financial economics from both a theoretical and practical point of view. The pioneering work of Black and Scholes [2] and Merton [25] laid the foundations of the field and stimulated important research in option pricing theory, its mathematical models and its computational techniques. A spread option is an option whose payoff depends on the price spread between two correlated underlying risky assets. If the asset prices are S_1 and S_2, then the payoff function to a spread put option with a strike K is $\max\{K - (S_1 - S_2), 0\}$. The spread options are traded either on an exchange or largely in the over-the-counter markets. The spread options are ubiquitous. The spread options are widely-used in various markets such as the currency and foreign exchange markets, the energy markets and the commodity futures markets (for instance, see Carmona and Durrleman [5] for an excellent survey). Some papers for pricing spread options

---

L. Chan (✉)
School of Mathematics and Statistics, University of New South Wales,
Sydney, NSW, 2052, Australia
e-mail: leung.chan@unsw.edu.au

under the Geometric Brownian motions (GBMs) include Ravindran [28], Shimko [29] and Venkatramanan and Alexander [31]. Even in the GBMs setting, there is no closed-form solution for European-style spread options except a special case with strike $K = 0$. This is the only case for which one has a solution in a closed form via Black-Scholes for the price of the spread option with zero strike (for instance, see [24]). Two approaches such as analytical approximations and numerical methods have been used for the pricing of spread options. For analytical approximations, see Kirk [21], Eydeland and Wolyniec [16] and Deng et al. [8]. For numerical methods, see Hurd and Zhou [20]. Beyond the GBMs framework, Dempster and Hong [7] priced spread options under a stochastic volatility model in order to capture volatility skews on the two underlying assets. Cheang and Chiarella [6] priced exchange options under jump-diffusion dynamics. Benth et al. [1] priced exchange options on a bivariate jump market and stability to model risk.

Despite its popularity the Black-Scholes-Merton model fails in various ways, such as the fact that implied volatility is not constant. During the past few decades many extensions to the Black-Scholes-Merton model have been introduced in the literature to provide more realistic descriptions for asset price dynamics. In particular, many models have been introduced to explain the empirical behavior of the implied volatility smile and smirk. Such models include the stochastic volatility models, jump-diffusion models, models driven by Lévy processes and regime switching models.

Maybe the simplest way to introduce additional randomness into the standard Black-Scholes-Merton model is to let the volatility and rate of return be functions of a finite state Markov chain. There has been considerable interest in applications of regime switching models driven by a Markov chain to various financial problems. Many papers in a regime switching framework include Elliott and van der Hoek [10], Guo [18], Elliott et al. [12, 13, 14, 15] and Buffington and Elliott [3, 4]. In addition, Siu et al. [30] priced the value of credit default swaps under a Markov-modulated Merton structural model. Yuen and Yang [32] proposed a recombined trinomial tree to price simple options and barrier options in a jump-diffusion model with regime switching. Yuen and Yang [33] used a trinomial tree method to price Asian options and equity-indexed annuities with regime switching. Zhu et al. [36] derived a closed-form solution for European options with a two-state regime switching model. Zhu and Chan [35] derived a closed-form solution for American options under a two-state regime switching model.

In this paper, we investigate the pricing of American spread options with strike $K = 0$ when the price dynamics of each underlying risky asset are assumed to follow a Geometric Brownian motion with a constant correlation. The spread options with strike $K = 0$ is called exchange options. We derive analytical solutions for American exchange options by means of the homotopy analysis method (HAM). HAM was initially suggested by Ortega and Rheinboldt [27]. The HAM consists of some standard methods such as Lyapunov's small artificial method, the δ-expansion method, Adomian decomposition method, and the Euler transform, so that it has the great generality. Consequently, the HAM provides us an useful tool to solve highly nonlinear problems in science and mathematical finance. For instance, the HAM has been successfully used to solve a number of heat transfer problems, see Liao and Zhu [23]. Zhu [34] proposed to adopt HAM to obtain an analytic pricing formula for American options in the Black-Scholes model. Gounden and O'Hara [17] extended the work of Zhu to pricing an American-style Asian option of floating strike type in the Black-Scholes model.

Obviously, it is important to guarantee the convergence of an approximation series. Unfortunately, near all previous analytic approximation methods such as perturbation methods, Lyapunov's artificial small parameter method, Adomian decomposition method, the δ-expansion method, cannot guarantee the convergence of approximation series, this is the reason why they are valid mainly for weakly nonlinear problems. However, based on the homotopy in topology, the HAM provides us a great freedom to introduce a non-zero auxiliary parameter, namely the convergence-control parameter h, into the so-called zeroth-order deformation equation. One can find some proper values of h to guarantee the convergence of the homotopy series. For more details, we refer to Liao [22]. In addition, Odibat [26] studied the convergence of the homotopy analysis method when applied to nonlinear problems. The sufficient condition for convergence of the method and the error estimate are presented in his work. Without the loss of generality, we choose $h = 1$ in this paper.

This paper is organized as follows. Section 9.2 describes the dynamics of the asset price under the Markov-modulated Geometric Brownian motion. Section 9.3 formulates the partial differential equations for the prices of European spread options and American spread options. In the case of an exchange option, the governing PDEs are reduced into two-dimensional PDEs which are just ordinary European or American options' PDEs. Section 9.4 presents an exact, closed-form solution for the American exchange options. The final section contains a conclusion.

9.2 Asset Price Dynamics

Consider a complete probability space $(\Omega, \mathscr{F}, \mathscr{P})$, where $\mathscr{P}$ is a real-world probability measure. Let $\mathscr{T}$ denote the time index set $[0, T]$ of the model. Write $\{W_t\}_{t \in \mathscr{T}}$ for a standard Brownian motion on $(\Omega, \mathscr{F}, \mathscr{P})$. Suppose the states of an economy are modelled by a finite state continuous-time Markov chain $\{X_t\}_{t \in \mathscr{T}}$ on $(\Omega, \mathscr{F}, \mathscr{P})$. Without loss of generality, we can identify the state space of $\{X_t\}_{t \in \mathscr{T}}$ with a finite set of unit vectors $\mathscr{X} := \{e_1, e_2, \ldots, e_N\}$, where $e_i = (0, \ldots, 1, \ldots, 0) \in \mathscr{R}^N$. We suppose that $\{X_t\}_{t \in \mathscr{T}}$ and $\{W_t\}_{t \in \mathscr{T}}$ are independent.

Let A be the generator $[a_{ij}]_{i,j=1,2,\ldots,N}$ of the Markov chain process. From Elliott et al. [11], we have the following semi-martingale representation theorem for $\{X_t\}_{t \in \mathscr{T}}$:

$$X_t = X_0 + \int_0^t AX_s ds + M_t, \qquad (9.1)$$

where $\{M_t\}_{t \in \mathscr{T}}$ is an $\mathscr{R}^N$-valued martingale increment process with respect to the filtration generated by $\{X_t\}_{t \in \mathscr{T}}$.

We consider a financial model with three primary traded assets, namely a money market account B and two risky assets or stocks with respective prices $S_1(t)$ and $S_2(t)$ at time t. Suppose the market is frictionless; the borrowing and lending interest rates are the same; the investors are price-takers.

The instantaneous market interest rate $\{r(t, X_t)\}_{t \in \mathscr{T}}$ of the bank account is given by:

$$r_t := r(t, X_t) = <r, X_t>, \tag{9.2}$$

where $r := (r_1, r_2, \ldots, r_N)$ with $r_i > 0$ for each $i = 1, 2, \ldots, N$ and $<\cdot, \cdot>$ denotes the inner product in $\mathscr{R}^N$.

We suppose the dynamics of the price process $\{B_t\}_{t \in \mathscr{T}}$ for the bank account are described by:

$$dB_t = r_t B_t dt, \quad B_0 = 1, \tag{9.3}$$

Suppose the stock appreciation rates $\{\mu_1(t)\}_{t \in \mathscr{T}}$, $\{\mu_2(t)\}_{t \in \mathscr{T}}$ and the volatilities $\{\sigma_1(t)\}_{t \in \mathscr{T}}$ and $\{\sigma_2(t)\}_{t \in \mathscr{T}}$ of $S_1(t)$ and $S_2(t)$ depend on $\{X_t\}_{t \in \mathscr{T}}$ and are described by:

$$\mu_1(t) := \mu_1(t, X_t) = <\mu_1, X_t>, \quad \mu_2(t) := \mu_2(t, X_t) = <\mu_2, X_t>,$$
$$\sigma_1(t) := \sigma_1(t, X_t) = <\sigma_1, X_t>, \quad \sigma_2(t) := \sigma_2(t, X_t) = <\sigma_2, X_t>, \tag{9.4}$$

where $\mu_1 := (\mu_1^1, \mu_1^2, \ldots, \mu_1^N)$, $\mu_2 := (\mu_2^1, \mu_2^2, \ldots, \mu_2^N)$, $\sigma_1 := (\sigma_1^1, \sigma_1^2, \ldots, \sigma_1^N)$, $\sigma_2 := (\sigma_2^1, \sigma_2^2, \ldots, \sigma_2^N)$ with $\sigma_1^i > 0$ and $\sigma_2^i > 0$ for each $i = 1, 2, \ldots, N$ and $<\cdot, \cdot>$ denotes the inner product in $\mathscr{R}^N$.

We assume that the dynamics of the underlying risky asset prices $S_1(t)$ and $S_2(t)$ over time are modelled by the following stochastic differential equations:

$$dS_1(t) = \mu_1 S_1(t) dt + \sigma_1 S_1(t) dW_t^1,$$
$$dS_2(t) = \mu_2 S_2(t) dt + \sigma_2 S_2(t) dW_t^2,$$
$$S_1(0) = s_1, \quad S_2(0) = s_2, \tag{9.5}$$

where $(W_t^1)_{t \in \mathscr{T}}$ and $(W_t^2)_{t \in \mathscr{T}}$ are Brownian motions with $E[dW_t^1 dW_t^2] = \rho dt$; $\rho \in (-1, 1)$ is a constant correlation coefficient. Note that W_t^2 can be decomposed over a basis of uncorrelated Brownian motions W_t^1, $(W_t^1)^\perp$:

$$W_t^2 = \rho W_t^1 + \sqrt{1 - \rho^2}(W_t^1)^\perp. \tag{9.6}$$

Setting $\hat{W}_t^1 = W_t^1$ and $\hat{W}_t^2 = (W_t^1)^\perp$, Eqs. (9.5) become

$$dS_1(t) = \mu_1 S_1(t) dt + \sigma_1 S_1(t) d\hat{W}_t^1,$$
$$dS_2(t) = \mu_2 S_2(t) dt + \sigma_2 \rho S_2(t) d\hat{W}_t^1 + \sigma_2 \sqrt{1 - \rho^2} S_2(t) d\hat{W}_t^2, \tag{9.7}$$

where $\hat{W}_t^1$ and $\hat{W}_t^2$ are independent standard Brownian motions. Let $\mathscr{F}_t$ be the σ-algebra generated by the random variables $\hat{W}_s^1$ and $\hat{W}_s^2$ for $s \leq t$. Then the processes $(\hat{W}_s^1)_{t \in \mathscr{T}}$ and $(\hat{W}_s^2)_{t \in \mathscr{T}}$ are $\mathscr{F}_t$-Brownian motions.

9.3 Problem Formulation

In this section, we formulate the partial differential equations for the prices of European spread options and American spread options.

It is assumed that under the risk-neutral measure Q the price dynamics of the underlying risky assets $S_1(t)$ and $S_2(t)$ satisfy

$$dS_1(t) = r_t S_1(t) dt + \sigma_1 S_1(t) d\hat{W}_t^1,$$
$$dS_2(t) = r_t S_2(t) dt + \sigma_2 \rho S_2(t) d\hat{W}_t^1 + \sigma_2 \sqrt{1-\rho^2} S_2(t) d\hat{W}_t^2. \tag{9.8}$$

Write $\tilde{\mathscr{G}}(t)$ for the σ-field $\mathscr{F}^W(t) \vee \mathscr{F}^{\mathbf{X}}(T)$, for each $t \in \mathscr{T}$. Then, given $\tilde{\mathscr{G}}(t)$, a conditional price of a European spread option V is:

$$V(t, S_1, S_2, \mathbf{X}) = E^Q[e^{-\int_t^T r_u du} V(S_1(T), S_2(T)) | \tilde{\mathscr{G}}(t)]. \tag{9.9}$$

Given $S_1(t) = s_1$, $S_2(t) = s_2$ and $\mathbf{X}(t) = \mathbf{x}$, a price of a European spread option V is:

$$V(t, s_1, s_2, \mathbf{x}) = E^Q[e^{-\int_t^T r_u du} V(S_1(T), S_2(T)) | S_1(t) = s_1, S_2(t) = s_2, \mathbf{X}(t) = \mathbf{x}]. \tag{9.10}$$

Let $V_i := V(t, s_1, s_2, \mathbf{e}_i)$, for each $i = 1, 2, \ldots, N$. Write $\mathbf{V} := (V_1, V_2, \ldots, V_N)'$, so $V(t, s_1, s_2, \mathbf{x}) = \langle \mathbf{V}, \mathbf{x} \rangle$. Applying the Feynman-Kac formula to Eq. (9.10), then $V := V(t, s_1, s_2, \mathbf{x})$ satisfies the system of partial differential equations (PDEs)

$$\frac{\partial V}{\partial t} + \frac{1}{2}\sigma_1^2 s_1^2 \frac{\partial^2 V}{\partial s_1^2} + \rho \sigma_1 \sigma_2 s_1 s_2 \frac{\partial^2 V}{\partial s_1 \partial s_2} + \frac{1}{2}\sigma_2^2 s_2^2 \frac{\partial^2 V}{\partial s_2^2} + r_t s_1 \frac{\partial V}{\partial s_1} + r_t s_2 \frac{\partial V}{\partial s_2}$$
$$- r_t V + \langle \mathbf{V}, \mathbf{A}\mathbf{x} \rangle = 0, \tag{9.11}$$

with terminal condition:

$$V(T, s_1, s_2, \mathbf{x}) = (K - (s_1 - s_2))^+, \tag{9.12}$$

and boundary condition:

$$\lim_{s_1 \to \infty} V(t, s_1, s_2, \mathbf{x}) = 0. \tag{9.13}$$

So, if $\mathbf{X}(t) := \mathbf{e}_i$ ($i = 1, 2, \ldots, N$),

$$\sigma_1(t) = \sigma_{1i}, \sigma_2(t) = \sigma_{2i}, \quad V(t, s_1, s_2, \mathbf{x}) = V(t, s_1, s_2, \mathbf{e}_i) := V_i, \quad (9.14)$$

and V_i ($i = 1, 2, \ldots, N$) satisfy the following system of PDEs:

$$\frac{\partial V_i}{\partial t} + \frac{1}{2}\sigma_{1i}^2 s_1^2 \frac{\partial^2 V_i}{\partial s_1^2} + \rho \sigma_{1i} \sigma_{2i} s_1 s_2 \frac{\partial^2 V_i}{\partial s_1 \partial s_2} + \frac{1}{2}\sigma_{2i}^2 s_2^2 \frac{\partial^2 V_i}{\partial s_2^2} + r_i s_1 \frac{\partial V_i}{\partial s_1} + r_i s_2 \frac{\partial V_i}{\partial s_2}$$
$$- r_i V_i + \langle \mathbf{V}, \mathbf{A} \mathbf{e}_i \rangle = 0, \quad (9.15)$$

with the terminal condition:

$$V(T, s_1, s_2, \mathbf{e}_i) = (K - (s_1 - s_2))^+, \quad i = 1, 2, \ldots, N, \quad (9.16)$$

and boundary condition:

$$\lim_{s_1 \to \infty} V(t, s_1, s_2, \mathbf{e}_i) = 0, \quad i = 1, 2, \ldots, N. \quad (9.17)$$

With $K = 0$, the price of an exchange option satisfies the following system of PDEs:

$$\frac{\partial V_i}{\partial t} + \frac{1}{2}\sigma_{1i}^2 s_1^2 \frac{\partial^2 V_i}{\partial s_1^2} + \rho \sigma_{1i} \sigma_{2i} s_1 s_2 \frac{\partial^2 V_i}{\partial s_1 \partial s_2} + \frac{1}{2}\sigma_{2i}^2 s_2^2 \frac{\partial^2 V_i}{\partial s_2^2} + r_i s_1 \frac{\partial V_i}{\partial s_1} + r_i s_2 \frac{\partial V_i}{\partial s_2}$$
$$- r_i V_i + \langle \mathbf{V}, \mathbf{A} \mathbf{e}_i \rangle = 0, \quad (9.18)$$

with the terminal condition:

$$V(T, s_1, s_2, \mathbf{e}_i) = (s_2 - s_1)^+, \quad i = 1, 2, \ldots, N, \quad (9.19)$$

and boundary condition:

$$\lim_{s_1 \to \infty} V(t, s_1, s_2, \mathbf{e}_i) = 0, \quad i = 1, 2, \ldots, N. \quad (9.20)$$

The three-dimensional PDEs for the price of an exchange option can be reduced into a two-dimensional PDEs either by the method of change measure or by a similarity reduction. We adopt the PDE approach to reduce the dimensionality. From the definition of an exchange option and the fact that $S_1(T)$ and $S_2(T)$ are linear in the initial conditions s_1 and s_2 respectively, V_i has a linear homogeneous property:

$$\lambda V_i(s_1, s_2) = V_i(\lambda s_1, \lambda s_2), \quad (9.21)$$

for any $\lambda > 0$, so taking $\lambda = \frac{1}{s_2}$, we have

$$V_i(t, s_1, s_2) = s_2 V_i'(t, \frac{s_1}{s_2}) \quad (9.22)$$

9 An Exact Formula for Pricing American Exchange Options with Regime Switching

where $V_i'(t,y) = V_i(t,y,1)$. The derivatives of V_i can be expressed in terms of those of V_i'. We have

$$\frac{\partial V_i}{\partial s_1} = \frac{\partial V_i'}{\partial y}, \quad \frac{\partial V_i}{\partial s_2} = V_i' - \frac{s_1}{s_2}\frac{\partial V_i'}{\partial y}.$$

We observe that $s_1 \frac{\partial V_i}{\partial s_1} + s_2 \frac{\partial V_i}{\partial s_2} = V_i$, so that $r_i s_1 \frac{\partial V_i}{\partial s_1} + r_i s_2 \frac{\partial V_i}{\partial s_2} - r_i V_i = 0$. After substitution and simple calculation, Eq. (9.18) is reduced to the following PDE for V_i'

$$\frac{\partial V_i'}{\partial t} + \frac{1}{2}y^2 \sigma_i^2 \frac{\partial^2 V_i'}{\partial y^2} + \langle \mathbf{V}', \mathbf{A}\mathbf{e}_i \rangle = 0 \tag{9.23}$$

where $\sigma_i = \sqrt{\sigma_{1i}^2 + \sigma_{2i}^2 - 2\rho \sigma_{1i}\sigma_{2i}}$. The boundary condition is

$$V_i'(T,y) = (1-y)^+. \tag{9.24}$$

Now consider an American exchange option. Given that $S_t = S$ and that $X_t = X$, the price of an American exchange put option at time t is given by:

$$V(t,T,S_1,S_2,X)$$
$$= \sup_{\tau \in [t,T]} E^Q\left[\exp\left(-\int_t^T r_u du\right)(S_2(T) - S_1(T))^+ \Big| S_1(t) = S_1, S_2(t) = S_2, X_t = X\right]$$
$$= \sup_{\tau \in [t,T]} E^Q\left[S_2(T)\exp\left(-\int_t^T r_u du\right)\left(1 - \frac{S_1(T)}{S_2(T)}\right)^+ \Big| S_1(t) = S_1, S_2(t) = S_2, X_t = X\right]$$
$$= \sup_{\tau \in [t,T]} s_2(0) E^{\tilde{Q}}\left[\exp\left(-\int_t^T r_u du\right)(1-y)^+ \Big| y(t) = y, X_t = X\right]$$
$$= s_2 V'(t,T,y,X) \tag{9.25}$$

where the supremum is taken over the set of stopping times τ taking values in $[t,T]$ and $\frac{d\tilde{Q}}{dQ} = \frac{1}{s_2(0)} S_2(T)$.

Let $\mathbf{V}' := \mathbf{V}'(t,T,y) = (V'(t,T,y,e_1),\ldots,V'(t,T,y,e_N)) = (V_1',\ldots,V_N')$. When $X_t = e_i$ ($i = 1,2,\ldots,N$), the continuation region is given by:

$$\mathscr{C}^i = \{(y,t) \in \mathscr{R}^+ \times [0,T] | V'(t,T,y,e_i) > (1-y)^+\}, \tag{9.26}$$

and the stopping region is given by:

$$\mathscr{S}^i = \{(y,t) \in \mathscr{R}^+ \times [0,T] | V'(t,T,y,e_i) = (1-y)^+\}. \tag{9.27}$$

As in Buffington and Elliott [4], we write $\mathscr{C}_t^i$ for the t-section of $\mathscr{C}^i$, for each $i = 1,2,\ldots,N$ and $t \in [0,T]$. Let $y_{f_i}(t) = y_f(t,e_i)$ denote the critical price of the American put when $X_t = e_i$ ($i = 1,2,\ldots,N$); that is, when the state $X_t = e_i$ and the price of the underlying risky asset at time t falls below $y_{f_i}(t)$, it is rational for the holder of the American put to exercise the option at time t. Note that $y_{f_i}(T) = K$, for each $i = 1, 2, \ldots, N$. Then, by Elliott and Kopp [9] and Buffington and Elliott [4],

$$\mathscr{C}_t^i = (y_{f_i}(t), \infty), \quad i = 1, 2, \ldots, N. \tag{9.28}$$

The price for an American exchange option $V := s_2 V' := s_2 V'(t, T, y, \mathbf{x})$ where $V'(t, T, y, \mathbf{x})$ satisfies the system of partial differential equations (PDEs)

$$\frac{\partial V_i'}{\partial t} + \frac{1}{2} y^2 \sigma_i^2 \frac{\partial^2 V_i'}{\partial y^2} + \langle \mathbf{V}', \mathbf{A} \mathbf{e}_i \rangle = 0. \tag{9.29}$$

The boundary conditions are

$$V_i'(t, T, y_{f_i}(t)) = 1 - y_{f_i}(t), \tag{9.30}$$

$$\frac{\partial V_i'}{\partial y}(t, T, y_{f_i}(t)) = -1, \tag{9.31}$$

$$V_i'(T, T, y) = (1 - y)^+. \tag{9.32}$$

Note that the above free boundary system of PDEs for the price of American exchange option is just an American put option with the strike $K = 1$ and the underlying process y.

Now, we restrict ourselves to a special case with the number of regimes N being 2 in order to simplify our discussion. In this two-state regime switching model, we have two regions: the common continuation region and the transition region. We first discuss the American put value function in the common continuation region where $y > y_{f_2}(t)$ and V_1', V_2' satisfy the equations:

$$\begin{cases} \frac{\partial V_1'}{\partial t} + \frac{1}{2} y^2 \sigma_1^2 \frac{\partial^2 V_1'}{\partial y^2} + \langle \mathbf{V}', \mathbf{A} \mathbf{e}_1 \rangle = 0 \\ V_1'(t, T, y_{f_1}(t)) = 1 - y_{f_1}(t) \\ \frac{\partial V_1'}{\partial y}(t, T, y_{f_1}(t)) = -1 \\ V_1'(T, T, y) = \max\{1 - y, 0\} \\ \lim_{y \to \infty} V_1'(t, T, y) = 0 \end{cases} \tag{9.33}$$

$$\begin{cases} \frac{\partial V_2'}{\partial t} + \frac{1}{2} y^2 \sigma_2^2 \frac{\partial^2 V_2'}{\partial y^2} + \langle \mathbf{V}', \mathbf{A} \mathbf{e}_2 \rangle = 0 \\ V_2'(t, T, y_{f_2}(t)) = 1 - y_{f_2}(t) \\ \frac{\partial V_2'}{\partial y}(t, T, y_{f_2}(t)) = -1 \\ V_2'(T, T, y) = \max\{1 - y, 0\} \\ \lim_{y \to \infty} V_2'(t, T, y) = 0. \end{cases} \tag{9.34}$$

Now consider the transition region between the two stopping curves:

$$\Gamma : \{(y, t) : y_{f_1}(t) \leq y \leq y_{f_2}(t)\}$$

9 An Exact Formula for Pricing American Exchange Options with Regime Switching

In this region $V'_2 = V'(t,T,y,e_2) = 1-y$ and V'_1 satisfies the Black-Scholes equation

$$\frac{\partial V'_1}{\partial t} + \frac{1}{2}y^2\sigma^2\frac{\partial^2 V'_1}{\partial y^2} + a_{11}V'_1 - a_{11}(1-y) = 0, \quad (9.35)$$

with a terminal condition

$$V'_1(T,T,y) = \max\{1-y, 0\}.$$

Also, continuity on $y_{f_1}(t)$ gives:

$$V'_1(t,T,y) = 1 - y_{f_1}(t)$$

and smoothness on $y_{f_1}(t)$ gives:

$$\frac{\partial V'_1}{\partial y}(t,T,y_{f_1}(t)) = -1. \quad (9.36)$$

9.4 A Closed-Form Formula

In this section, we modify the result of Zhu and Chan [35] to obtain a closed-form formula for the price of an American exchange option under a two-state regime switching model by means of the homotopy analysis method. The pricing American exchange option PDEs is just the American option PDEs. The closed-form formula for the price of an American option under a two-state regime switching model is given by Zhu and Chan [35]. We state the following closed-form solution for the price of the American exchange option under a two-state regime switching model which was due to Zhu and Chan [35].

Lemma 9.1. *Let* $\tau_i := \frac{\sigma_i^2}{2}(T-t)$, $\lambda_i = \frac{2a_{ii}}{\sigma_i^2}$ *and* $x_i = \ln\left(\frac{y}{y_{f_i}(\tau_i)}\right)$. *Then the American exchange put price* $V_i = s_2 V'_i, i = 1,2$ *in the common continuation region is given as follows:*

$$V_i = s_2 V'_i = s_2 \sum_{n=0}^{\infty} \frac{\bar{V}_i^n(\tau_i, x_i)}{n!}, \quad i = 1,2 \quad (9.37)$$

with the initial guess $\bar{V}_i^0(\tau_i, x_i)$ is given by a closed-form solution of a European put option with a two-state regime switching (for instance, see Zhu et al. [36]):

$V_i^0(t,y)$

$$= e^{-r_i(T-t)} + \frac{1}{4\pi\sqrt{2}}\sqrt{y}e^{-\frac{1}{2}(r_i+a_{21}+a_{12}+\frac{\sigma_1^2+\sigma_2^2}{8})(T-t)} \int_0^\infty \frac{(-1)^{i-1}2\hat{f}_1(\rho)(a_{21}+a_{12})}{M(\rho)(\rho^4+\frac{1}{16})(\sigma_1^2-\sigma_2^2)}$$

$$\times \left\{ e^{X_i(\rho)}\left[(2\rho^2-\frac{1}{2})\sin(\hat{f}_2(\rho)+\theta(\rho)-Y_i(\rho)) - (2\rho^2+\frac{1}{2})\cos(\hat{f}_2(\rho)+\theta(\rho)-Y_i(\rho))\right] \right.$$

$$\left. - e^{-X_i(\rho)}\left[(2\rho^2-\frac{1}{2})\sin(\hat{f}_2(\rho)+\theta(\rho)+Y_i(\rho)) - (2\rho^2+\frac{1}{2})\cos(\hat{f}_2(\rho)+\theta(\rho)+Y_i(\rho))\right] \right\}$$

$$+ \frac{2\hat{f}_1(\rho)}{M(\rho)}\left\{ e^{X_i(\rho)}\left[\sin(\hat{f}_2(\rho)+\theta(\rho)-Y_i(\rho)) + \cos(\hat{f}_2(\rho)+\theta(\rho)-Y_i(\rho))\right] \right.$$

$$\left. - e^{-X_i(\rho)}\left[\sin(\hat{f}_2(\rho)+\theta(\rho)+Y_i(\rho)) + \cos(\hat{f}_2(\rho)+\theta(\rho)+Y_i(\rho))\right] \right\}$$

$$+ \frac{\hat{f}_1(\rho)}{\rho^4+\frac{1}{16}}\left\{ e^{X_i(\rho)}\left[(2\rho^2-\frac{1}{2})\sin(\hat{f}_2(\rho)-Y_i(\rho)) - (2\rho^2+\frac{1}{2})\cos(\hat{f}_2(\rho)-Y_i(\rho))\right] \right.$$

$$\left. + e^{-X_i(\rho)}\left[(2\rho^2-\frac{1}{2})\sin(\hat{f}_2(\rho)+Y_i(\rho)) - (2\rho^2+\frac{1}{2})\cos(\hat{f}_2(\rho)+Y_i(\rho))\right] \right\} d\rho,$$

(9.38)

for $i = 1,2$, where

$$\tau = \frac{\sigma_1^2-\sigma_2^2}{4}(T-t), \quad \alpha = \frac{2(a_{12}-a_{21})}{\sigma_1^2-\sigma_2^2}, \quad \mu^2 = \frac{4a_{12}a_{21}}{(\sigma_1^2-\sigma_2^2)^2},$$

$$M(\rho) = \left\{ [(\tfrac{1}{4}+\alpha)^2 - \rho^4 + \mu^2]^2 + 4\rho^4(\tfrac{1}{4}+\alpha)^2 \right\}^{\frac{1}{4}},$$

$$\theta(\rho) = \frac{1}{2}\tan^{-1}\left[\frac{2\rho^2(\tfrac{1}{4}+\alpha)}{(\tfrac{1}{4}+\alpha)^2-\rho^4+\mu^2}\right],$$

$$X_i(\rho) = (-1)^{i-1}M(\rho)\tau\cos\theta(\rho), \quad Y_i(\rho) = (-1)^{i-1}M(\rho)\tau\sin\theta(\rho)$$

and

$$\hat{f}_1(\rho) = e^{-\frac{\rho}{\sqrt{2}}|\ln(y)+r_i(T-t)|}, \quad \hat{f}_2(\rho) = \frac{\rho^2}{4}(\sigma_1^2+\sigma_2^2)(T-t) - \frac{\rho}{\sqrt{2}}|\ln(y)+r_i(T-t)|.$$

Then each $\bar{V}_i^n(\tau_i,x_i), n = 1,2,\ldots$, is recursively given by

$$\bar{V}_1^n(\tau_1,x_1) = -\frac{2}{\sqrt{\pi}}\int_0^\infty \int_{(x_1+\eta_1)/2\sqrt{\tau_1}}^\infty \phi_1^n\left(\tau_1 - \frac{(x_1+\eta_1)^2}{4\xi_1^2}\right)$$

$$\times e^{-[(x_1+\eta_1)/\xi_1]^2 - \xi_1^2} d\xi_1 d\eta_1 + \int_0^{\tau_1}\left\{\frac{e^{\eta_1}}{\sqrt{\pi}}\left[e^{x_1}\right.\right.$$

9 An Exact Formula for Pricing American Exchange Options with Regime Switching   221

$$\times \int_{-x_1/2\sqrt{\tau_1-\eta_1}}^{x_1/2\sqrt{\tau_1-\eta_1}} f_1^n(\eta_1, 2\sqrt{\tau_1-\eta_1}\xi_1 + x_1)e^{\sqrt{\tau_1-\eta_1}\xi_1 - \xi_1^2} d\xi_1$$

$$+ \int_{x_1/2\sqrt{\tau_1-\eta_1}}^{\infty} \left[e^{x_1} f_1^n(\eta_1, 2\sqrt{\tau_1-\eta_1}\xi_1 + x_1) \right.$$

$$\left. + e^{-x_1} f_1^n(\eta_1, 2\sqrt{\tau_1-\eta_1}\xi_1 - x_1) \right] \times e^{\sqrt{\tau_1-\eta_1}\xi_1 - \xi_1^2} d\xi_1 \Bigg]$$

$$- 2\sqrt{\tau_1-\eta_1} e^{\tau_1} \int_{x_1/2\sqrt{\tau_1-\eta_1}}^{\infty} f_1^n(\eta_1, 2\sqrt{\tau_1-\eta_1}\xi_1 - x_1)$$

$$\times e^{2\sqrt{\tau_1-\eta_1}\xi_1} \operatorname{erfc}\left(\xi_1 + \sqrt{\tau_1-\eta_1}\right) d\xi_1 \Bigg\} d\eta_1, \tag{9.39}$$

$$\bar{V}_2^n(\tau_2, x_2) = -\frac{2}{\sqrt{\pi}} \int_0^{\infty} \int_{(x_2+\eta_2)/2\sqrt{\tau_2}}^{\infty} \phi_2^n\left(\tau_2 - \frac{(x_2+\eta_2)^2}{4\xi_2^2}\right)$$

$$\times e^{-[(x_2+\eta_2)/\xi_2]^2 - \xi_2^2} d\xi_2 d\eta_2 + \int_0^{\tau_2} \left\{ \frac{e^{\eta_2}}{\sqrt{\pi}} \left[e^{x_2} \right. \right.$$

$$\times \int_{-x_2/2\sqrt{\tau_2-\eta_2}}^{x_2/2\sqrt{\tau_2-\eta_2}} f_2^n(\eta_2, 2\sqrt{\tau_2-\eta_2}\xi_2 + x_2)e^{\sqrt{\tau_2-\eta_2}\xi_2 - \xi_2^2} d\xi_2$$

$$+ \int_{x_2/2\sqrt{\tau_2-\eta_2}}^{\infty} \left[e^{x_2} f_2^n(\eta_2, 2\sqrt{\tau_2-\eta_2}\xi_2 + x_2) \right.$$

$$\left. + e^{-x_2} f_2^n(\eta_2, 2\sqrt{\tau_2-\eta_2}\xi_2 - x_2) \right] \times e^{\sqrt{\tau_2-\eta_2}\xi_2 - \xi_2^2} d\xi_2 \Bigg]$$

$$- 2\sqrt{\tau_2-\eta_2} e^{\tau_2} \int_{x_2/2\sqrt{\tau_2-\eta_2}}^{\infty} f_2^n(\eta_2, 2\sqrt{\tau_2-\eta_2}\xi_2 - x_2)$$

$$\times e^{2\sqrt{\tau_2-\eta_2}\xi_2} \operatorname{erfc}\left(\xi_2 + \sqrt{\tau_2-\eta_2}\right) d\xi_2 \Bigg\} d\eta_2, \tag{9.40}$$

where

$$f_i^n(\tau_i, x_i) = \begin{cases} -\mathcal{L}_i[\bar{V}_i^0(\tau_i, x_i)] + \tilde{\mathcal{A}}_i(\tau_i, x_i, 0), & \text{if } n = 1 \\ n \frac{\partial^{n-1} \tilde{\mathcal{A}}_i}{\partial p^{n-1}}\big|_{p=0}, & \text{if } n \geq 2, \end{cases} \tag{9.41}$$

$$\phi_i^n(\tau_i) = \begin{cases} \bar{V}_i^0(\tau_i, 0) - \frac{\partial \bar{V}_i^0}{\partial x_i}(\tau_i, 0) - 1, & \text{if } n = 1 \\ 0, & \text{if } n \geq 2. \end{cases} \tag{9.42}$$

Here $\mathcal{L}_i$ is a differential operator defined as

$$\mathcal{L}_i = \frac{\partial}{\partial \tau_i} - \frac{\partial^2}{\partial x_i^2}, \tag{9.43}$$

$\tilde{\mathscr{A}}_i = \mathscr{L}_i[\bar{V}_i] - \mathscr{A}_i, i = 1, 2$ and $\mathscr{A}_i$ is a functional defined as

$$\mathscr{A}_1[\bar{V}_1(\tau_1, x_1, p), \bar{y}_{f_1}(\tau_1, p)]$$
$$= \mathscr{L}_1(\bar{V}_1) - \lambda_1(\bar{V}_1 - \bar{V}_2) - \frac{1}{\bar{y}_{f_1}(\tau_1, p)} \frac{d\bar{y}_{f_1}(\tau_1, p)}{d\tau_1} \frac{\partial \bar{V}_1}{\partial x_1}(\tau_1, x_1, p) \quad (9.44)$$

$$\mathscr{A}_2[\bar{V}_2(\tau_2, x_2, p), \bar{y}_{f_2}(\tau_2, p)]$$
$$= \mathscr{L}_2(\bar{V}_2) - \lambda_2(\bar{V}_2 - \bar{V}_1) - \frac{1}{\bar{y}_{f_2}(\tau_2, p)} \frac{d\bar{y}_{f_2}(\tau_2, p)}{d\tau_2} \frac{\partial \bar{V}_2}{\partial x_2}(\tau_2, x_2, p). \quad (9.45)$$

We also obtain the exact, explicit and closed-form formula for the optimal exercise boundaries $y_{f_i}(\tau_i), i = 1, 2$ in the form:

$$y_{f_1}(\tau_1) = \frac{2}{\sqrt{\pi}} \sum_{n=0}^{\infty} \frac{1}{n!} \left\{ e^{\tau_1} \int_0^{\infty} e^{-\eta_1} \int_{\eta_1/2\sqrt{\tau_1}}^{\infty} \phi_1^n \left(\tau_1 - \frac{\eta_1^2}{4\xi_1^2} \right) \right.$$
$$\times e^{-(\eta_1/\xi_1)^2 - \xi_1^2} d\xi_1 d\eta_1$$
$$+ \int_0^{\tau_1} [e^{\eta_1} \int_0^{\infty} f_1^n(\eta_1, 2\sqrt{\tau_1 - \eta_1}\xi_1) e^{\sqrt{\tau_1 - \eta_1}\xi_1 - \xi_1^2} d\xi_1$$
$$- \sqrt{\pi}\sqrt{\tau_1 - \eta_1} e^{\tau_1} \int_0^{\infty} f_1^n(\eta_1, 2\sqrt{\tau_1 - \eta_1}\xi_1)$$
$$\left. \times e^{2\sqrt{\tau_1 - \eta_1}\xi_1} \text{erfc}\left(\xi_1 + \sqrt{\tau_1 - \eta_1} \right) d\xi_1] d\eta_1 \right\}, \quad (9.46)$$

and

$$y_{f_2}(\tau_2) = \frac{2}{\sqrt{\pi}} \sum_{n=0}^{\infty} \frac{1}{n!} \left\{ e^{\tau_2} \int_0^{\infty} e^{-\eta_2} \int_{\eta_2/2\sqrt{\tau_2}}^{\infty} \phi_2^n \left(\tau_2 - \frac{\eta_2^2}{4\xi_2^2} \right) \right.$$
$$\times e^{-(\eta_2/\xi_2)^2 - \xi_2^2} d\xi_2 d\eta_2$$
$$+ \int_0^{\tau_2} [e^{\eta_2} \int_0^{\infty} f_2^n(\eta_2, 2\sqrt{\tau_2 - \eta_2}\xi_2) e^{\sqrt{\tau_2 - \eta_2}\xi_2 - \xi_2^2} d\xi_2$$
$$- \sqrt{\pi}\sqrt{\tau_2 - \eta_2} e^{\tau_2} \int_0^{\infty} f_2^n(\eta_2, 2\sqrt{\tau_2 - \eta_2}\xi_2)$$
$$\left. \times e^{2\sqrt{\tau_2 - \eta_2}\xi_2} \text{erfc}\left(\xi_2 + \sqrt{\tau_2 - \eta_2} \right) d\xi_2] d\eta_2 \right\}. \quad (9.47)$$

Lemma 9.2. *Let $\tau_1 := \frac{\sigma_1^2}{2}(T-t)$, $\lambda_1 = \frac{2a_{11}}{\sigma_1^2}$ and $x_1 = \ln\left(\frac{y}{y_{f_1}(\tau_1)}\right)$. Then the American exchange put price $V_1 = s_2 V_1'$, in the transition region is given as follows:*

$$V_1 = s_2 V_1' = s_2 \sum_{n=0}^{\infty} \frac{\bar{V}_1^n(\tau_1, x_1)}{n!}, \quad (9.48)$$

and $V_2 = s_2 V_1' = s_2(1-y)$.

The final analytic solution at each order is then:

$$\bar{V}_1^n(\tau_1, x_1) = \int_0^\infty \psi_1^n(\xi_1) G(x_1, \xi_1, \tau_1) d\xi_1 - \int_0^{\tau_1} \phi_1^n(\eta_1) G(x_1, 0, \tau_1 - \eta_1) d\eta_1$$

$$= \int_0^{\tau_1} \int_0^\infty \tilde{f}_1^n(\eta_1, \xi_1) G(x_1, \xi_1, \tau_1 - \eta_1) d\xi_1 d\eta_1, \tag{9.49}$$

where

$$G(x_1, \xi_1, \tau_1) = \frac{1}{2\sqrt{\pi \tau_1}} \left\{ \exp\left[\frac{-(x_1 - \xi_1)^2}{4\tau_1}\right] + \exp\left[\frac{-(x_1 + \xi_1)^2}{4\tau_1}\right] \right.$$
$$\left. - 2\sqrt{\pi \tau_1} \exp\left[\tau_1 + x_1 + \xi_1\right] \right.$$
$$\left. \times \text{erfc}\left(\frac{x_1 + \xi_1}{2\sqrt{\tau_1}} + \sqrt{\tau_1}\right) \right\}, \tag{9.50}$$

$$\tilde{f}_1^n(\tau_1, x_1) = \begin{cases} -\mathcal{L}_1[\bar{V}_1^0(\tau_1, x_1)] + \bar{\mathcal{A}}_1(\tau_1, x_1, 0), & \text{if } n = 1 \\ n \frac{\partial^{n-1} \bar{\mathcal{A}}_1}{\partial p^{n-1}}|_{p=0}, & \text{if } n \geq 2, \end{cases} \tag{9.51}$$

$$\phi_1^n(\tau_1) = \begin{cases} \bar{V}_1^0(\tau_1, 0) - \frac{\partial \bar{V}_1^0}{\partial x_1}(\tau_1, 0) - 1, & \text{if } n = 1 \\ 0, & \text{if } n \geq 2. \end{cases} \tag{9.52}$$

Here $\mathcal{L}_1$ is a differential operator defined as

$$\mathcal{L}_1 = \frac{\partial}{\partial \tau_1} - \frac{\partial^2}{\partial x_1^2}, \tag{9.53}$$

$\hat{\mathcal{A}}_1$ is a functional defined as

$$\hat{\mathcal{A}}_1[\bar{V}_1(\tau_1, x_1, p), \bar{y}_{f_1}(\tau_1, p)]$$
$$= \mathcal{L}_1(\bar{V}_1) - \lambda_1(\bar{V}_1 + e^{x_1}\bar{y}_{f_1}(\tau_1, p) - 1) - \frac{1}{\bar{y}_{f_1}(\tau_1, p)} \frac{d\bar{y}_{f_1}(\tau_1, p)}{d\tau_1} \frac{\partial \bar{V}_1}{\partial x_1}(\tau_1, x_1, p) \tag{9.54}$$

and $\bar{\mathcal{A}}_1 = \mathcal{L}[\bar{V}_1] - \hat{\mathcal{A}}_1$.

The optimal exercise boundary $y_{f_1}(\tau_1)$ is given by

$$y_{f_1}(\tau_1) = \sum_{n=0}^\infty \frac{\bar{y}^n(\tau_1)}{n!}, \tag{9.55}$$

where

$$\bar{y}^n(\tau_1) = -\bar{V}_1^n(\tau_1, 0). \tag{9.56}$$

For exact solutions in the form of infinite series, quite often it is difficult to theoretically show the convergence of the series. Although sometimes the convergency is assumed (e.g. see Hildebrand [19]), it is more reasonable to at least show the convergency through numerical evidence of the computed results of the solution. Zhu [34] was the first who applied the HAM to the American put option under the Black-Scholes environment. Using the Landau transform and the HAM, Zhu [34] gave a solution in the form of infinite recursive series involving double integrals. With a 30th-order approximation through numerical integration, Zhu [34] numerically demonstrated the convergence of his results. At the same line of Zhu [34], we shall consider the numerical examples in our future research.

9.5 Conclusion

We consider the pricing of an American exchange option in a two-state regime switching model. We show that the price of an American exchange option can be reduced to the price of an American option. Then, we modify the result of Zhu and Chan [35], a closed-form analytical pricing formula for the American exchange option is presented.

Acknowledgements We wish to thank the referees for helpful comments.

References

1. Benth, F.E., Di Nunno, G., Khedher, A., Schmeck, M.D.: Pricing of spread options on a bivariate jump market and stability to model risk. Working paper, University of Oslo (2012)
2. Black, F., Scholes, M.: The pricing of options and corporate liabilities. J. Politi. Econ. **81**, 637–659 (1973)
3. Buffington, J., Elliott, R.J.: Regime switching and European options. In: Pasik-Duncan B. (eds.) Stochastic Theory and Control, Proceedings of a Workshop, Lawrence, pp. 73–81. Springer, Berlin (2002)
4. Buffington, J., Elliott R.J.: American options with regime switching. Int. J. Theor. Appl. Financ. **5**, 497–514 (2002)
5. Carmona, R., Durrleman, V.: Pricing and hedging spread options. SIAM Rev. **45**(4), 627–685 (2003)
6. Cheang, G.H.L., Chiarella, C.: Exchange options under jump-diffusion dynamics. Appl. Math. Financ. **18**(3), 245–276 (2011)

7. Dempster, M., Hong, S.: Spread option valuation and the fast Fourier transform. In: Geman, H., Madan, D., Pliska, S.R., Vorst, T. (eds.) Mathematical Finance, Bachelier Congress, vol. 1, pp. 203–220. Springer, Berlin (2000)
8. Deng, S., Li, M., Zhou, J.: Closed-form approximations for spread option prices and greeks. J. Deriv. **16**(4), 58–80 (2008)
9. Elliott, R.J., Kopp, P.E.: Mathematics of Financial Markets. Springer, Berlin/Heidelberg/New York (1999)
10. Elliott, R.J., van der Hoek, J.: An application of hidden Markov models to asset allocation problems. Financ. Stoch. **3**, 229–238 (1997)
11. Elliott, R.J., Aggoun, L., Moore J.B.: Hidden Markov Models: Estimation and Control. Springer, Berlin/Heidelberg/New York (1994)
12. Elliott, R.J., Hunter, W.C., Jamieson, B.M.: Financial signal processing. Int. J. Theor. Appl. Financ. **4**, 567–584 (2001)
13. Elliott, R.J., Malcolm, W.P., Tsoi, A.H.: Robust parameter estimation for asset price models with Markov modulated volatilities. J. Econ. Dyn. Control **27**(8), 1391–1409 (2003)
14. Elliott, R.J., Chan, L.L., Siu, T.K.: Option pricing and Esscher transform under regime switching. Ann. Financ. **1**(4), 423–432 (2005)
15. Elliott, R.J., Chan, L., Siu, T.K.: Option valuation under a regime-switching constant elasticity of variance process. Appl. Math. Comput. **219**(9), 4434–4443 (2013)
16. Eydeland, A., Wolyniec, K.: Energy and Power Risk Management: New Developments in Modeling, Pricing and Hedging. Wiley, Hoboken (2003)
17. Gounden, S., O'Hara, J.G.: An analytic formula for the price of an American-style Asian option of floating strike type. Appl. Math. Comput. **217**, 2923–2936 (2010)
18. Guo, X.: Information and option pricings. Quant. Financ. **1**, 38–44 (2001)
19. Hildebrand, F.B.: Advanced Calculus and Applications. Prentice Hall, Englewood Cliffs (1976)
20. Hurd, T., Zhou, Z.: A Fourier transform method for spread option pricing. SIAM J. Financ. Math. **1**, 142–157 (2009)
21. Kirk, E.: Correlation in energy markets. In: Kaminski, V. (ed.) Managing Energy Price Risk, pp. 71–78. Risk Publications, London (1996)
22. Liao, S.J.: Homotopy Analysis Method in Nonlinear Differential Equations. Springer, Heidelberg/New York/Higher Education Press, Beijing (2012)
23. Liao, S.J., Zhu, S.P.: Solving the Liouville equation with the general boundary element method approach. Bound. Elem. Technol. **XIII**, 407–416 (1999)
24. Margrabe, W.: The value of an option to exchange one asset for another. J. Financ. **33**(1), 177–186 (1978)
25. Merton, R.: Theory of rational option pricing. Bell J. Econ. Manag. Sci. **4**, 141–183 (1973)
26. Odibat, Z.M.: A study on the convergence of homotopy analysis method. Appl. Math. Comput. **217**, 782–789 (2010)
27. Ortega, J.M., Rheinboldt, W.C.: Iterative Solution of Nonlinear Equations in Several Variables. Academic, New York (1970)

28. Ravindran, K.: Low-fat spreads. Risk **6**(10), 56–57 (1993)
29. Shimko, D.: Options on futures spreads: hedging, speculation and valuation. J. Futures Mark. **14**(2), 183–213 (1994)
30. Siu, T.K., Erlwein, C., Mamon, R.: The pricing of credit default swaps under a Markov-modulated Mertons structural model. North Am. Actuar. J. **12**(1), 19–46 (2008)
31. Venkatramanan, A., Alexander, C.: Closed form approximations for spread options. Appl. Math. Financ. **18**(5), 447–472 (2011)
32. Yuen, F.L., Yang, H.: Option pricing in a jump-diffusion model with regime-switching. ASTIN Bull. **39**(2), 515–539 (2009)
33. Yuen, F.L., Yang, H.: Pricing Asian options and equity-indexed annuities with regime switching by the trinomial tree method. North Am. Actuar. J. **14**(2), 256–277 (2010)
34. Zhu, S.P.: An exact and explicit solution for the valuation of American put options. Quant. Financ. **6**(3), 229–242 (2006)
35. Zhu, S.P., Chan, L.: An analytic formula for pricing American options with regime switching. Submitted for publication (2012)
36. Zhu, S.P., Badran, A., Lu, X.: A new exact solution for pricing European options in a two-state regime-switching economy. Comput. Math. Appl. **64**, 2744–2755 (2012)

Chapter 10
Parameter Estimation in a Weak Hidden Markov Model with Independent Drift and Volatility

Xiaojing Xi and Rogemar S. Mamon

Abstract We develop a multivariate higher-order Markov model, also known as weak hidden Markov model (WHMM), for the evolution of asset prices. The means and volatilities of asset's log-returns are governed by a second-order Markov chain in discrete time. WHMM enriches the usual HMM by incorporating more information from the past thereby capturing presence of memory in the underlying market state. A filtering technique in conjunction with the Expectation-Maximisation algorithm is adopted to develop the optimal estimates of model parameters. To ensure that the errors between the "true" parameters and estimated parameters are due only to the estimation method and not from model uncertainty, recursive filtering algorithms are implemented to a simulated dataset. Using goodness-of-fit metrics, we show that the WHMM-based filtering techniques are able to recover the "true" underlying parameters.

10.1 Introduction

We propose a general extension of the WHMM-modulated model for asset price dynamics by allowing the drift and volatility components to be driven by two independent WMCs not necessarily having the same number of states. That is, for instance, the drift may have three states whilst the volatility has only two states and the asset

---

X. Xi
Department of Applied Mathematics, University of Western Ontario, London, ON N6A 5B7, Canada

R.S. Mamon (✉)
Department of Statistical & Actuarial Sciences, University of Western Ontario, 1151 Richmond street, London, ON N6A 5B7, Canada

Department of Applied Mathematics, University of Western Ontario, London, ON N6A 5B7, Canada
e-mail: rmamon@stats.uwo.ca

returns components are driven by different Markov chains. In recent years, Markovian regime-switching models have received considerable interests in economics, finance and actuarial science. In a pioneering work, Hamilton [10] proposed a class of discrete-time Markov-switching autoregressive time series models. In Hamilton's model, the parameters are governed by a discrete-time, finite-state Markov chain so that the parameters have different values according to the state of the chain in a given period. This idea gives a natural and simple way to model the cyclical behaviour or the impact of changes in the financial market on the dynamics of financial time series. Many empirical studies reveal that Markovian regime-switching model can provide a better description of economic and financial series then that from a single regime model. Hardy [11] proposed a Markov regime-switching lognormal model for stock returns, and implemented the model on S&P500 and the Toronto Stock Exchange 300 indices; it was found that the regime-switching model performed better than the GARCH model in terms of goodness of fit. A survey of regime switching frameworks in econometric time series modelling and the differences between observation switching and Markov switching are given in Lange and Rohbek [13]. Ang and Timmermann [1] discussed the impact of regime switching on equilibrium asset prices and suggested that regimes exist in a variety of financial series such as fixed income, equity and currency markets. A three-state model in which the asset prices are highest in the "good" regime and lowest in the "bad" regime was considered in Veronesi [23].

Calvet and Fisher [3] suggested that a regime-switching mechanism can account for time-varying state and asymmetric reaction of equilibrium stock prices to news arrivals. A well-known class of such models is the HMM, in which a hidden Markov chain is employed to describe the random transition of the hidden state of an economy. HMM can provide a reasonably realistic description of some important empirical features such as heavy-tails of the returns distribution and time-varying conditional volatility. Due to their empirical successes, HMMs have been widely adopted in modelling financial time series. Rydén, et al. [20] considered an HMM for modelling daily return series, and investigated the capability of HMM to capture the series' stylised facts. By applying the model to S&P500 daily returns, the results suggest that the HMM can describe most of the observed features of the time series data except for the slowly decaying autocorrelation function of the absolute return. Early studies of HMM for financial time series include Tyssedal and Tjotheim [22], Pagan and Schwert [19] and Sola and Leroux [14]. The monograph by Elliott, et al. [5] provides a comprehensive discussion of parameter estimation under the HMM setup using filtering techniques. Since then, many researchers apply some of these techniques to finance and economics. Elliott et al. [6] developed robust filtering equations for a continuous-time HMM to estimate the volatility of a risky asset. Mamon et al. [18] derived and implemented the filters on logreturn of commodity prices, and compared the HMM to ARCH and GARCH models with respect to the prices' predictability. The HMM filtering method is applied to many other financial problems, for example, asset allocation [7, 9], interest rate modelling [8, 27], option pricing [15], and so on.

Although quite popular, the simple homogeneous Markov switching model is memoryless, which seems inadequate for real-world data. It is well-known in the economic and finance literature that the states of an economy and financial time series possess long-range dependence property, which can be verified through a simple plot of an empirical autocorrelation function. Motivated by this empirical phenomenon, we consider a WHMM as more flexible and generalises the HMM. The basic idea of an nth-order HMM is that the behaviour of the underlying Markov chain at present time depends on its behaviour in the past n time steps. WHMM is popular in speech and text recognition, but is rather new in probing the finer structures of the financial market. Bulla and Bulla [2] examined the fit of WHMMs to 18 daily sector return series and suggested that the stylised fact of slowly decaying autocorrelation can be described better by WHMMs. Yu et al. [26] explored the use of WHMMs to capture long-range dependence; they derived the recursive formula for the autocovarriance function over different time scales and the estimator of the Hurst parameter. Their empirical results demonstrated that WHMM captures long-range dependence if one state is heavy-tailed distributed. By weakening the Markov assumption, WHMMs provide a simple and flexible way to describe the duration dependence through their dependence on backward time recurrence. This has led some authors to investigate applications of WHMMs in the fields of financial derivatives, for example, risk management [21], option pricing [4], interest rate modelling [12], and asset return modelling [25].

Mamon and Jalen [17] proposed a method based on tensors to transform two independent chains into one Markov chain so that the regular filtering technique can be applied. This method is then implemented on the Dow Jones and NASDAQ indices. In this paper, we investigate a WHMM under the situation when the drift and volatility of the given data are assumed driven by two independent weak Markov chains (WMCs). In particular, we suppose that the rate of return of a risky asset is governed by a WHMM with two underlying WMCs. The tensor-based transformation is adopted and the filtering technique for WHMMs under one chain is subsequently applied. Numerical study based on simulated observation data is carried out to demonstrate the effectiveness of this method. We also provide error analyses for different combination of states through the h-step ahead prediction performance.

This article is organised as follows. In Sect. 10.2, we present the modelling framework of WHMM. The dynamics of a risky asset price under the WHMM extension is described. By utilising a measure-change method, we derive recursive filters for the state of the WMC and other processes of interests. Parameter estimation based on EM algorithm is established in Sect. 10.3. In Sect. 10.4, we implement the filters under the proposed extended set-up on simulated data. The method is examined using different algorithm starting values. We generate one- and five-step ahead forecasts for different models and compare the forecasting performance via four error metrics. We give a conclusion in Sect. 10.5.

10.2 Modelling Background

We present a WHMM for modelling asset prices, where the drift and volatility have independent probability behaviour. In the sequel, all vectors are denoted by bold English or Greek letters in lowercase and all matrices are denoted by bold letters in uppercase. Fix a complete probability space $(\Omega, \mathscr{F}, P)$, where P is a real-world probability measure. Define a discrete-time WMC $\{\mathbf{x}_k\}$, $k \geq 0$ on $(\Omega, \mathscr{F}, P)$ with a finite-state space $\mathscr{S} = \{s_1, s_2, \ldots, s_N\}$. The states of the MC represent different states of the economy. Without loss of generality, the points in $\mathscr{S}$ can be identified with the canonical basis $\{\mathbf{e}_1, \mathbf{e}_2, \ldots, \mathbf{e}_N\} \subset \mathbf{R}^N$, where $\mathbf{e}_i = (0, \ldots, 0, 1, 0, \ldots, 0)^\top$ and $\top$ denotes the transpose of a vector. In particular, $\langle \mathbf{x}_k, \mathbf{e}_i \rangle$ stands for the event that the economy is in state i at time t and $\langle , \rangle$ stands for the inner product in $\mathbf{R}^N$.

In this article, we concentrate on a second-order WMC to simplify the discussion and present a complete characterisation of the parameter estimation. The probability of the next time step for a second-order WMC depends on the information on current and previous time steps. Let $\mathbf{A} \in \mathbf{R}^{N \times N^2}$ be the transition probability matrix of $\mathbf{x}_k$. Each entry $a_{lmv} := P(\mathbf{x}_{k+1} = \mathbf{e}_l | \mathbf{x}_k = \mathbf{e}_m, \mathbf{x}_{k-1} = \mathbf{e}_v)$, $l, m, v \in 1, \ldots, N$ is the transition probability that $\mathbf{x}$ enters state l given that the current and previous states were states m and v, respectively. The salient idea in the filtering method for WHMM is that a second-order Markov chain is transformed into a first-order Markov chain through a mapping ξ, and then we may apply the regular filtering method. The mapping ξ is defined by

$$\xi(\mathbf{e}_r, \mathbf{e}_s) = \mathbf{e}_{rs}, \text{ for } 1 \leq r, s \leq N,$$

where $\mathbf{e}_{rs}$ is an $\mathbf{R}^{N^2}$–unit vector with unity in its $((r-1)N+s)th$ position. The identification of the new first-order Markov chain with the canonical basis is given by

$$\langle \xi(\mathbf{x}_k, \mathbf{x}_{k-1}), \mathbf{e}_{rs} \rangle = \langle \mathbf{x}_k, \mathbf{e}_r \rangle \langle \mathbf{x}_{k-1}, \mathbf{e}_s \rangle.$$

We further assume that the new Markov chain has a new transition probability matrix, $\mathbf{\Pi} \in \mathbf{R}^{N^2 \times N^2}$, given by

$$\pi_{ij} = \begin{cases} a_{lmv} & \text{if } i = (l-1)N + m, \ j = (m-1)N + v \\ 0 & \text{otherwise.} \end{cases}$$

Each non-zero element π_{ij} represents the probability

$$\pi_{ij} = a_{lmv} = P(\mathbf{x}_k = \mathbf{e}_l | \mathbf{x}_{k-1} = \mathbf{e}_m, \mathbf{x}_{k-2} = \mathbf{e}_v),$$

and each zero represents an impossible transition. Following the arguments in Elliott et al. [5], the new Markov chain $\xi(\mathbf{x}_k, \mathbf{x}_{k-1})$ has the semi-martingale representation

$$\xi(\mathbf{x}_k, \mathbf{x}_{k-1}) = \mathbf{\Pi} \xi(\mathbf{x}_{k-1}, \mathbf{x}_{k-2}) + \mathbf{v}_k, \tag{10.1}$$

where $\{\mathbf{v}_k\}_{k \geq 1}$ is a sequence of $\mathbf{R}^{N^2}$ martingale increments.

Let S_k, $k \geq 1$, be the asset price at time k and y_k denote the logarithmic increments. Xi and Mamon [25] examined the case where the drift and volatility of y_k are governed by the same hidden Markov chain. In particular, y_k is assumed to have the dynamics

$$y_{k+1} = f(\mathbf{x}_k) + \sigma(\mathbf{x}_k) z_{k+1} = \langle \mathbf{f}, \mathbf{x}_k \rangle + \langle \boldsymbol{\sigma}, \mathbf{x}_k \rangle z_{k+1}. \tag{10.2}$$

The sequence $\{z_k\}$ is a sequence of $N(0,1)$ IID random variables, which are independent of the $\mathbf{x}$-process. In this study, we consider the case when the drift and volatility have independent states and probabilistic behaviour. That is, we assume y_k has the dynamics

$$y_{k+1} = \langle \mathbf{f}, \mathbf{x}_k^1 \rangle + \langle \boldsymbol{\sigma}, \mathbf{x}_k^2 \rangle z_{k+1}, \tag{10.3}$$

where $\mathbf{x}^i$ is an N_i-state WMC on state space $\mathscr{S}_i$ with transition matrix $\mathbf{A}_i \in \mathbf{R}^{N_i \times N_i^2}$. Suppose the drift and volatility have the form $\mathbf{f} = (f_1, f_2, \ldots, f_{N_1}) \in \mathbf{R}^{N_1}$ and $\boldsymbol{\sigma} = (\sigma_1, \sigma_2, \ldots, \sigma_{N_2}) \in \mathbf{R}^{N_2}$ respectively. In order to apply the regular WHMM filtering technique, we aim to re-formulate the hidden WMCs so that the dynamics of y_k have the same form as in Eq. (10.2). Let $\otimes$ denote the tensor product. Following the idea in [17], we transform the two chains, $\mathbf{x}_k^1$ and $\mathbf{x}_k^2$, into a new WMC $\mathbf{x}_k$ using Kronecker product or tensor product, i.e., $\mathbf{x}_k = \mathbf{x}_k^1 \otimes \mathbf{x}_k^2$. Then $\mathbf{x}_k$ is an $N_1 N_2$-state WMC with transition matrix $\mathbf{A} = \mathbf{A}_1 \otimes \mathbf{A}_2$. Write $\mathbf{1}_N$ for the vector $(1, 1, \ldots, 1) \in \mathbf{R}^N$. The reformulated drift and volatility are given by

$$\boldsymbol{\alpha} = \mathbf{f} \otimes \mathbf{1}_{N_2},$$
$$\boldsymbol{\eta} = \mathbf{1}_{N_1} \otimes \boldsymbol{\sigma}.$$

Therefore the dynamics of y_k in Eq. (10.3) can be rewritten as

$$y_{k+1} = \langle \boldsymbol{\alpha}, \mathbf{x}_k \rangle + \langle \boldsymbol{\eta}, \mathbf{x}_k \rangle z_{k+1}. \tag{10.4}$$

We demonstrate the workings of this transformation through a numerical example in Sect. 10.4.

10.3 Filters and Parameter Estimation

Under the real-world measure P, we cannot observe the hidden state of the economy $\mathbf{x}_k$ directly. Instead, we are given market observations y_k, which contain information about $\mathbf{x}_k$. Certainly, the unknown drift and volatility are highly dependent on the WMC. Thus, the estimation of parameters is tantamount to "filtering" the noise out of the observations to recover the hidden WMC. However, the derivation of filters under P is complicated. Exploiting the Kolmogorov's Extension theorem, we note that there exists a reference probability measure $\bar{P}$ under which

- y_k's are $N(0,1)$ IID random variables and
- $\mathbf{x}$ is a finite state WMC satisfying (10.1) and $\bar{E}[\mathbf{v}_k | \mathscr{Y}_k] = 0$.

Under $\bar{P}$, y_k does not depend on $\mathbf{x}_k$, and therefore it is more convenient to evaluate the filtered estimates. The calculation starts with $\bar{P}$ and then we perform a measure change to construct the real-world measure P equivalent to $\bar{P}$. Consider a $\mathscr{Y}_k$-adapted process Λ_k, $k \geq 0$ defined by

$$\lambda_l = \frac{\phi\left(\sigma(\mathbf{x}_{l-1})^{-1}(y_l - f(\mathbf{x}_{l-1}))\right)}{\sigma(\mathbf{x}_{l-1})\phi(y_l)}, \tag{10.5}$$

$$\Lambda_k = \prod_{l=1}^{k} \lambda_l, \ k \geq 1, \ \Lambda_0 = 1, \tag{10.6}$$

where $\phi(z)$ is the probability density function of a standard normal random variable Z.

Define the Radon-Nikodŷm derivative of P with respect to $\bar{P}$ by

$$\left.\frac{dP}{d\bar{P}}\right|_{\mathscr{Y}_k} := \Lambda_k. \tag{10.7}$$

Suppose X_k is a $\mathscr{Y}_k$-adapted process. Write $\hat{X}_k := E[X_k|\mathscr{Y}_k]$ and $\gamma(X_k) := \bar{E}[\Lambda_k X_k|\mathscr{Y}_k]$. Then by Bayes' theorem for conditional expectation (see, for example, p. 22 of Elliott et al. [5]), we have

$$\hat{X}_k = \frac{\bar{E}[\Lambda_k X_k|\mathscr{Y}_k]}{\bar{E}[\Lambda_k|\mathscr{Y}_k]} = \frac{\gamma(X_k)}{\gamma(1)}. \tag{10.8}$$

Let us derive the conditional expectation of $\xi(\mathbf{x}_k, \mathbf{x}_{k-1})$ given $\mathscr{Y}_k$ under P. Write

$$p_k^{ij} := P(\mathbf{x}_k = \mathbf{e}_i, \mathbf{x}_{k-1} = \mathbf{e}_j|\mathscr{Y}_k) = E[\langle \xi(\mathbf{x}_k, \mathbf{x}_{k-1}), \mathbf{e}_{ij}\rangle|\mathscr{Y}_k] \tag{10.9}$$

with $\mathbf{p}_k = (p_k^{11}, \ldots, p_k^{ij}, \ldots, p_k^{NN}) \in \mathbf{R}^{N^2}$. Hence, by Bayes' theorem for conditional expectation,

$$\mathbf{p}_k = E[\xi(\mathbf{x}_k, \mathbf{x}_{k-1})|\mathscr{Y}_k] = \frac{\gamma(\xi(\mathbf{x}_k, \mathbf{x}_{k-1}))}{\gamma(1)}. \tag{10.10}$$

Note that

$$\sum_{i,j=1}^{N} \langle \xi(\mathbf{x}_k, \mathbf{x}_{k-1}), \mathbf{e}_{ij}\rangle = \langle \xi(\mathbf{x}_k, \mathbf{x}_{k-1}), \mathbf{1}_{N^2}\rangle = 1. \tag{10.11}$$

Let $\mathbf{q}_k = \gamma(\xi(\mathbf{x}_k, \mathbf{x}_{k-1}))$ so that

$$\langle \mathbf{q}_k, \mathbf{1}_{N^2}\rangle = \bar{E}[\Lambda_k \langle \xi(\mathbf{x}_k, \mathbf{x}_{k-1}), \mathbf{1}_{N^2}\rangle|\mathscr{Y}_k] = \gamma(1). \tag{10.12}$$

From Eqs. (10.10) and (10.12), we get the conditional distribution of $\xi(\mathbf{x}_k, \mathbf{x}_{k-1})$ under P as

$$\mathbf{p}_k = \frac{\mathbf{q}_k}{\langle \mathbf{q}_k, \mathbf{1}_{N^2}\rangle}. \tag{10.13}$$

In order to estimate the state process $\xi(\mathbf{x}_k, \mathbf{x}_{k-1})$, we shall establish the recursion for the process $\mathbf{q}_k$. Define the diagonal matrix $\mathbf{B}_k \in \mathbf{R}^{N^2 \times N^2}$ by

$$\mathbf{B}_k = \mathrm{diag}(b_k^1, \ldots, b_k^N, \ldots, b_k^1, \ldots, b_k^N) \qquad (10.14)$$

where $\mathrm{diag}(\mathbf{v})$ is a diagonal matrix whose diagonal entries are the components of the vector $\mathbf{v}$ and

$$b_k^i = \frac{\phi((y_k - f_i)/\sigma_i)}{\sigma_i \phi(y_k)}. \qquad (10.15)$$

To estimate the parameters of the model, we first present the set of quantities that are pertinent to the derivation. Three of these processes are related to the state process and one is related to both the state and observation processes. For $r, s, t = 1, \ldots, N$, these quantities are defined by

- J_k^{rst}, the number of jumps from $(\mathbf{e}_s, \mathbf{e}_t)$ to state $\mathbf{e}_r$ up to time k,

$$J_k^{rst} = \sum_{l=1}^{k} \langle \mathbf{x}_l, \mathbf{e}_r \rangle \langle \mathbf{x}_{l-1}, \mathbf{e}_s \rangle \langle \mathbf{x}_{l-2}, \mathbf{e}_t \rangle \qquad (10.16)$$

- O_k^{rs}, the occupation time of $\mathbf{x}$ spent in state $(\mathbf{e}_r, \mathbf{e}_s)$ up to time k,

$$O_k^{rs} = \sum_{l=1}^{k} \langle \mathbf{x}_{l-1}, \mathbf{e}_r \rangle \langle \mathbf{x}_{l-2}, \mathbf{e}_s \rangle \qquad (10.17)$$

- O_k^r, the occupation time spent by $\mathbf{x}$ in state $\mathbf{e}_r$ up to time k,

$$O_k^r = \sum_{l=1}^{k} \langle \mathbf{x}_{l-1}, \mathbf{e}_r \rangle \qquad (10.18)$$

- $T_k^r(g)$, the level sum for the state $\mathbf{e}_r$,

$$T_k^r(g) = \sum_{l=1}^{k} g(y_l) \langle \mathbf{x}_{l-1}, \mathbf{e}_r \rangle. \qquad (10.19)$$

Here, g is a function with the form $g(y) = y$ or $g(y) = y^2$, for $1 \leq l \leq k$.

The two propositions that follow are straightforward extensions of the main results in Xi and Mamon [25], where $N = N^1 \times N^2$, and thus their proofs are immediate. They provide the recursive formulae for the vectors $\gamma(J_k^{rst} \xi(\mathbf{x}_k, \mathbf{x}_{k-1}))$, $\gamma(O_k^{rs} \xi(\mathbf{x}_k, \mathbf{x}_{k-1}))$, $\gamma(O_k^r \xi(\mathbf{x}_k, \mathbf{x}_{k-1}))$ and $\gamma(T_k^r(g) \xi(\mathbf{x}_k, \mathbf{x}_{k-1}))$, which are the unnormalised filtered estimates for J_k^{rst}, O_k^{rs}, O_k^r and $T_k^r(g)$, respectively.

Proposition 10.1. *Let $\mathbf{V}_r, 1 \leq r \leq N$ be an $N^2 \times N^2$ matrix such that the $((i-1)N + r)$th column of $\mathbf{V}_r$ is $\mathbf{e}_{ir}$ for $i = 1 \ldots N$ and zero elsewhere. If $\mathbf{B}$ is the diagonal matrix defined in Eq. (10.14) then*

$$\mathbf{q}_{k+1} = \mathbf{B}_{k+1} \Pi \mathbf{q}_k \qquad (10.20)$$

and

$$\gamma(J^{rst}_{k+1}\xi(\mathbf{x}_{k+1},\mathbf{x}_k)) = \mathbf{B}_{k+1}\boldsymbol{\Pi}\gamma(J^{rst}_k\xi(\mathbf{x}_k,\mathbf{x}_{k-1})) + b^r_{k+1}\langle\boldsymbol{\Pi}\mathbf{e}_{st},\mathbf{e}_{rs}\rangle\langle\mathbf{q}_k,\mathbf{e}_{st}\rangle\mathbf{e}_{rs}, \tag{10.21}$$

$$\gamma(O^{rs}_{k+1}\xi(\mathbf{x}_{k+1},\mathbf{x}_k)) = \mathbf{B}_{k+1}\boldsymbol{\Pi}\gamma(O^{rs}_k\xi(\mathbf{x}_k,\mathbf{x}_{k-1})) + b^r_{k+1}\langle\mathbf{q}_k,\mathbf{e}_{rs}\rangle\boldsymbol{\Pi}\mathbf{e}_{rs}, \tag{10.22}$$

$$\gamma(O^r_{k+1}\xi(\mathbf{x}_{k+1},\mathbf{x}_k)) = \mathbf{B}_{k+1}\boldsymbol{\Pi}\gamma(O^r_k\xi(\mathbf{x}_k,\mathbf{x}_{k-1})) + b^r_{k+1}\mathbf{V}_r\boldsymbol{\Pi}\mathbf{q}_k, \tag{10.23}$$

$$\gamma(T^r_{k+1}(g)\xi(\mathbf{x}_{k+1},\mathbf{x}_k)) = \mathbf{B}_{k+1}\boldsymbol{\Pi}\gamma(T^r_k(g)\xi(\mathbf{x}_k,\mathbf{x}_{k-1})) + g(y_{k+1})b^r_{k+1}\mathbf{V}_r\boldsymbol{\Pi}\mathbf{q}_k. \tag{10.24}$$

Similar to Eq. (10.11), the unnormalised filtered estimates for $\gamma(J^{rst}_k)$, $\gamma(O^{rs}_k)$, $\gamma(O^r_k)$ and $\gamma(T^r_k(g))$ can be determined by taking the inner products with $\mathbf{1}_{N^2}$. For example,

$$\gamma(J^{rst}_k) = \langle\gamma(J^{rst}_k\xi(\mathbf{x}_k,\mathbf{x}_{k-1})),\mathbf{1}_{N^2}\rangle.$$

The normalised estimates are obtained by dividing $\gamma(J^{rst}_k)$ by $\gamma(1)$, that is, $\hat{J}^{rst}_k = \gamma(J^{rst}_k)/\gamma(1)$.

We now briefly illustrate the EM algorithm for estimating the optimal parameters using the filters in Proposition 10.1. The parameters in our model are given by the set

$$\theta = \{a_{rst}, f_r, \sigma_r. \ 1 \le r,s,t \le N\}.$$

The EM method is a two-stage estimation technique; these stages are the expectation and maximisation steps. The algorithm proceeds by initially selecting any set of parameters, denoted by θ_0, for the model. The change to a new set of parameters is described by a change of probability measure from P_0 to P_θ. This means that the likelihood function for estimating the parameter θ based on the given information $\mathscr{Y}$ is

$$L(\theta) = E_0\left[\frac{dP_\theta}{dP_0}\bigg|\mathscr{Y}\right].$$

The logarithm of the Radon-Nykodŷm derivative of the new measure with respect to the old measure is then calculated. A set of parameters that maximise the conditional log-likelihood is then determined. It is shown in [24] that the sequence of the estimated log-likelihood is monotonically increasing and the sequence of estimates converges to a local maximum of the expectation of the estimated likelihood function. This method yields a self-tuning approximation of the maximum likelihood estimate. As an example, let us consider the case of estimating the transition matrix. Note that the non-zero entries of $\boldsymbol{\Pi}$ are the same as the entries of $\mathbf{A}$. We estimate the matrix $\mathbf{A}$ then construct $\boldsymbol{\Pi}$ for the calculation of filters. We first perform a change of measure from P_θ to $P_{\hat{\theta}}$ for this method. Under P_θ, $\mathbf{x}$ is a WMC with transition matrix $\mathbf{A} = (a_{rst})$. In [16], it is proved that under $P_{\hat{\theta}}$, $\mathbf{x}$ is still a WMC and the transition matrix is $\hat{\mathbf{A}} = (\hat{a}_{rst})$. To replace $\mathbf{A}$ by $\hat{\mathbf{A}}$, the likelihood function

$$\frac{dP_\theta}{dP_0}\bigg|_{\mathscr{Y}_k} = \Gamma^A_k,$$

$$\Gamma_k^A = \prod_{l=2}^{k} \prod_{r,s,t=1}^{N} \left(\frac{\hat{a}_{rst}}{a_{rst}}\right)^{\langle \mathbf{x}_l, \mathbf{e}_r \rangle \langle \mathbf{x}_{l-1}, \mathbf{e}_s \rangle \langle \mathbf{x}_{l-2}, \mathbf{e}_t \rangle}$$

is considered. In case $a_{rst} = 0$, take $\hat{a}_{rst} = 0$ and $\hat{a}_{rst}/a_{rst} = 1$. The maximum likelihood estimates are given in the following proposition.

Proposition 10.2. *Suppose the observation is d-dimensional and the set of parameters $\{\hat{a}_{rst}, \hat{\mathbf{f}}_r, \hat{\sigma}_r\}$ determines the dynamics of y_k, $k \geq 1$. Then the EM estimates for these parameters are given by*

$$\hat{a}_{rst} = \frac{\hat{J}_k^{rst}}{\hat{O}_k^{st}} = \frac{\gamma(J_k^{rst})}{\gamma(O_k^{st})}, \forall \, pairs\,(r,s),\, r \neq s, \quad (10.25)$$

$$\hat{f}_r = \frac{\hat{T}_k^r}{\hat{O}_k^r} = \frac{\gamma(T_k^r(y))}{\gamma(O_k^r)}, \quad (10.26)$$

$$\hat{\sigma}_r = \sqrt{\frac{\hat{T}_k^r(y^2) - 2\hat{f}_r \hat{T}_k^r(y) + \hat{f}_r^2 \hat{O}_k^r}{\hat{O}_k^r}}. \quad (10.27)$$

The recursions in Eqs. (10.21)–(10.24) can be used to obtain dynamic parameter estimates given in Proposition 10.2. In other words, every time a set of financial time series observations is available up to time k, new parameters $\hat{a}_{rst}(k)$, $\hat{f}_r(k)$, $\hat{\sigma}_r(k)$, $1 \leq r,s,t \leq N$ are obtained from Eqs. (10.25) to (10.27) through filtering recursions. The parameter estimation is then self-tuning as recursive filters for the unobserved Markov chain and related processes in Proposition 10.1 can easily get updated every time new information arrives.

10.4 Numerical Implementation

In this section, we demonstrate the implementation of the filtering technique in the estimation of model parameters using simulated data. Suppose we are given a set of data generated from a process with a 3-state drift and 2-state volatility: $\mathbf{f} = (f_1, f_2, f_3)$ and $\boldsymbol{\sigma} = (\sigma_1, \sigma_2)$. The re-formulated drift and volatility vectors are

$$\boldsymbol{\alpha} = \mathbf{f} \otimes \mathbf{1}_2 = (f_1, f_1, f_2, f_2, f_3, f_3)$$
$$\boldsymbol{\eta} = \mathbf{1}_3 \otimes \boldsymbol{\sigma} = (\sigma_1, \sigma_2, \sigma_1, \sigma_2, \sigma_1, \sigma_2).$$

Note that the new WMC $\mathbf{x}$ has six states. Instead of estimating all values in $\boldsymbol{\alpha}$ and $\boldsymbol{\eta}$, we only estimate $\mathbf{f}$ and $\boldsymbol{\sigma}$, then reformulate $\boldsymbol{\alpha}$ and $\boldsymbol{\eta}$ for the recursive filters. In this way, the algorithm estimates fewer parameters than the actual 6-state model, but it is rich enough to capture all information.

The steps of the algorithm are as follows:

1. Initialise $\mathbf{f}$, $\boldsymbol{\sigma}$, $\mathbf{A}_1$ and $\mathbf{A}_2$.
2. Construct $\boldsymbol{\alpha}$, $\boldsymbol{\eta}$, $\mathbf{A}$ and $\boldsymbol{\Pi}$.

3. Calculate the filters in Proposition 10.1 using $\boldsymbol{\alpha}$, $\boldsymbol{\eta}$, $\mathbf{A}$ and $\boldsymbol{\Pi}$.
4. Using a new batch of y_k values, compute new estimates of $\mathbf{f}$, $\boldsymbol{\sigma}$ and $\mathbf{A}$ using the recursive filters in Proposition 10.2.
5. Construct the new $\boldsymbol{\alpha}$, $\boldsymbol{\eta}$ and $\boldsymbol{\Pi}$, and use these estimates as the initial values in the processing of the next batch of data points. Repeat from step. 3

Clearly, the algorithm allows us to generate new estimates when new information arrives.

We illustrate the proposed scheme using simulated data. For our simulated observation data, two sets of 1,000-point WMCs were generated with the following parameter "true" values:

$$\mathbf{A}_1 = \begin{pmatrix} 0.8 & 0.8 & 0.8 & 0.1 & 0.1 & 0.1 & 0.1 & 0.1 & 0.1 \\ 0.1 & 0.1 & 0.1 & 0.8 & 0.8 & 0.8 & 0.1 & 0.1 & 0.1 \\ 0.1 & 0.1 & 0.1 & 0.1 & 0.1 & 0.1 & 0.8 & 0.8 & 0.8 \end{pmatrix},$$

$$\mathbf{A}_2 = \begin{pmatrix} 0.7 & 0.7 & 0.3 & 0.3 \\ 0.3 & 0.3 & 0.7 & 0.7 \end{pmatrix}.$$

The initial states of both WMCs are state 1. The "true" values of the drift and volatility components are $\mathbf{f} = (0.04, 0, -0.02)$ and $\boldsymbol{\sigma} = (0.02, 0.05)$, respectively. Using the dynamics in Eq. (10.3), 1,000 simulated observation points are obtained, which can be considered as daily returns of an asset price. We run our on-line algorithm in batches consisting of 10 data points per batch this produces a set of updated parameter estimates and therefore, the parameters were updated every 2 weeks. We call the processing of a batch of data as one complete algorithm pass or step. In our simulation study, there are 100 algorithm steps.

An important aspect to consider when implementing the EM algorithm is to determine the number of states. Erlwein and Mamon [8] determined the optimal number of regimes using the Akaike information criteria and found that a two-state HMM outperforms other multi-state HMMs in capturing the dynamics of the Canadian short rates proxied by the 30-day T-bill yields. Indeed, the number of states can be any reasonable value indicated by the data. In our case, we run the algorithm using different choices for the number of states to obtain the parameter estimates. In order to compare the performance of various WHMM model candidates, we compare their corresponding 1-step ahead forecasts under different goodness-of-fit measures.

Figure 10.1 displays the evolution of $\mathbf{f}$ and $\boldsymbol{\sigma}$ estimates under various WHMM settings. Initial values of $\mathbf{f}$ and $\boldsymbol{\sigma}$ are indicated below each plot, which were based on the values chosen for the "true" parameter values. All initial entries of the transition matrix, $\mathbf{A}$, are set to $1/N$. Parameter estimates achieve stability after approximately five steps as shown by the plots in Fig. 10.1a–c. In plot in Fig. 10.1d, convergence is attained after 25 steps. Our experiment shows that the convergence can be achieved with other reasonable starting values. However, the choice of initial parameter values and the model setting affect the speed of convergence.

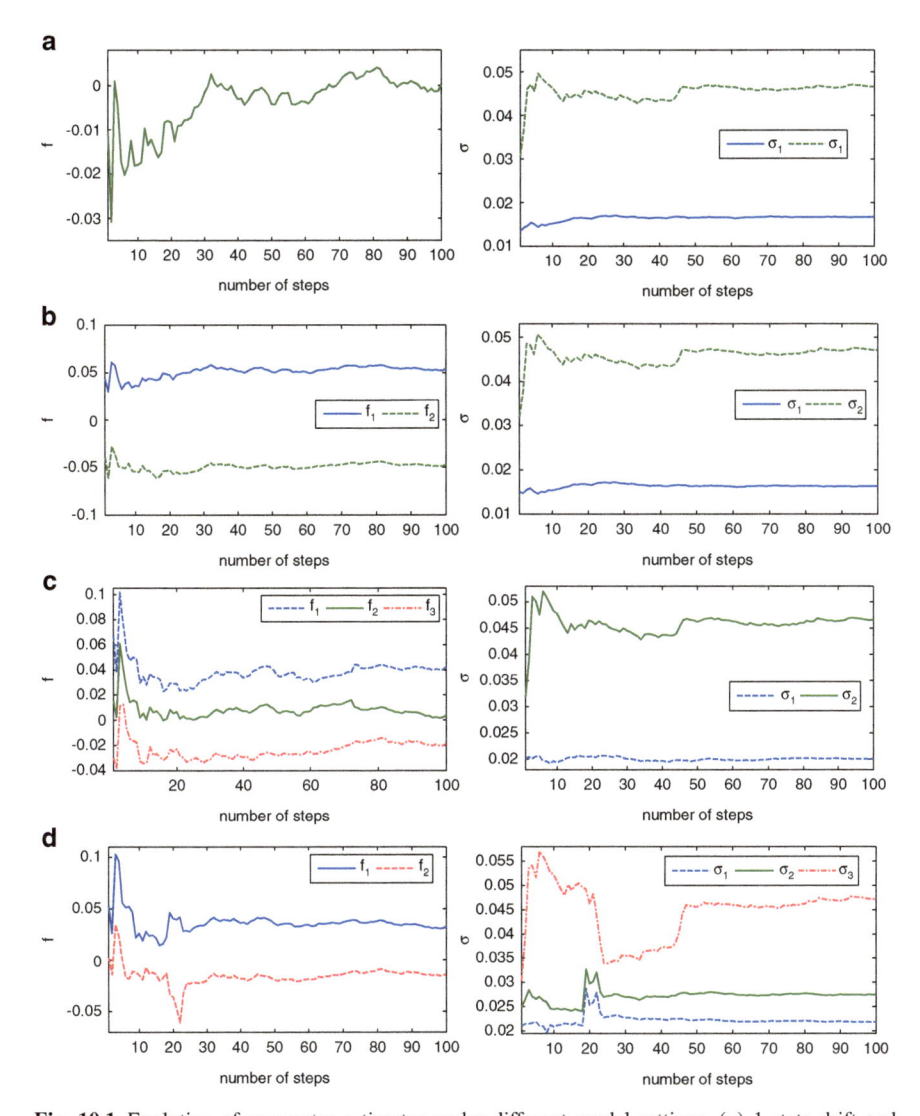

Fig. 10.1 Evolution of parameter estimates under different model settings. (**a**) 1-state drift and 2-state volatility with initial values $f = 0$ and $\boldsymbol{\sigma} = (0.01, 0.04)$. (**b**) 2-state drift and 2-state volatility with initial values $\mathbf{f} = (0.03, -0.03)$ and $\boldsymbol{\sigma} = (0.01, 0.03)$. (**c**) 3-state drift and 2-state volatility with initial values $\mathbf{f} = (0.03, 0, -0.03)$ and $\boldsymbol{\sigma} = (0.01, 0.03)$. (**d**) 2-state drift and 3-state volatility with initial values $\mathbf{f} = (0.03, 0)$ and $\boldsymbol{\sigma} = (0.01, 0.02, 0.03)$

The semi-martingale representation of a WMC in Eq. (10.1) and the definition of **A** lead to
$$E[\mathbf{x}_{k+h}|\mathcal{Y}_k] = \mathbf{A}\boldsymbol{\Pi}^{h-1}\mathbf{p}_k, \ h \geq 1.$$

Table 10.1 Comparison of 1-step ahead forecast errors

f	σ	RMSE	AME	RAE	APE
1-state	1-state	1.3911	1.0233	0.1010	0.0440
1-state	2-state	1.0780	0.7001	0.0691	0.0282
2-state	3-state	1.0842	0.7010	0.0690	0.0819
2-state	1-state	1.0787	0.7006	0.0692	0.0820
2-state	2-state	1.0824	0.7088	0.0700	0.0285
3-state	2-state	1.0742	0.6854	0.0687	0.0279

Table 10.2 Comparison of 5-step ahead forecast errors

f	σ	RMSE	AME	RAE	APE
1-state	1-state	2.6274	1.8770	0.1854	0.0775
1-state	2-state	2.4653	1.7449	0.1723	0.0705
2-state	3-state	2.4717	1.7477	0.1726	0.0706
2-state	1-state	2.4653	1.7450	0.1723	0.0705
2-state	2-state	2.4689	1.7470	0.1725	0.0706
3-state	2-state	2.4648	1.7442	0.1722	0.0704

The h-step ahead forecasts of the logarithmic increment y_k is

$$E[y_{k+h}|\mathscr{Y}_k] = \langle \mathbf{f}, \mathbf{A}\boldsymbol{\Pi}^{h-1}\mathbf{p}_k \rangle,$$

and the conditional variance of y_{k+h} is

$$\text{Var}[y_{k+h}|\mathscr{Y}_k] = \mathbf{f}^\top \text{diag}(\mathbf{A}\boldsymbol{\Pi}^{h-1}\mathbf{p}_k)\mathbf{f} + \boldsymbol{\sigma}^\top \text{diag}(\mathbf{A}\boldsymbol{\Pi}^{h-1}\mathbf{p}_k)\boldsymbol{\sigma} - \langle \mathbf{f}, \mathbf{A}\boldsymbol{\Pi}^{h-1}\mathbf{p}_k \rangle^2.$$

We compare the forecasting performance of different WHMM settings. To assess the goodness-of-fit of the h-step ahead forecasts, we evaluate the root mean square (RMSE), absolute mean error (AME), relative absolute error (RAE) and absolute percentage error (APE) for $h = 1$ and $h = 5$. These errors are reported in Tables 10.1 and 10.2. In both cases of $h = 1$ and $h = 5$, the model with 3-state drift and 2-state volatility gives the best fit in terms of lowest forecasting errors. Hence, the estimation method works very well given that it provides the best results consistent with the data-generating process.

Note that the 1-state drift and volatility model produces the highest errors under all metrics. It is apparent that the single-regime model is not able to capture the dynamics of the data series. Furthermore, the model with 2-state drift and 3-state volatility produces higher errors than those from the other 2-state volatility models. Since the data were simulated using 2-state volatility, a model with more states will lead to an overestimation of parameters and consequently generates large errors.

10.5 Conclusion

We extended a WHMM framework to model the evolution of a risky asset by considering the drift and volatility of the logarithmic increments governed by two independent hidden Markov chains. A tensor-based technique was employed to convert the two independent WMCs into a single WMC. Filtered estimates of the drift, volatility and the state of the new WMC were derived based on EM algorithm. Numerical examples on simulated data were given. Other than the true number of states, we also considered other combinations of states for drift and volatility, and analysed the resulting prediction errors. Our empirical results suggest that using the number of states and the initial values indicated by the observation data will make the model perform better. Indeed, our study confirms that starting values do directly affect the algorithm performance. The determination at the outset of the correct number of states for each parameter via error analysis or other methods improves the estimation approach's performance. It is definitely worth exploring the effectiveness and efficiency of the proposed filtering and parameter estimation technique on other models in finance and economics.

References

1. Ang, A., Timmermann, A.: Regime changes and financial markets. Ann. Rev. Financ. Econ. **4**, 313–337 (2012)
2. Bulla, J., Bulla, I.: Stylised facts of financial time series and hidden semi-Markov models. Comput. Stat. Data Anal. **51**, 2192–2209 (2006)
3. Calvet, L.E., Fisher, A.J.: Multifrequency news and stock returns. J. Financ. Econ. **86**, 178–212 (2007)
4. Ching, W.K., Siu, T.K., Li, L.M.: Pricing exotic options under a higher-order Markovian regime. J. Appl. Math. Decis. Sci. 1–15 (2007) Article ID 18014
5. Elliott, R.J., Moore, J., Aggoun, L.: Hidden Markov Models: Estimation and Control. Springer, New York (1995)
6. Elliott, R.J., Malcolm, W.P., Tsoi, A.H.: Robust parameter estimation for asset price models with Markov modulated volatilities. J. Econ. Dyn. Control **27**, 1391–1409 (2003)
7. Elliott, R.J., Siu, T.K., Badescu. A.: On mean-variance portfolio selection under a hidden Markovian regime-switching model. Econ. Model. **27**, 678–686 (2010)
8. Erlwein, C., Mamon, R.: An online estimation scheme for a Hull-White model with HMM-driven parameters. Stat. Methods Appl. **18**(1), 87–107 (2009)
9. Erlwein, C., Mamon, R., Davison, M.: An examination of HMM-based investment strategies for asset allocation. Appl. Stoch. Model Bus. Ind. **27**, 204–221 (2011)
10. Hamilton, J.: A new approach to the economic analysis of nonstationary time series and the business cycle. Econometrica **57**(2), 357–384 (1989)

11. Hardy. M.R.: A regime-switching model of long-term stock returns. North Am. Actuar. J. **5**, 41–53 (2001)
12. Hunt, J., Devolder, P.: Semi-Markov regime switching interest rate models and minimal entropy measure. Phys. A Stat. Mech. Appl. **390**, 3767–3781 (2011)
13. Lange, T., Rahbek, A.: An introduction to regime switching time series model. In: Lange, T., Rahbek, A. (eds.) Handbook of Financial Time Series, pp. 871–887. Springer, Berlin/Heidelberg (2009)
14. Leroux. B.G.: Maximum-likelihood estimation for hidden Markov models. Stoch. Process. Appl. **40**, 127–143 (1992)
15. Liew, C.C, Siu, T.K.: A hidden Markov-switching model for option valuation. Insur. Math. Econ. **47**, 374–384 (2010)
16. Luo, S., Tsoi, A.H.: Filtering of hidden weak Markov chain-discrete range observation. In: Mamon, R.S., Elliott, R.J. (eds.) Hidden Markov Models in Finance, pp. 106–119. Springer, New York (2007)
17. Mamon, R.S., Jalen. L.: Parameter estimation in a regime-switching model when the drift and volatility are independent. In: Proceedings of 5th International Conference on Dynamic Systems and Applications, pp. 291–298. Dynamic Publishers, Atlanta (2008)
18. Mamon, R.S., Erlwein, C., Gopaluni, R.B.: Adaptive signal processing of asset price dynamics with predictability analysis. Inform. Sci. **178**, 203–219 (2008)
19. Pagan, A.R., Schwert, G.W.: Alternative models for conditional stock volatility. J. Econ. **45**, 267–290 (1990)
20. Rydén, T., Terasvirta, T., Asbrink, S.: Stylized facts of daily return series and the hidden Markov model. J. Appl. Econ. **13**, 217–244 (1998)
21. Siu, T., Ching, W., Fung, E., Ng, M. Li, X.: A high-order Markov-switching model for risk measurement. Comput. Math. Appl. **58**, 1–10 (2009)
22. Tyssedal, J.S., Tjostheim, D.: An autoregressive model with suddenly changing parameters and an application to stock market prices. Appl. Stat. **37**, 353–369 (1988)
23. Veronesi, P.: How does information affect stock returns? J. Financ. **55**, 807–837 (2009)
24. Wu, C.: On the convergence properties of the EM algorithm. Ann. Stat. **11**, 95–103 (1983)
25. Xi, X., Mamon, R.S.: Parameter estimation of an asset price model driven by a weak hidden Markov chain. Econ. Model. **28**, 36–46 (2011)
26. Yu, S., Liu, Z., Squillante, M.S., Xia, C.H., Zhang, L.: A hidden semi-Markov model for web workload self-similarity. In: Proceedings of 21st IEEE International Performance, Computing, and Communications Conference, pp. 65–72, Phoenix (2002)
27. Zhou, N., Mamon, R.: An accessible implementation of interest rate models with Markov-switching. Expert Syst. Appl. **39**(5), 4679–4689 (2012)

Chapter 11
Parameter Estimation in a Regime-Switching Model with Non-normal Noise

Luka Jalen and Rogemar S. Mamon

Abstract This paper deals with the estimation of a Markov-modulated regime-switching model for asset prices, where the noise term is assumed non-normal consistent with the well-known observed market phenomena that log-return distributions exhibit heavy tails. Hence, the proposed model augments the flexibility of the current Markov-switching models with normal perturbation whilst still achieving dynamic calibration of parameters. In particular, under the setting where the model's noise term follows a t-distribution, we employ the method of change of reference probability measure to provide recursive filters for the estimate of the state and transition probabilities of the Markov chain. Although recursive filters are no longer available for the maximum likelihood estimation of the model's drift and volatility components under the current extension, we show that such estimation is tantamount to solving numerically a manageable system of nonlinear equations. Practical applications with the use of simulated and real-market data are included to demonstrate the implementation of our proposed algorithms.

11.1 Introduction

Various authors previously considered a discrete-time, finite-state Markov chain which is observed through a real-valued function whose values are corrupted by noise assumed to be independent and identically distributed (IID) normals. See for example, Mamon and Elliott [5], Erlwein and Mamon [3], Erlwein et al. [4], and

---

L. Jalen
MVG Equity, Nomura International plc, 1 Angel Lane, London, EC4R 3AB, UK
e-mail: Luka.Jalen@nomura.com

R.S. Mamon (✉)
Department of Statistical and Actuarial Sciences, University of Western Ontario,
1151 Richmond Street, London, ON, N6A 5B7 Canada
e-mail: rmamon@stats.uwo.ca

Date et al. [1], amongst others. It is, however, an accepted fact that the noise distorting the true state of the observable processes does not necessarily follow the normal distribution. In financial time series modeling, assumptions of normal IID noise term results in tails thinner than those observed in the market.

Moving away from the normality assumption, i.e., employing non-normal noise, complicates considerably the algebra involved in the calculation of filters. Despite this inherent complexity, this paper attempts to relax the assumption on the noise term and explore the implications. The noise term is assumed to have a general distribution, and even depend on the state of the underlying Markov chain. With this flexibility and extension, it is in general not possible to come up with all the recursive filtering equations for the parameter re-estimation just like in the current literature. Despite this drawback, recursive filters for estimating the transition probabilities could be derived without regard to the distribution of the noise. However, numerical methods are necessary in order to approximate the remaining parameters, namely, the drift vector $\boldsymbol{\alpha}$ and volatility vector $\boldsymbol{\beta}$ of the observation process in the proposed modeling framework of this paper.

This chapter is structured as follows. In Sect. 11.2, the modeling set up is introduced giving the dynamics of the state, which is captured by the Markov chain, and observation processes. The change of reference probability technique, as it applies to our modelling formulation, is discussed in Sect. 11.3. We establish the filtering equations in Sect. 11.4. In Sect. 11.5, we apply the Expectation-Maximisation algorithm to find the representation of the optimal estimates for the transition probabilities in terms of the filters. A system of equations is given whose solutions provide the optimal estimates of the drift and volatility. Section 11.6 contains several examples illustrating parameter estimation using simulated and observed market data. Model validation is performed via error analysis. This chapter is concluded in Sect. 11.7.

11.2 Model Set Up

It is assumed that all processes are defined on a complete probability space $(\Omega, \mathscr{F}, P)$. Suppose $\mathbf{x}_k$ is a homogenous Markov chain with a finite state space. Without loss of generality, we can assume that the state space of $\mathbf{x}_k$ is associated with the canonical basis $\{\mathbf{e}_1, \ldots, \mathbf{e}_N\} \in \mathbb{R}^N$ and $\mathbf{e}_i = (0, \ldots, 0, 1, 0, \ldots, 0)^\top$, where $\top$ denotes the transpose of a vector. Further, assume that $\mathbf{x}_0$ is given, or its distribution is known and $\Pi = (\pi_{ji})$ is the transition probability matrix with $\pi_{ji} = P(\mathbf{x}_{k+1} = \mathbf{e}_j \mid \mathbf{x}_k = \mathbf{e}_i)$, $1 \leq i, j \leq N$. Additionally, $\mathbf{x}_k$ is not observed directly, rather there is an observation process

$$y_{k+1} = \langle \boldsymbol{\alpha}, \mathbf{x}_k \rangle + \langle \boldsymbol{\beta}, \mathbf{x}_k \rangle z_{k+1}(\mathbf{x}_k), \quad (11.1)$$

where $\{z_{k+1}(\mathbf{x}_k)\}$ is a sequence of independent random variables with a distribution function $\phi_{\mathbf{x}_k}(\cdot)$, possibly dependent on the state of the Markov chain. It is a known result (cf. Elliott et al. [2]) that $\mathbf{x}_k$ has a semimartingale representation given by

$$\mathbf{x}_{k+1} = \Pi \mathbf{x}_k + \mathbf{v}_{k+1}, \quad (11.2)$$

where $\mathbf{v}_{k+1}$ is a martingale increment, and $\boldsymbol{\alpha} = (\alpha_1, \alpha_2, \ldots, \alpha_N)^\top$ and $\boldsymbol{\beta} = (\beta_1, \beta_2, \ldots, \beta_N)^\top$ are the respective drift and volatility parameters.

The filtrations or histories associated with our processes are: (i) $\mathscr{F}_k$, the complete filtration generated by $\mathbf{x}_0, \mathbf{x}_1, \ldots, \mathbf{x}_k$; (ii) $\mathscr{Y}_k$, the complete filtration generated by $y_0, y_1, \ldots, y_k$; and (iii) $\mathscr{H}_k := \mathscr{F}_k \vee \mathscr{Y}_k$.

11.3 Reference Probability Measure

Adopting the idea of change-of-measure-based filtering approach in discrete time, we consider a real-world measure P under which the signal model with the real-valued y process on $(\Omega, \mathscr{F}, P)$ has the dynamics

$$\mathbf{x}_{k+1} = \Pi \mathbf{x}_k + \mathbf{v}_{k+1}$$
$$y_{k+1} = \langle \boldsymbol{\alpha}, \mathbf{x}_k \rangle + \langle \boldsymbol{\beta}, \mathbf{x}_k \rangle z_{k+1}(\mathbf{x}_k).$$

We introduce a new probability measure Q via a Radon-Nikodým derivative

$$\left. \frac{dP}{dQ} \right|_{\mathscr{H}_k} = \Lambda_k,$$

such that under Q the random variable y_{k+1} has density $\phi_{\mathbf{x}_k}(\cdot)$. Write

$$\lambda_l := \sum_{i=1}^N \langle \mathbf{x}_{l-1}, \mathbf{e}_i \rangle \frac{\phi_{\mathbf{e}_i}\left(\frac{y_l - \alpha_i}{\beta_i}\right)}{\beta_i \phi_{\mathbf{e}_i}(y_l)},$$

$$\Lambda_k := \prod_{l=1}^k \lambda_l, \qquad \Lambda_0 = 1. \tag{11.3}$$

Lemma 11.1. *Write* $\Phi(\cdot) := \sum_{i=1}^N \langle \mathbf{x}_k, \mathbf{e}_i \rangle \phi_{\mathbf{e}_i}(\cdot)$. *Under Q, the y_k's are IID random variables following* $\Phi(\cdot)$, *which is a mixture of* $\phi_{\mathbf{e}_i}(\cdot)$ *distributions.*

Proof. Using Bayes' theorem for conditional expectation, we have

$$Q(y_{k+1} \leq t \mid \mathscr{H}_k) = \mathbb{E}^Q[I(y_{k+1} \leq t) \mid \mathscr{H}_k]$$
$$= \frac{\mathbb{E}[\lambda_{k+1}^{-1} I(y_{k+1} \leq t) \mid \mathscr{H}_k]}{\mathbb{E}[\lambda_{k+1}^{-1} \mid \mathscr{H}_k]}. \tag{11.4}$$

Now,

$$\mathbb{E}[\lambda_{k+1} \mid \mathscr{H}_k] = \int_{-\infty}^{\infty} \frac{\langle \boldsymbol{\beta}, \mathbf{x}_k \rangle \Phi(y_{k+1})}{\Phi(z_{k+1})} \Phi(z_{k+1}) dz_{k+1} = 1. \tag{11.5}$$

Therefore, we can write

$$Q(y_{k+1} \le t \mid \mathcal{H}_k) = \mathbb{E}^Q\left[I(y_{k+1} \le t) \mid \mathcal{H}_k\right]$$
$$= \int_{-\infty}^{\infty} \frac{\langle \boldsymbol{\beta}, \mathbf{x}_k \rangle \Phi(y_{k+1})}{\Phi(z_{k+1})} \Phi(z_{k+1}) I(y_{k+1} \le t) dz_{k+1}$$
$$= \int_{-\infty}^{t} \Phi(y_{k+1}) dy_{k+1} = Q(y_{k+1} \le t), \qquad (11.6)$$

and the result follows. □

Remark 11.1. The converse of Lemma 11.1 may be proved similarly. That is, if we start with a probability measure Q on $(\Omega, \mathcal{F})$ such that under Q, $\mathbf{x}_k$ is a Markov chain with dynamics $\mathbf{x}_{k+1} = \boldsymbol{\Pi} \mathbf{x}_k + \mathbf{v}_{k+1}$ and $\{y_k\}$ is a sequence of IID random variables with a distribution function $\Phi(\cdot)$ then under P, z_k is a sequence of IID random variables following the distribution $\Phi(\cdot)$ on $(\Omega, \mathcal{F})$.

Our aim is to estimate $\mathbf{x}$ given the observation under P, the real-world probability. Under P, even the recursive formulae for $\mathbf{x}_k$ are not linear in $\mathbf{x}_k$, and this undoubtedly leads to complications to their evaluation. So, calculations will be performed under Q due to the convenience of the IID assumptions for y_k's; see Elliott et al. [2] for more details.

Write $\boldsymbol{\xi}_k := \mathbb{E}^Q[\Lambda_k \mathbf{x}_k \mid \mathcal{Y}_k]$. Observing that $\sum_{i=1}^{N} \langle \mathbf{x}_k, \mathbf{e}_i \rangle = 1$, we have

$$\sum_{i=1}^{N} \mathbb{E}^Q\left[\langle \Lambda_k \mathbf{x}_k, \mathbf{e}_i \rangle \mid \mathcal{Y}_k\right] = \sum_{i=1}^{N} \langle \mathbb{E}^Q[\Lambda_k \mathbf{x}_k \mid \mathcal{Y}_k], \mathbf{e}_i \rangle = \sum_{i=1}^{N} \langle \boldsymbol{\xi}_k, \mathbf{e}_i \rangle. \qquad (11.7)$$

In addition, letting

$$\hat{p}_k^i = \mathbb{P}(\mathbf{x}_k = \mathbf{e}_i \mid \mathcal{Y}_k) \quad \text{and} \quad \hat{\mathbf{p}}_k = \left(\hat{p}_k^1, \ldots, \hat{p}_k^N\right), \qquad (11.8)$$

we get an explicit expression for the conditional distribution of $\mathbf{x}_k$ under P given $\mathcal{Y}_k$ given by

$$\hat{\mathbf{p}}_k = \frac{\boldsymbol{\xi}_k}{\sum_{i=1}^{N} \langle \boldsymbol{\xi}_k, \mathbf{e}_i \rangle}. \qquad (11.9)$$

Equation (11.9) provides the optimal estimate for the state of the Markov chain. Its update as new information becomes available is given in the next section.

11.4 Recursive Estimation

We present a recursive filter for the vector process $\boldsymbol{\xi}_k$ as well as derive the recursive filtering relations and formulae which will be used in turn for the estimation of transition probabilities. Write $\mathbf{D}$ for the diagonal matrix whose i-th element on the diagonal is

$$\frac{\phi_{e_i}\left(\frac{y_l - \alpha_i}{\beta_i}\right)}{\beta_i \phi_{e_i}(y_l)}. \tag{11.10}$$

Lemma 11.2. *If* $\boldsymbol{\xi}_k := \mathbb{E}^Q[\Lambda_k \mathbf{x}_k \mid \mathscr{Y}_k]$, $\mathbf{D}$ *is a matrix with diagonal elements of the form* (11.10) *and* $\boldsymbol{\Pi}$ *a transition matrix corresponding to the Markov chain* $\mathbf{x}_k$ *then*

$$\boldsymbol{\xi}_{k+1} = \boldsymbol{\Pi} \mathbf{D} \boldsymbol{\xi}_k. \tag{11.11}$$

Proof. From the definition of $\boldsymbol{\xi}_k$, we have

$$\boldsymbol{\xi}_{k+1} := \mathbb{E}^Q[\Lambda_{k+1} \mathbf{x}_{k+1} \mid \mathscr{Y}_{k+1}] \tag{11.12}$$

and therefore,

$$\begin{aligned}
\boldsymbol{\xi}_{k+1} &= \mathbb{E}^Q[\Lambda_{k+1} \mathbf{x}_{k+1} \mid \mathscr{Y}_{k+1}] = \mathbb{E}^Q[\Lambda_k \lambda_{k+1}(\boldsymbol{\Pi} \mathbf{x}_k + \mathbf{v}_{k+1}) \mid \mathscr{Y}_{k+1}] \\
&= \mathbb{E}^Q\left[\Lambda_k \left(\sum_{i=1}^N \langle \mathbf{x}_k, \mathbf{e}_i \rangle \frac{\phi_{e_i}\left(\frac{y_{k+1} - \alpha_i}{\beta_i}\right)}{\beta_i \phi_{e_i}(y_{k+1})}\right) \boldsymbol{\Pi} \mathbf{x}_k \,\bigg|\, \mathscr{Y}_{k+1}\right] \\
&= \sum_{i=1}^N \mathbb{E}^Q[\Lambda_k \langle \mathbf{x}_k, \mathbf{e}_i \rangle \mid \mathscr{Y}_{k+1}] \frac{\phi_{e_i}\left(\frac{y_{k+1} - \alpha_i}{\beta_i}\right)}{\beta_i \phi_{e_i}(y_{k+1})} \boldsymbol{\Pi} \mathbf{e}_i = \boldsymbol{\Pi} \mathbf{D} \boldsymbol{\xi}_k. \tag{11.13}
\end{aligned}$$

$\square$

In addition to the optimal state estimation presented in Lemma 11.2, we also aim to estimate the parameters of the model given in Eqs. (11.1) and (11.2), i.e., estimate the vectors $\boldsymbol{\alpha}$ and $\boldsymbol{\beta}$, and transition matrix $\boldsymbol{\Pi}$. To do this, we define the processes, $\mathscr{O}_k^r$, $\mathscr{J}_k^{sr}$, and $\mathscr{T}_k^r$ for the respective occupation time, number of jumps of $\mathbf{x}$ from r to s up to time k, and related auxiliary processes, as follows:

$$\mathscr{O}_{k+1}^r = \sum_{i=1}^{k+1} \langle \mathbf{x}_i, \mathbf{e}_r \rangle \tag{11.14}$$

$$\mathscr{J}_{k+1}^{rs} = \sum_{i=1}^{k+1} \langle \mathbf{x}_i, \mathbf{e}_r \rangle \langle \mathbf{x}_i, \mathbf{e}_s \rangle \tag{11.15}$$

$$\mathscr{T}_{k+1}^r(g) = \sum_{i=1}^{k+1} \langle \mathbf{x}_i, \mathbf{e}_r \rangle g(y), \tag{11.16}$$

for some function $g(y)$ in the estimation of α_i, and being replaced by $h(y)$ in the estimation of β_i.

Notation: For any $\mathscr{Y}$-adapted process C, we shall use the notation $\gamma(C)_k := \mathbb{E}^Q[\Lambda_k C_k \mid \mathscr{Y}_k]$.

Theorem 11.1. *If* $\mathbf{D}$ *is the diagonal matrix as defined in* (11.10), *the recursive relations for the vector processes* $\gamma(\mathscr{J}^{sr}\mathbf{x})_k$, $\gamma(\mathscr{O}^r\mathbf{x})_k$ *and* $\gamma(\mathscr{T}^r\mathbf{x})_k$ *are*

$$\gamma(\mathscr{J}^{sr}\mathbf{x})_k = \mathbf{\Pi D}(y_k)\gamma(\mathscr{J}^{sr}\mathbf{x})_{k-1} + \langle \boldsymbol{\xi}_{k-1}, \mathbf{e}_r \rangle \frac{\phi_{\mathbf{e}_r}\left(\frac{y_k-\alpha_r}{\beta_r}\right)}{\beta_r \phi_{\mathbf{e}_r}(y_k)} \pi_{sr}\mathbf{e}_s \quad (11.17)$$

$$\gamma(\mathscr{O}^r\mathbf{x})_k = \mathbf{\Pi D}(y_k)\gamma(\mathscr{O}^r\mathbf{x})_{k-1} + \langle \boldsymbol{\xi}_{k-1}, \mathbf{e}_r \rangle \frac{\phi_{\mathbf{e}_r}\left(\frac{y_k-\alpha_r}{\beta_r}\right)}{\beta_r \phi_{\mathbf{e}_r}(y_k)} \mathbf{\Pi e}_r \quad (11.18)$$

$$\gamma(\mathscr{T}^r(g)\mathbf{x})_k = \mathbf{\Pi D}(y_k)\gamma(\mathscr{T}^r(g)\mathbf{x})_{k-1} + \langle \boldsymbol{\xi}_{k-1}, \mathbf{e}_r \rangle \frac{\phi_{\mathbf{e}_r}\left(\frac{y_k-\alpha_r}{\beta_r}\right)}{\beta_r \phi_{\mathbf{e}_r}(y_k)} g(y_k)\mathbf{\Pi e}_r. \quad (11.19)$$

Proof.

$$\begin{aligned}
\gamma(\mathscr{J}^{sr}\mathbf{x})_k &= \mathbb{E}^Q\left[\Lambda_k \mathscr{J}_k^{sr} \mathbf{x}_k \mid \mathscr{Y}_k\right] \\
&= \mathbb{E}^Q\left[\Lambda_{k-1}\lambda_k(\mathscr{J}_{k-1}^{sr} + \langle \mathbf{x}_{k-1}, \mathbf{e}_r\rangle\langle \mathbf{x}_k, \mathbf{e}_s\rangle)\mathbf{x}_k \mid \mathscr{Y}_k\right] \\
&= \sum_{i=1}^N \mathbb{E}^Q\left[\Lambda_{k-1}\langle \mathbf{x}_{k-1}, \mathbf{e}_r\rangle \mathscr{J}_{k-1}^{sr} \mid \mathscr{Y}_k\right]\left[\sum_{r=1}^N \langle \mathbf{x}_{k-1}, \mathbf{e}_r\rangle \frac{\phi_{\mathbf{e}_r}(y_k - \alpha_i)}{\phi_{\mathbf{e}_r}(y_k)}\right] \\
&\quad + \mathbb{E}^Q\left[\Lambda_{k-1}\langle \mathbf{x}_{k-1}, \mathbf{e}_r\rangle \mid \mathscr{Y}_k\right] \frac{\phi_{\mathbf{e}_r}(y_k - \alpha_r)}{\phi_{\mathbf{e}_r}(y_k)} \pi_{sr}\mathbf{e}_s \\
&= \mathbf{\Pi D}(y_k)\gamma(\mathscr{J}^{sr}\mathbf{x})_{k-1} + \langle \boldsymbol{\xi}_k, \mathbf{e}_r\rangle \frac{\phi_{\mathbf{e}_r}(y_k - \alpha_r)}{\phi_{\mathbf{e}_r}(y_k)} \pi_{sr}\mathbf{e}_s.
\end{aligned}$$

The proofs of the two other recursive formulae, namely $\gamma(\mathscr{O}^r\mathbf{x})_k$ and $\gamma(\mathscr{T}^r\mathbf{x})_k$ follow similar arguments and are thus omitted. □

To make the results of Theorem 11.1 usable in the optimal estimation of a scalar quantity, we note that

$$\gamma(C)_k = \gamma(C_k\langle \mathbf{x}_k, \mathbf{1}\rangle) = \gamma(\langle C_k\mathbf{x}_k, \mathbf{1}\rangle) = \langle \gamma(C_k\mathbf{x}_k), \mathbf{1}\rangle.$$

11.5 Parameter Estimation

In order to estimate the parameters of the model, the Expectation-Maximisation (EM) algorithm is employed to determine the optimal approximation for each parameter in the set $\boldsymbol{\theta}$. Initial values for the EM algorithm are assumed to be given. Updated parameter approximations are then carried out based on the maximisation of the conditional expected log-likelihoods. Recursive filters for the occupation time process and jump process can be used for the dynamic optimal estimation of transition probabilities.

11.5.1 EM Algorithm and the Estimation of Transition Probabilities

The EM algorithm entails the change of measure from P^θ to $P^{\hat\theta}$. Under P^θ, **x** is a Markov chain with transition matrix $\mathbf{\Pi}$. Under $P^{\hat\theta}$, **x** is still a Markov chain with transition matrix $\hat{\mathbf{\Pi}}$ and the set of optimal parameters is obtained by maximising $\mathbb{E}^\theta \left[\frac{dP^{\hat\theta}}{dP^\theta} \,\middle|\, \mathscr{Y} \right]$ with respect to the set of parameters $\hat{\boldsymbol{\theta}}$, where

$$\left. \frac{dP^{\hat\theta}}{dP} \right|_{\mathscr{H}_k} = \prod_{l=1}^{k} \left(\prod_{s,r=1}^{N} \left(\frac{\hat\pi_{sr}}{\pi_{sr}} \right)^{\langle \mathbf{x}_l, \mathbf{e}_s \rangle \langle \mathbf{x}_{l-1}, \mathbf{e}_r \rangle} \right).$$

Theorem 11.2. *If at time k the sequence of observations $y_1, \ldots, y_k$ is available and the parameter set $\boldsymbol{\theta} = \{\pi_{sr}, \alpha_r, \beta_r\}$ determine the model then the EM estimates for the transition probabilities are*

$$\hat\pi_{sr} = \frac{\gamma(\mathscr{J}^{sr}\mathbf{x})_k}{\gamma(\mathscr{O}^r\mathbf{x})_k}. \tag{11.20}$$

Proof. Using the Radon-Nikodým derivative of $P^{\hat\theta}$ with respect to P^θ we have

$$\begin{aligned}
\log \frac{dP^{\hat\theta}}{dP^\theta} &= \sum_{l=1}^{k} \log \left(\prod_{s,r=1}^{N} \left(\frac{\hat\pi_{sr}}{\pi_{sr}} \right)^{\langle \mathbf{x}_l, \mathbf{e}_s \rangle \langle \mathbf{x}_{l-1}, \mathbf{e}_r \rangle} \right) \\
&= \sum_{l=1}^{k} \sum_{s,r=1}^{N} (\log \hat\pi_{sr} - \log \pi_{sr}) \langle \mathbf{x}_l, \mathbf{e}_s \rangle \langle \mathbf{x}_{l-1}, \mathbf{e}_r \rangle \\
&= \sum_{s,r=1}^{N} \mathscr{J}_k^{sr} \log \hat\pi_{sr} + R,
\end{aligned} \tag{11.21}$$

where the R does not involve $\hat\pi_{sr}$. Observe that $\sum_{s=1}^{N} \mathscr{J}_k^{sr} = \mathscr{O}_k^r$, hence

$$\sum_{s=1}^{N} \hat{\mathscr{J}}_k^{sr} = \hat{\mathscr{O}}_k^r. \tag{11.22}$$

The $\hat\pi_{sr}$'s optimal estimate is the value that maximises the log-likelihood function in (11.21) subject to the constraint $\sum_{s=1}^{N} \hat\pi_{sr} = 1$.

Introducing the Lagrange multiplier δ, we consider the function

$$L(\hat\pi, \delta) = \sum_{r,s=1}^{N} \hat{\mathscr{J}}_k^{sr} \log \hat\pi_{sr} + \delta \left(\sum_{s=1}^{N} \hat\pi_{sr} - 1 \right) + R(\pi_{sr}). \tag{11.23}$$

Differentiating (11.23) with respect to $\hat\pi_{sr}$ and δ and equating the derivatives to 0, we have

$$\frac{1}{\hat{\pi}_{sr}}\hat{\mathscr{J}}_k^{sr} + \delta = 0. \qquad (11.24)$$

Equation (11.24), however, can be re-written as

$$\hat{\pi}_{sr} = \frac{\hat{\mathscr{J}}_k^{sr}}{-\delta}. \qquad (11.25)$$

Therefore,

$$\sum_{s=1}^{N} \hat{\pi}_{sr} = \frac{\sum_{s=1}^{N} \hat{\mathscr{J}}_k^{sr}}{-\delta}. \qquad (11.26)$$

Considering $\sum_{s=1}^{N} \hat{\pi}_{sr} = 1$ together with (11.22), Eq. (11.26) simplifies to

$$\delta = -\hat{\mathscr{O}}_k^r.$$

Thus, from Eq. (11.25), the optimal estimate for $\hat{\pi}_{sr}$ is

$$\hat{\pi}_{sr} = \frac{\hat{\mathscr{J}}_k^{sr}}{\hat{\mathscr{O}}_k^r} = \frac{\gamma_k(\mathscr{J}_k^{sr})}{\gamma_k(\mathscr{O}_k^r)}. \qquad \square$$

In the completely general case, i.e., making no assumptions concerning the distribution of the noise term, it is not possible to derive the formulae for the re-estimation of the model parameters. Again, in order to use the EM algorithm one needs to change the measure from P^θ to $P^{\hat{\theta}}$ which depends on the specific distribution function of the noise assumed for a model.

Furthermore, it is not possible to obtain recursive filtering formulae that could provide updates straightforwardly for the model parameters apart from the transition probabilities in the general case. One needs to resort to numerical methods in evaluating the maximum likelihood. As will be shown in the next subsection for the case where noise follows the Student's t-distribution, we can reduce the estimation problem to finding a zero of a function. This is then a relatively simple numerical problem that can be solved quickly using modern computers.

11.5.2 Student's t-Distributed Noise Term

In this subsection, we focus on the model in (11.1) when a Student's t-distribution with v degrees of freedom governs the noise term. In general, the noise term can be a function of the state of the underlying Markov chain $\mathbf{x}_k$. Write $\mathbf{v} = (v_1, \ldots, v_n)^\top$ for the vector of degrees of freedom and consider again

$$y_{k+1} = \langle \boldsymbol{\alpha}, \mathbf{x}_k \rangle + \langle \boldsymbol{\beta}, \mathbf{x}_k \rangle z_{k+1}(\langle \mathbf{v}, \mathbf{x}_k \rangle),$$

where $\{z_{k+1}(v_i)\}$ is a sequence of independent random variables following the Student's t-distribution with v_i degrees of freedom. For simplicity, we assume that v_i is assigned and not estimated from the data.

Theorem 11.3. *Let $y_1,\ldots,y_k$ be a sequence of observations available at time k and let the parameter set $\boldsymbol{\theta} = \{\pi_{sr},\alpha_r,\beta_r\}$ determine the model. Then the EM estimates for $\{\hat{\boldsymbol{\alpha}}_r,\hat{\boldsymbol{\beta}}_r\}$ are solutions to the following system of equations:*

$$(v_i+1)\gamma(\mathscr{T}^{(i)}(g_{\hat{\alpha}_i})\mathbf{x})_k = 0,$$

$$\text{where} \quad g_{\hat{\alpha}_i} : y \longmapsto \frac{\hat{\alpha}_i - y}{\beta_i v_i + (\hat{\alpha}_i - y)^2}, \quad \text{and}$$

$$(v_i+1)\left(\gamma(\mathscr{T}^{(i)}(h_{\hat{\beta}_i})\mathbf{x})_k - \frac{\gamma(\mathscr{O}^{(i)}\mathbf{x})_k}{\hat{\beta}_i}\right) = 0,$$

$$\text{where} \quad h_{\hat{\beta}_i} : y \longmapsto \frac{\hat{\beta}_i v_i}{\hat{\beta}_i v_i + (y - \alpha_i)^2}.$$

Proof. Consider the parameters α_i, $i \in \{1,2,\ldots,N\}$. To get the updates for parameters $\hat{\alpha}_i$ from α_i, we first consider the factors λ_{k+1}'s in the likelihood function. We have

$$\lambda_{k+1}(\mathbf{x}_k,y_{k+1}) = \frac{\phi_{\langle \mathbf{x}_k,\mathbf{v}\rangle}\left(\frac{y_{k+1}-\langle \mathbf{x}_k,\hat{\boldsymbol{\alpha}}\rangle}{\langle \mathbf{x}_k,\boldsymbol{\beta}\rangle}\right)}{\phi_{\langle \mathbf{x}_k,\mathbf{v}\rangle}\left(\frac{y_{k+1}-\langle \mathbf{x}_k,\boldsymbol{\alpha}\rangle}{\langle \mathbf{x}_k,\boldsymbol{\beta}\rangle}\right)}$$

$$= \frac{\left(1+\frac{(y_{k+1}-\langle \mathbf{x}_k,\hat{\boldsymbol{\alpha}}\rangle)^2}{\langle \mathbf{x}_k,\boldsymbol{\beta}\rangle^2\langle \mathbf{x}_k,\mathbf{v}\rangle}\right)^{-\frac{\langle \mathbf{x}_k,\mathbf{v}\rangle+1}{2}}}{\left(1+\frac{(y_{k+1}-\langle \mathbf{x}_k,\boldsymbol{\alpha}\rangle)^2}{\langle \mathbf{x}_k,\boldsymbol{\beta}\rangle^2\langle \mathbf{x}_k,\mathbf{v}\rangle}\right)^{-\frac{\langle \mathbf{x}_k,\mathbf{v}\rangle+1}{2}}}$$

$$= \left(\frac{\langle \mathbf{x}_k,\boldsymbol{\beta}\rangle^2\langle \mathbf{x}_k,\mathbf{v}\rangle + y_{k+1}^2 - 2y_{k+1}\langle \mathbf{x}_k,\hat{\boldsymbol{\alpha}}\rangle + \langle \mathbf{x}_k,\hat{\boldsymbol{\alpha}}\rangle^2}{\langle \mathbf{x}_k,\boldsymbol{\beta}\rangle^2\langle \mathbf{x}_k,\mathbf{v}\rangle + y_{k+1}^2 - 2y_{k+1}\langle \mathbf{x}_k,\boldsymbol{\alpha}\rangle + \langle \mathbf{x}_k,\boldsymbol{\alpha}\rangle^2}\right)^{-\frac{\langle \mathbf{x}_k,\mathbf{v}\rangle+1}{2}} \quad (11.27)$$

Write $\Lambda_{k+1}^*(\mathbf{x}_k,y_{k+1}) := \prod_{l=1}^{k}\lambda_{l+1}^*(\mathbf{x}_l,y_{l+1})$, and introduce a new measure P^* defined by

$$\left.\frac{dP^*}{dP}\right|_{\mathscr{H}_k} = \Lambda_{k+1}(\mathbf{x}_k,y_{k+1}). \quad (11.28)$$

Now,

$$\log \Lambda_{k+1}^* = -\sum_{l=1}^{k}\frac{\langle \mathbf{x}_l,\mathbf{v}\rangle+1}{2}\log\left(\frac{\langle \mathbf{x}_k,\boldsymbol{\beta}\rangle^2\langle \mathbf{x}_l,\mathbf{v}\rangle + y_{l+1}^2 - 2y_{l+1}\langle \mathbf{x}_l,\hat{\boldsymbol{\alpha}}\rangle + \langle \mathbf{x},\hat{\boldsymbol{\alpha}}\rangle^2}{\langle \mathbf{x}_k,\boldsymbol{\beta}\rangle^2\langle \mathbf{x}_l,\mathbf{v}\rangle + y_{l+1}^2 - 2y_{l+1}\langle \mathbf{x}_l,\boldsymbol{\alpha}\rangle + \langle \mathbf{x},\boldsymbol{\alpha}\rangle^2}\right)$$

$$= -\sum_{l=1}^{k}\sum_{i=1}^{n}\langle \mathbf{x}_l,\mathbf{e}_i\rangle\frac{v_i+1}{2}\log\left(\beta_i v_i + y_{l+1}^2 - 2y_{l+1}\hat{\alpha}_i + \hat{\alpha}_i^2\right) + R, \quad (11.29)$$

where R is independent of $\hat{\alpha}$.

We wish to find the maximum of

$$\mathbb{E}\left[\frac{dP^*}{dP} \mid \mathcal{Y}_k\right] \quad (11.30)$$

at $\hat{\boldsymbol{\alpha}}$. To do this, we differentiate the expected value in (11.30) with respect to $\hat{\alpha}_i$ and equate the resulting derivative to zero.

We have

$$\frac{\partial}{\partial \hat{\alpha}_i}\mathbb{E}\left[\log \Lambda_k^* \mid \mathcal{Y}_k\right] = \mathbb{E}\left[\frac{\partial}{\partial \hat{\alpha}_i}\log \Lambda_k^* \mid \mathcal{Y}_k\right]$$

$$= \mathbb{E}\left[(v_i+1)\sum_{l=1}^{k}\langle \mathbf{x}_{l-1}, \mathbf{e}_i\rangle \frac{\hat{\alpha}_i - y_l}{\beta_i v_i + (\hat{\alpha}_i - y_l)^2} \mid \mathcal{Y}_k\right]. \quad (11.31)$$

It can be seen from (11.31) that in order to maximise (11.30) one needs to find $\hat{\alpha}_i$ which makes the weighted sum of the differences $y_l - \hat{\alpha}_i$ equal to zero. Given this structure of the equation, a recursive estimation procedure for $\hat{\alpha}_i$ is not feasible. We find

$$\frac{\partial}{\partial \hat{\alpha}_i}\mathbb{E}\left[\log \Lambda_k^* \mid \mathcal{Y}_k\right] = (v_i+1)\mathbb{E}\left[\sum_{l=1}^{k}\langle \mathbf{x}_{l-1}, \mathbf{e}_i\rangle \frac{\hat{\alpha}_i - y_l}{\beta_i v_i + (\hat{\alpha}_i - y_l)^2} \mid \mathcal{Y}_k\right]$$

$$= (v_i+1)\gamma(\mathcal{T}^{(i)}(g_{\hat{\alpha}_i})\mathbf{x})_k \quad (11.32)$$

$$\text{where} \quad g_{\hat{\alpha}_i} : y \longmapsto \frac{\hat{\alpha}_i - y}{\beta_i v_i + (\hat{\alpha}_i - y)^2},$$

which finishes the proof of the first part.

In order to get the updated parameter $\widehat{\beta}_i$ from β_i, we need to consider the factors

$$\lambda_{k+1}^{**}(\mathbf{x}_k, y_{k+1}) = \frac{\phi_{\langle \mathbf{x}_k, \mathbf{v}\rangle}\left(\frac{y_{k+1}-\langle \mathbf{x}_k, \boldsymbol{\alpha}\rangle}{\langle \mathbf{x}_k, \hat{\boldsymbol{\beta}}\rangle}\right)}{\phi_{\langle \mathbf{x}_k, \mathbf{v}\rangle}\left(\frac{y_{k+1}-\langle \mathbf{x}_k, \boldsymbol{\alpha}\rangle}{\langle \mathbf{x}_k, \boldsymbol{\beta}\rangle}\right)}$$

$$= \frac{\left(1 + \frac{(y_{k+1}-\langle \mathbf{x}_k, \boldsymbol{\alpha}\rangle)^2}{\langle \mathbf{x}_k, \hat{\boldsymbol{\beta}}\rangle^2 \langle \mathbf{x}_k, \mathbf{v}\rangle}\right)^{-\frac{\langle \mathbf{x}_k, \mathbf{v}\rangle+1}{2}}}{\left(1 + \frac{(y_{k+1}-\langle \mathbf{x}_k, \boldsymbol{\alpha}\rangle)^2}{\langle \mathbf{x}_k, \boldsymbol{\beta}\rangle^2 \langle \mathbf{x}_k, \mathbf{v}\rangle}\right)^{-\frac{\langle \mathbf{x}_k, \mathbf{v}\rangle+1}{2}}}$$

$$= \left(\frac{\langle \mathbf{x}_k, \boldsymbol{\beta}\rangle^2 \langle \mathbf{x}_k, \mathbf{v}\rangle \langle \mathbf{x}_k, \hat{\boldsymbol{\beta}}\rangle^2 + \langle \mathbf{x}_k, \boldsymbol{\beta}\rangle^2 (y_{k+1}-\langle \mathbf{x}_k, \boldsymbol{\alpha}\rangle)^2}{\langle \mathbf{x}_k, \boldsymbol{\beta}\rangle^2 \langle \mathbf{x}_k, \mathbf{v}\rangle \langle \mathbf{x}_k, \hat{\boldsymbol{\beta}}\rangle^2 + \langle \mathbf{x}_k, \hat{\boldsymbol{\beta}}\rangle^2 (y_{k+1}-\langle \mathbf{x}_k, \boldsymbol{\alpha}\rangle)^2}\right)^{-\frac{\langle \mathbf{x}_k, \mathbf{v}\rangle+1}{2}} \quad (11.33)$$

Write $\Lambda_{k+1}^{**}(\mathbf{x}_k, y_{k+1}) = \prod_{l=1}^{k}\lambda_{l+1}^{**}(\mathbf{x}_l, y_{l+1})$ and consider a new measure P^* defined via

$$\left.\frac{dP^*}{dP}\right|_{\mathcal{H}_k} = \Lambda_{k+1}(\mathbf{x}_k, y_{k+1}). \quad (11.34)$$

11 Parameter Estimation in a Regime-Switching Model with Non-normal Noise

Now,

$$\log \Lambda_{k+1}^{**} = -\sum_{l=1}^{k} \frac{\langle \mathbf{x}_l, \mathbf{v} \rangle + 1}{2} \Big(\log \big(\langle \mathbf{x}_k, \boldsymbol{\beta} \rangle^2 \langle \mathbf{x}_k, \mathbf{v} \rangle \langle \mathbf{x}_k, \hat{\boldsymbol{\beta}} \rangle^2 \\ + \langle \mathbf{x}_k, \boldsymbol{\beta} \rangle^2 \big(y_{k+1} - \langle \mathbf{x}_k, \boldsymbol{\alpha} \rangle \big)^2 \big) \\ - \log \big(\langle \mathbf{x}_k, \boldsymbol{\beta} \rangle^2 \langle \mathbf{x}_k, \mathbf{v} \rangle \langle \mathbf{x}_k, \hat{\boldsymbol{\beta}} \rangle^2 + \langle \mathbf{x}_k, \hat{\boldsymbol{\beta}} \rangle^2 \big(y_{k+1} - \langle \mathbf{x}_k, \boldsymbol{\alpha} \rangle \big)^2 \big) \Big). \qquad (11.35)$$

Similar to the situation of calculating the EM estimates for $\hat{\alpha}_i$, it is not possible to get a recursive estimation procedure for $\hat{\beta}_i$, unless a standard normal noise term is assumed. However, it is still possible to re-estimate $\hat{\beta}_i$ numerically.

Following similar principles employed in deriving Eq. (11.31), we need to differentiate (11.35) with respect to $\hat{\beta}_i$ and equate the resulting derivative to zero. Doing this, we have

$$\frac{\partial}{\partial \hat{\beta}_i} \mathbb{E}\big[\log \Lambda_k^{**} \mid \mathscr{Y}_k\big] = \mathbb{E}\Big[\frac{\partial}{\partial \hat{\beta}_i} \log \Lambda_k^{**} \mid \mathscr{Y}_k\Big]$$

$$= \frac{v_i+1}{2} \mathbb{E}\Big[\sum_{l=1}^{k} \langle \mathbf{x}_{l-1}, \mathbf{e}_i \rangle \Big(\frac{2 \hat{\beta}_i v_i \beta_i^2}{\hat{\beta}_i^2 v_i \beta_i^2 + \beta_i^2 (y_l - \alpha_i)^2} \\ + \frac{2 \hat{\beta}_i v_i \beta_i^2 + 2 \hat{\beta}_i (y_l - \alpha_i)^2}{\hat{\beta}_i^2 v_i \beta_i^2 + \hat{\beta}_i^2 (y_l - \alpha_i)^2} \Big) \mid \mathscr{Y}_k \Big]$$

$$= \frac{v_i+1}{2} \mathbb{E}\Big[\sum_{l=1}^{k} \langle \mathbf{x}_{l-1}, \mathbf{e}_i \rangle \Big(\frac{2 \hat{\beta}_i v_i}{\hat{\beta}_i v_i + (y_l - \alpha_i)^2} \\ - \frac{1}{\hat{\beta}_i} \frac{2 \beta_i^2 v_i + 2(y_l - \alpha_i)^2}{\beta_i^2 v_i + (y_l - \alpha_i)^2} \Big) \mid \mathscr{Y}_k \Big]$$

$$= (v_i+1) \mathbb{E}\Big[\sum_{l=1}^{k} \langle \mathbf{x}_{l-1}, \mathbf{e}_i \rangle \Big(\frac{\hat{\beta}_i v_i}{\hat{\beta}_i v_i + (y_l - \alpha_i)^2} - \frac{1}{\hat{\beta}_i} \Big) \mid \mathscr{Y}_k \Big]$$

$$= (v_i+1) \Big(\gamma(\mathscr{T}^{(i)}(h_{\hat{\beta}_i}) \mathbf{x})_k - \frac{\gamma(\mathscr{O}^i \mathbf{x})_k}{\hat{\beta}_i} \Big) \qquad (11.36)$$

where $h_{\hat{\beta}_i} : y \longmapsto \dfrac{\hat{\beta}_i v_i}{\hat{\beta}_i v_i + (y - \alpha_i)^2}.$

This completes the proof. □

Remark 11.2. Performing further manipulations of Eq. (11.31), we have

$$\frac{\partial}{\partial \hat{\alpha}_i} \mathbb{E}\big[\log(\Lambda_k) \mid \mathcal{Y}_k\big]$$

$$= \mathbb{E}\left[(v_i+1) \frac{\sum_{l=1}^{k}\langle \mathbf{x}_{l-1}, \mathbf{e}_i\rangle(\hat{\alpha}_i - y_l) \prod_{j=1, j\neq l}^{k}\left(\beta_i v_i + (\hat{\alpha}_i - y_j)^2\right)}{\prod_{l=1}^{k}\left(\beta_i v_i + (\hat{\alpha}_i - y_l)^2\right)} \,\bigg|\, \mathcal{Y}_k\right]$$

$$= \mathbb{E}\left[\frac{\sum_{l=1}^{k}\langle \mathbf{x}_{l-1}, \mathbf{e}_i\rangle(\hat{\alpha}_i - y_l)(v_i+1) \prod_{j=1, j\neq l}^{k}\left(\beta_i v_i + (\hat{\alpha}_i - y_j)^2\right)}{\prod_{l=1}^{k}\left(\beta_i v_i + (\hat{\alpha}_i - y_l)^2\right)} \,\bigg|\, \mathcal{Y}_k\right]$$

$$= \mathbb{E}\left[\frac{\sum_{l=1}^{k}\langle \mathbf{x}_{l-1}, \mathbf{e}_i\rangle(\hat{\alpha}_i - y_l)\left(\beta_i v_i^k + \mathcal{O}(v_i^{k-1})\right)}{\beta_i v_i^k + \mathcal{O}(v_i^{k-1})} \,\bigg|\, \mathcal{Y}_k\right]. \qquad (11.37)$$

Consider the case of normally distributed noise terms by letting $v_i \to \infty$. It is apparent from the above that we would recover the known EM parameter estimate for α_i, derived for example in Eq. 18 of Mamon et al. [6]. Hence, it can be updated via the recursive filters known in the current literature.

The result in Remark 11.2 can be obtained by observing that

$$\lim_{v_i \to \infty} \left[\frac{\partial}{\partial \hat{\alpha}_i}\mathbb{E}\big[\log(\Lambda_k) \mid \mathcal{Y}_k\big]\right] = \mathbb{E}\left[\sum_{l=1}^{k}\langle \mathbf{x}_{l-1}, \mathbf{e}_i\rangle(\hat{\alpha}_i - y_l)\right]$$

$$= \hat{\alpha}_i \mathbb{E}\left[\sum_{l=1}^{k}\langle \mathbf{x}_{l-1}, \mathbf{e}_i\rangle\right] - \mathbb{E}\left[\sum_{l=1}^{k}\langle \mathbf{x}_{l-1}, \mathbf{e}_i\rangle y_l\right]$$

$$= \hat{\alpha}_i \gamma(\mathcal{O}^i \mathbf{x})_k - \gamma(\mathcal{T}^{(i)}(y)\mathbf{x})_k. \qquad (11.38)$$

Since Eq. (11.38) has to be zero, we get the optimal estimate for the drift, which is the ratio of $\mathcal{T}$ to $\mathcal{O}$, under the assumption of normally distributed noise terms.

11.6 Numerical Application of the Filters

11.6.1 Filtering Using Simulated Data

We provide a numerical demonstration of the results presented in the previous section. We investigate the performance of the filtering recursions on a simulated data set to concentrate on the accuracy of the estimation approach and thus, do not have to deal with the issue of model uncertainty. The filtering algorithm is tested on three sets of simulated data generated from a Markov chain process with two, three and four states. In each of the three examples presented below, 200 data points were generated by simulation in accordance with the model

$$y_{k+1} = \langle \boldsymbol{\alpha}, \mathbf{x}_k\rangle + \langle \boldsymbol{\beta}, \mathbf{x}_k\rangle z_{k+1}(\langle \mathbf{v}, \mathbf{x}_k\rangle), \qquad (11.39)$$

where $\{z_k(v_i)\}$ is a sequence of independent random variables following a Student's t-distribution with $v_i = 3$ degrees of freedom.

The filtering procedure is applied to the simulated data for three different numbers of states of the underlying Markov chain, with the re-estimation period containing 50 data points. In other words, the parameters are re-estimated upon the revelation of 50 new data points. All calculations were performed in Matlab, on a 1.83 Ghz dual core processor. The results of running the filtering algorithm with the computing times for the three examples are given. We also present the graphs of the simulated data ("true" data) versus the estimated values. For all graphs, the estimated values are calculated as $\boldsymbol{\alpha}^\top \hat{\mathbf{x}}_l$, where $\hat{\mathbf{x}}_l$ is the filtered state vector after the l-th data point is processed, and $\hat{\mathbf{x}}_k = E[\mathbf{x}_k] = \hat{p}_k$ is calculated using Eq. (11.9). In each example, we state the errors of the estimated parameters, calculated as the second norm of the difference between the filtered parameters and the ones used to simulate the data ("true" parameters underlying the data). The error metrics are

$$||\boldsymbol{\alpha} - \hat{\boldsymbol{\alpha}}||_2 = \sqrt{(\alpha_1 - \hat{\alpha}_1)^2 + \ldots + (\alpha_N - \hat{\alpha}_N)^2},$$

$$||\boldsymbol{\beta} - \hat{\boldsymbol{\beta}}||_2 = \sqrt{(\beta_1 - \hat{\beta}_1)^2 + \ldots + (\beta_N - \hat{\beta}_N)^2},$$

$$||\boldsymbol{\Pi} - \hat{\boldsymbol{\Pi}}||_2 = \sqrt{\text{eig}_{\max}((\boldsymbol{\Pi} - \hat{\boldsymbol{\Pi}})^\top (\boldsymbol{\Pi} - \hat{\boldsymbol{\Pi}}))},$$

where $\text{eig}_{\max}$ denotes the largest eigenvalue of a matrix.

The error is calculated after each re-estimation of the parameter values and decreasing errors depict the improvement in filtering as more data is processed. Due to the nature of the EM algorithm behind the parameter re-estimation procedure, a good guess of initial values in filtering is required. In all three examples presented below, we are able to set initial values for the transition matrix $\boldsymbol{\Pi}$ in a random manner but with some heuristic structure. The elements of the matrix are drawn from a uniform distribution over $(0,1)$ and then normalized to ensure the columns of $\boldsymbol{\Pi}$ sum up to one.

Here, we are not interested in trying to calculate the one-step (or more) ahead predictions; we are simply examining the performance of the filtering itself. Applications of the filtering algorithm to observed data are endeavours that require more in-depth analysis and are considered in Sect. 11.6.2.

Example 11.1. The values used to simulate the data for the case of a two-state Markov chain are reported in Table 11.1 and the initial values for α and β used in filtering are displayed in Table 11.2. The transition matrix $\boldsymbol{\Pi}$ used as an initial guess for filtering was random, i.e., its elements were drawn from the uniformly distributed random numbers on interval $(0,1)$. The calculated final values of the parameters are reported in Table 11.3 and the errors on the parameter estimates are displayed in Table 11.4. The graph of the simulated data (blue) together with the estimated value is shown in Fig. 11.1 and the total calculation time is 24.8 s.

Example 11.2. For the three-state Markov chain, the initial values for the data simulation are reported in Table 11.5. The values for α and β used as initial guesses

Table 11.1 Values of parameters (Π, α, β) used in the simulation for a two-state Markov chain

$$\Pi = \begin{bmatrix} 0.8 & 0.3 \\ 0.2 & 0.7 \end{bmatrix}, \quad \alpha = \begin{bmatrix} 1 \\ -1 \end{bmatrix} \quad \text{and} \quad \beta = \begin{bmatrix} 0.09 \\ 0.10 \end{bmatrix}$$

Table 11.2 Initial values of parameters (α, β) used in filtering for a two-state Markov chain

$$\alpha = \begin{bmatrix} 0.90 \\ -0.90 \end{bmatrix} \quad \text{and} \quad \beta = \begin{bmatrix} 0.10 \\ 0.09 \end{bmatrix}$$

Table 11.3 Final values of parameters (Π, α, β) calculated from the simulated data for a two-state Markov chain

$$\hat{P} = \begin{bmatrix} 0.7590 & 0.2924 \\ 0.2421 & 0.7091 \end{bmatrix}, \quad \hat{\alpha} = \begin{bmatrix} 1.0205 \\ -1.0046 \end{bmatrix} \quad \text{and} \quad \hat{\beta} = \begin{bmatrix} 0.1073 \\ 0.1281 \end{bmatrix}$$

Table 11.4 Errors of the estimated parameter values in the case of a two-state Markov chain

Re-estimation number	Errors		
	α	β	Π
1	0.0556	0.0396	0.1367
2	0.0375	0.0329	0.0763
3	0.0247	0.0330	0.0645
4	0.0210	0.0330	0.0599

in the filtering process can be found in Table 11.6 whilst the initial guesses for the transition matrix are random, its elements were drawn from uniformly distributed random numbers on interval $(0, 1)$. The outputs of the filtering algorithm are reported in Table 11.7 with the errors reported in Table 11.8. The graph of the simulated data (blue) versus the estimated values is shown in Fig. 11.2. Finally, the entire calculation took 56.3 s.

Table 11.5 Values of parameters (Π, α, β) used to simulation for a three-state Markov chain

$$\Pi = \begin{bmatrix} 0.8 & 0.2 & 0.05 \\ 0.1 & 0.7 & 0.15 \\ 0.1 & 0.1 & 0.80 \end{bmatrix}, \quad \alpha = \begin{bmatrix} 0 \\ 1 \\ -1 \end{bmatrix} \quad \text{and} \quad \beta = \begin{bmatrix} 0.08 \\ 0.09 \\ 0.10 \end{bmatrix}$$

Table 11.6 Initial values of parameters (α, β) used in filtering for a three-state Markov chain

$$\alpha = \begin{bmatrix} 0.01 \\ 0.90 \\ -0.90 \end{bmatrix} \quad \text{and} \quad \beta = \begin{bmatrix} 0.1 \\ 0.1 \\ 0.1 \end{bmatrix}$$

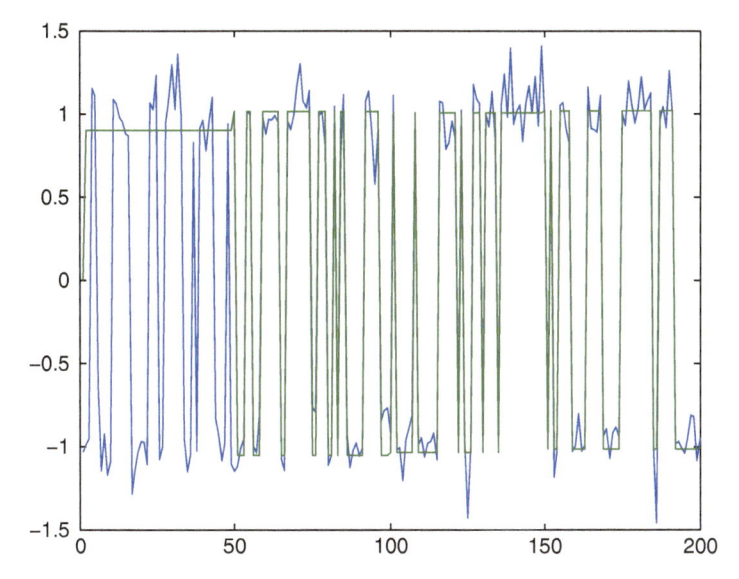

11 Parameter Estimation in a Regime-Switching Model with Non-normal Noise 255

Fig. 11.1 Simulated data (*blue*) with the estimated values (*green-fine dotted lines*)

Table 11.7 Final values of parameters (Π, α, β) calculated from the simulated data for a three-state Markov chain

$$\hat{\Pi} = \begin{bmatrix} 0.8157 & 0.1641 & 0.0465 \\ 0.0982 & 0.7238 & 0.0618 \\ 0.0944 & 0.1168 & 0.8937 \end{bmatrix}, \quad \hat{\alpha} = \begin{bmatrix} 0.01335 \\ 1.0156 \\ -1.0063 \end{bmatrix} \quad \text{and} \quad \hat{\beta} = \begin{bmatrix} 0.1471 \\ 0.0855 \\ 0.0957 \end{bmatrix}$$

Table 11.8 Errors of the estimated parameter values in the case of a three-state Markov chain

Re-estimation number	Errors		
	α	β	Π
1	0.0607	0.0191	0.6159
2	0.0212	0.0574	0.2058
3	0.0192	0.0711	0.1066
4	0.0215	0.0674	0.1288

Example 11.3. Finally, we present the results for the case of a four-state Markov chain driving the observation process. The values used for data simulation are exhibited in Table 11.9 and the initial guesses for the filtering algorithm in Table 11.10 with the transition matrix are random as in the previous two examples. The calculation time for the implementation under the four-state Markov chain is 120.1 seconds. The estimated parameter values for the vectors α and β can be found in Table 11.11 whilst the graph of the simulated data and the estimated values is displayed in Fig. 11.3. The errors of the parameters for each re-estimation are given in Table 11.12.

As illustrated in the above examples, we can still use partially the filtering-based estimation technique despite departure from the assumption of normally distributed

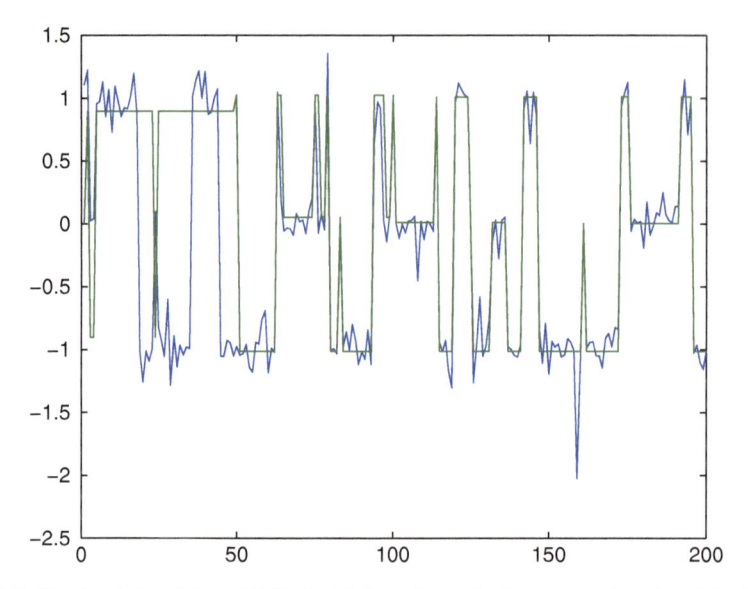

Fig. 11.2 Simulated data (*blue solid line*) with the estimated values (*green fine-dotted line*)

Table 11.9 Values of parameters (Π, α, β) used in the simulation for a four-state Markov chain

$$\Pi = \begin{bmatrix} 0.80 & 0.15 & 0.05 & 0.05 \\ 0.10 & 0.70 & 0.05 & 0.05 \\ 0.05 & 0.10 & 0.80 & 0.10 \\ 0.05 & 0.05 & 0.10 & 0.80 \end{bmatrix}, \quad \alpha = \begin{bmatrix} 0.0 \\ 0.5 \\ -0.5 \\ -1.0 \end{bmatrix} \quad \text{and} \quad \beta = \begin{bmatrix} 0.06 \\ 0.07 \\ 0.08 \\ 0.09 \end{bmatrix}$$

Table 11.10 Initial values of parameters (α, β) used in filtering for a four-state Markov chain

$$\alpha = \begin{bmatrix} 0.01 \\ 0.40 \\ -0.40 \\ -1.20 \end{bmatrix} \quad \text{and} \quad \beta = \begin{bmatrix} 0.1 \\ 0.1 \\ 0.1 \\ 0.1 \end{bmatrix}$$

Table 11.11 Final values of parameters (Π, α, β) calculated from the simulated data for a four-state Markov chain

$$\hat{\Pi} = \begin{bmatrix} 0.8184 & 0.1043 & 0.1299 & 0.0478 \\ 0.0995 & 0.7379 & 0.0898 & 0.1168 \\ 0.0441 & 0.0338 & 0.7745 & 0.0545 \\ 0.0442 & 0.1298 & 0.0151 & 0.7869 \end{bmatrix}, \quad \hat{\alpha} = \begin{bmatrix} -0.0115 \\ 0.5479 \\ -0.4837 \\ -1.0201 \end{bmatrix} \quad \text{and} \quad \hat{\beta} = \begin{bmatrix} 0.0935 \\ 0.0563 \\ 0.0762 \\ 0.0974 \end{bmatrix}$$

noise term. The calculations are, nonetheless, more demanding due to the fact that one needs to resort to numerical methods in the estimation of the drift and volatility parameters. Hence, the total computation time is longer. Considering the ability

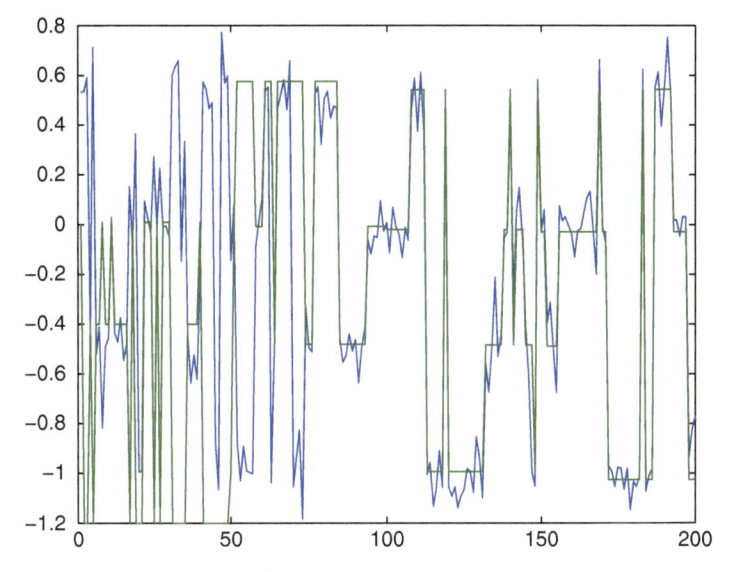

11 Parameter Estimation in a Regime-Switching Model with Non-normal Noise

Fig. 11.3 Simulated data (*blue solid line*) with the estimated values (*green fine-dotted line*)

Table 11.12 Errors of the estimated parameter values in the case of a four-state Markov chain

Re-estimation number	Errors		
	α	β	Π
1	0.0901	0.0435	0.5193
2	0.0594	0.0473	0.1752
3	0.0462	0.0461	0.1575
4	0.0389	0.0371	0.1976

to employ recursive formulae for re-estimating the transition probabilities and state of the Markov chain, it is still worth going through the procedure of changing the measure along with re-estimating the remaining parameters. It is also evident from Figs. 11.1 to 11.3 that the estimated values follow very closely the state of the underlying Markov chain after the first parameter re-estimation. Parameters were re-estimated after processing 50 data points, however there is no notable difference in the goodness of fit after the second and third parameter update. Therefore, assuming the dynamics of the observation data do not change much, there is no need to increase the frequency of re-estimations, i.e., lengthen the data window for each pass. An algorithm pass or step in this case comprises of 50 data points.

11.6.2 Application of the Filters to Observed Market Data

In the previous subsection, we showed that the derived filters were successfully used to estimate the parameters of the model on a simulated data set. In this section, we illustrate that the filters can also be applied on a larger data set of observed market data. We consider both the NASDAQ and DOW JONES data sets for the period 28 February 2003–16 February 2007.

Suppose S_k is a sequence of asset prices. Then, we can observe the logarithmic increments

$$y_k = \ln S_k - \ln S_{k-1} = \ln \frac{S_k}{S_{k-1}}$$

or

$$S_k = S_{k-1} \exp(y_{k-1}).$$

The logarithmic increments are assumed driven by a function f of the underlying Markov chain and some noise term, that is, $y_k = f(\mathbf{x}_k, z_{k+1})$. Here, $\{z_k\}$ is a sequence of IID random variables following a Student's t-distribution with 3 degrees of freedom.

Table 11.13 Summary statistics for the NASDAQ and DOW JONES logarithmic returns for the period 28/02/2003–16/02/2007

Statistic	NASDAQ data	DOW JONES data
Mean	3.998×10^{-4}	4.819×10^{-4}
Median	-4.834×10^{-4}	4.585×10^{-4}
Standard deviation	0.008	0.007
Skewness	0.446	0.124
Kurtosis	4.532	4.881
Range	0.063	0.072
Minimum	−0.026	−0.036
Maximum	0.036	0.035
Count	1,000	1,000

Table 11.14 Comparison of RMSEs and computational time in seconds for the DOW JONES and NASDAQ data

Data set	Number of MC states	RMSE	Computational time (s)
NASDAQ	2	3.2258×10^{-3}	86.3
	3	3.1522×10^{-3}	166.4
	4	3.1163×10^{-3}	303.3
DOW JONES	2	1.1243×10^{-3}	81.0
	3	1.0983×10^{-3}	153.9
	4	1.0859×10^{-3}	252.06

The filters from the previous section are applied to both data sets whose summary statistics are given in Table 11.13.

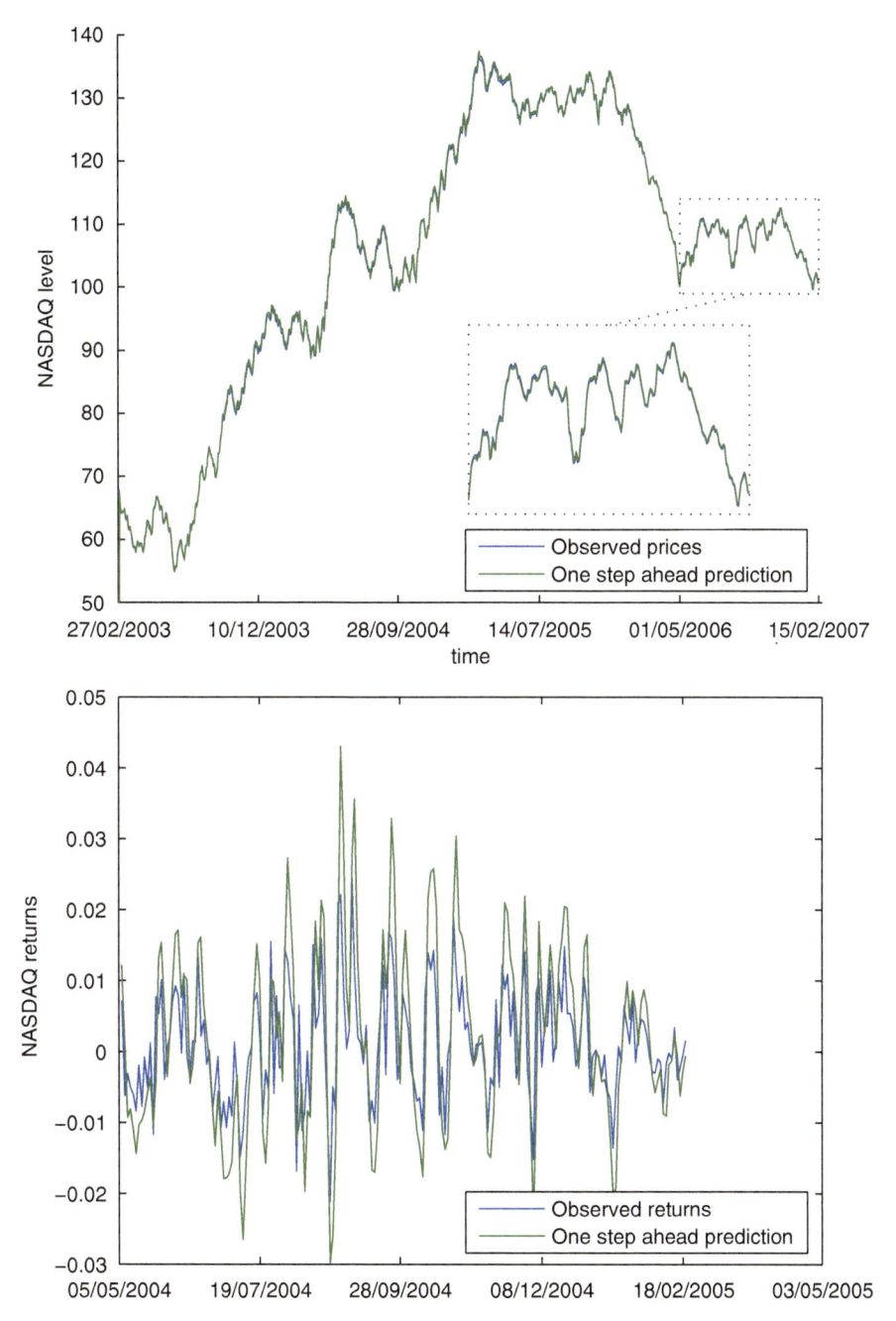

Fig. 11.4 NASDAQ actual returns series (*blue solid line*) and one-step ahead predictions (*green fine-dotted line*)

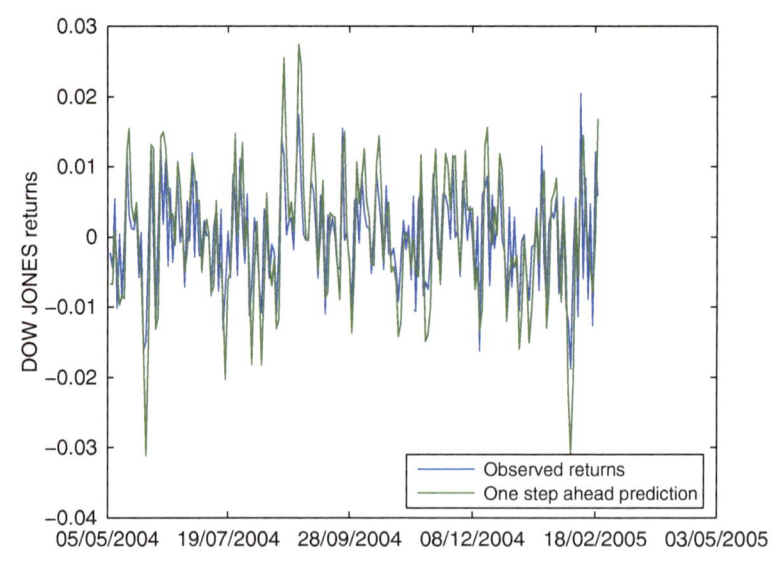

Fig. 11.5 DOW JONES actual returns series (*blue solid line*) and one-step ahead predictions (*green fine-dotted line*)

The data were processed in batches of 50 data points and the parameters were re-estimated using the results of Theorems 11.2 and 11.3. Table 11.14 depicts the fitting errors (RMSEs) and the computational time in seconds needed to complete the calculations under the assumption of student's t-distributed noise term. Compared with the naive, no change model $E[y_{k+1} \mid y_k] = y_k$, which has the RMSEs of 7.2682×10^{-3} and 8.3338×10^{-3} for the NASDAQ and DOW JONES data sets respectively, the HMM based filters perform very well. In Figs. 11.4 and 11.5, we present the plots of returns of the actual data and the one step-ahead predictions for the period 05 May 2004–18 February 2005. The respective plots of the NASDAQ and DOW JONES returns as well as the actual observations vis-à-vis the one step ahead predictions for the entire period of our investigation are displayed in Figs. 11.4 and 11.5. These figures are generated using a three-state Markov chain in conjunction with the filtering and estimation procedures put forward in this paper.

11.7 Conclusions

We revisited the estimation techniques from HMM filtering theory and extended the framework that allows non-normal noise term. Recursive procedure was obtained for the re-estimation of transition probabilities of the underlying Markov chain. We provided a system of non-linear equations to enable the re-estimation of the drift and volatility parameters. Concentrating specifically on the noise following the Student's t-distribution, we gave examples outlining the implementation of the filters and re-estimation of model parameters as well as the optimal state of the Markov chain. Our results can be adopted to capture the dynamics of other financial variables for the purpose of financial modeling. In particular, applications to pricing, risk management and portfolio optimization with a view of testing the model's performance using historical financial data are a natural course for future research exploration.

References

1. Date, P., Mamon, R., Tenyakov, A.: Filtering and forecasting commodity futures prices under an HMM framework. Energy Economics **40**, 1001–1013 (2013)
2. Elliott, R., Moore, J., Aggoun, L.: Hidden Markov Models: Estimation and Control. Springer, New York (1995)
3. Erlwein, C., Mamon, R.: An online estimation scheme for a Hull-White model with HMM-driven parameters. Stat. Methods Appl. **18**(1), 87–107 (2009)
4. Erlwein, C., Mamon, R., Davison, M.: An examination of HMM-based investment strategies for asset allocation. Appl. Stoch. Models Bus. Ind. **27**, 204–221 (2011)
5. Mamon, R., Elliott, R.: Hidden Markov Models in Finance. International Series in Operations Research and Management Science. Springer, New York (2007)
6. Mamon, R., Erlwein, C., Gopaluni, R.B.: Adaptive signal processing of asset price dynamics with predictability analysis. Inf. Sci. **178**, 203–219 (2008)